P9-CJF-951

CCD Astronomy

Construction and Use of an Astronomical CCD Camera

Christian Buil

Translated and Adapted from the French by Emmanuel and Barbara Davoust

Published by:
Willmann–Bell, Inc.
P.O. Box 35025
Richmond, Virginia 23235

Published by Willmann-Bell, Inc.
P.O. Box 35025, Richmond, Virginia 23235

First English Edition

First published in French as *Astronomie CCD, Construction et Utilisation des Cameras CCD en Astronomie Amateur* Copyright ©1989 by Christian Buil

Printed in the United States of America

Library of Congress Cataloging-in-Publication Data.

Buil, Christian.
[Astronomie CCD. English]
CCD astronomy : construction and use of an astronomical CCD camera
Christian Buil : translated and adapted from the French by
Emmanuel and Barbara Davoust. – 1st English ed.
p. cm.
Translation of: Astronomie CCD.
Includes bibliographical references (p.) and index.
ISBN 0-943396-29-8
1. Astronomical photography. 2. Imaging systems in astronomy.
3. Charge coupled devices. I. Title.
QB121.B8513 1991 90-25474
522'.63–dc20 CIP

95 96 97 98 9 8 7 6 5 4 3

Foreword

In the search to understand the universe, the astronomer seeks ever more powerful observing instruments. Today progress in optics (thin mirrors), materials (light and rigid structures), and computer science (image quality improvements in real-time) allows giant telescopes to flourish around the world.

Compared to their ancestors of just 20 years ago, new telescopes are often located without regard to onsite living amenities. Rather, the controlling factors are the maximum number of clear nights *and* a calm, transparent atmosphere. The astronomer is often found half a world away sitting comfortably in front of a computer screen from which he controls the telescope and collects the data via satellite link.

Throughout the history of the telescope the sensor at the focal point has been recognized as vital. Today it is even more so as ever larger instruments come only at enormous costs. Nobody can afford to waste the few rare and precious photons gathered by expensive telescopes because of an inadequate sensor.

For most of history the practical sensor has been the human eye. However, its sensitivity, limited by physiological considerations and human subjectivity, placed serious limits on what could be discovered. The advent of photography in the last century was a monumental step forward.

A fundamental advantage of the photographic plate was its ability to record "unseeable" faint stars with long exposure times. During the early days of astrophotography, some astronomers made exposures lasting several dozen hours spread over several nights. Photographic technology has evolved over the years with ever more sensitive emulsions and treatment techniques. Most amateur astronomers are familiar with one of the most recent techniques—gas hypersensitizing.

Despite these improvements, the efficiency of a photographic plate remains relatively low. For every 100 photons that strike the plate, at best only 3 or 4 react with the silver "sensors" in the film's emulsion to create the image. This means that a one meter telescope using photographic techniques is no better than a 20 cm telescope equipped with a 100% efficient sensor! We can therefore understand why the astronomer continues to seek a better sensor than the 3 to 4% efficient photographic plate.

The micro-electronic revolution of the last several decades has not been ignored by astronomy. The Charge Coupled Device (CCD) working at better than 20% efficiency has quickly won over the astronomical community. Besides its increased efficiency in collecting photons, its sensitivity extends into the infrared spectrum. Further, unlike photography its response is a linear function of incident flux and exposure time—there is no reciprocity failure as encountered in long duration astrophotograpy.

A Story of Buckets

The outward appearance of a CCD resembles the integrated circuit now found in all modern electronic equipment: a flat case several tens of millimeters long, several millimeters thick and surrounded by "pins" that make electric connections to the associated circuitry. However, a more careful examination reveals something different: one of the long sides of the circuit is covered with a transparent glass window. Looking through it we can see a little black square—the circuit's silicon chip. This chip viewed through a microscope is a fascinating sight.

Almost the whole surface of the chip is covered with numerous, tiny, square cells that almost touch. These cells measure about 20 μm on a side (0.02mm). Since the silicon chip is about 10 millimeters on a side, we can easily calculate that there are 250,000 cells in this tiny space.

Each of these cells (which we will call *photosites* or *pixel*—short for "picture element"—from now on) has the ability to convert the light that strikes it into an electric signal. The chip is therefore the photosensitive detector, and for our purposes we will place it, rather than a photographic plate, at the focus of our telescope to record the sky's images.

To illustrate how a CCD image sensor works, we will compare the organization of the photoelements to a large number of buckets tightly arranged in an open field. Further we will say that the rain that falls on this field is similar to the particles of light (photons) that "rain" down from the sky.

The field is so big that in certain places the rain is heavier than in others. Therefore, as time passes, the buckets fill unevenly. Once the rain stops, we measure the depth of the water in each bucket to make a map of the precipitation reaching the surface of the field. Very soon we discover that this work is tedious and disagreeable.

We give up for the time being, but in anticipation of the next rainfall, we invent a clever device that will allow us to pour, on demand and at the same time, the contents of the second line of buckets in the first line, the third in the second and so on. Then all we will have to do is to measure the level in the first line of buckets between each *transfer* to reconstruct line after line the *image* of the precipitation on the field. Of course, before transferring the liquid from one line into another, we will have to empty the contents of the first line to avoid mixing. The system works perfectly and we can draw a map of the precipitation, line by line.

However, even this reduced work load is tiresome—it's too much to constantly go up and down the line measuring the water level of each bucket. Therefore, after further thought we modify our device so that it transfers the liquid down the line from bucket to bucket so that we can sit in front of this last bucket measuring water levels to make a point-by-point map of the field.

The analogy between our field and a CCD is not gratuitous. When an image is formed on the surface of the silicon chip, each photoelement is the origin of an electric signal which is proportional to the local luminous flux and to the exposure time (the length of the rainfall). The electric signal materializes in the form of packets of electric charges (the contents of the buckets), i.e., electrons. The CCD's organizational structure allows the transfer of stored charges at each photosite—similar to our rain measuring bucket device—first in lines, then in columns. Finally, we record this information sequentially at a single output pin as electric charges (traceable back to a specific photoelement in the CCD). Knowing where the charge occurred in the device and its amount, we can recreate an electronic map of the optical image.

The CCD in Action

The production of an electronic image with a CCD is carried out in two steps. First, the optical image is integrated in such a way that a significant number of charges is placed in the photosites. Secondly, at the end of the integration, the photocharge packets are transferred one behind the other toward an output stage where their contents can be measured. The measurement can either be analog, the image being reconstituted on a video screen (the camcorder principle), or numerical to be visualized later on a computer screen. The latter is the only method used in astronomy because a maximum amount of information can be obtained thanks to mathematical processing software.

In the term *CCD camera*, we include everything making up the CCD: the adjoining mechanical components, the electronic sequence for signal processing, and the computer hardware and software. But even more may be required—for example long exposures require the CCD to be cooled. The fact is that a CCD camera optimized for astronomical observation can be intimidating at first sight. The interconnecting cables and pipes jutting out at all angles are quite a sight! But once the camera is made, all the technical aspects are quickly forgotten and the CCD camera becomes at least as easy to use as an ordinary camera.

Once he gets used to it, an astrophotographer would not feel out of place with a CCD camera. Like the film in astrophotography, the sensitive surface of the CCD must be very precisely placed at the telescope's focus. However, with a CCD, precise focus cannot be found on a ground glass or with a knife edge (Foucault) test. CCD focusing is based on criteria of sharpness of stellar images which is just as precise and much faster.

One has to be more careful pointing a telescope outfitted with a CCD than with a photographic camera because of the small sensitive surface of the former (several millimeters long for the CCD compared to several dozen millimeters for photography). This difficulty is partly compensated by the CCD's high sensitivity since, in an exposure lasting a few seconds on a medium-sized amateur telescope (20 cm diameter), it is possible to record 15 or 16 magnitude objects. A quick trial exposure will confirm that the desired field will be recorded during the longer exposure. One then only needs to select a guide star and track it during an exposure that can last several minutes. The CCD's high sensitivity still does not preclude long exposures despite the fact that in a few minutes one can record as much information as in an hour-long exposure on photographic film. Even with a CCD, exposures of several dozen minutes are common to get the faintest objects. The limiting factors are usually the telescope and the quality of the observing site and not the CCD camera.

Once the exposure is finished, the electronic image recorded by the CCD can be immediately visualized on the computer screen. This is much better than photography and is just as pleasant as visual observation. It is fascinating and exciting to be able to see almost in real time the results of a long exposure and to be able to redo it if a problem such as poor guiding or incorrect framing arises. The frustration of working blindly as in photography is finished.

CCD imaging is very different from photography when we come to the problem of analysis. On a CCD image we can easily and precisely measure many parameters of the objects (brightness, dimensions, position, etc.). We cannot say the same for photographs that are more often contemplated than studied. In other words, it is much simpler to obtain scientifically valuable results with a CCD than with a photographic camera.

Of course, the CCD is not a panacea. There are areas in which photography remains the best choice: sky surveys with the wide field Schmidt telescopes to detect new comets, new asteroids, or novae and the study of extended low surface brightness objects.

Costs

CCD cameras are like telescopes. Some people buy theirs completely ready-to-use from a manufacturer; others build theirs from beginning to end. The former want to have the fewest problems and to use them as soon as possible. The latter are put off by the high prices, want an instrument that fulfils an exact need, and want to get personal satisfaction from their achievement.

At present it is possible to buy cameras especially made for astronomical observation. Unfortunately, their price is much too high for most amateurs, about $13,000 for a good complete system. This alone is a good reason for building one's own camera. But even if it is built at home, the cost of developing a CCD camera is high. Following is a typical budget: CCD chip $350, electronics $350, cooling $350, vacuum pump $1,200, mechanical components $500, and Computer $5,000 for a total of $7,750. Should one already own a computer, the costs will be proportionally less.

One of the forces motivating the builder and user of a CCD camera should be the realization that in astronomy the sensor is as important as the telescope. This is true no matter what the sensor is; an astronomer equipped with a modest telescope but working precisely and with experienced eyes will always get better results than a poor, sloppy observer using a bigger telescope. Therefore it is only common sense to build a detection system that is *at least* as good as the telescope itself.

The Arrangement of This Book

The meeting, in 1981, between the amateur astronomer that I am and the magical electronic component that is the CCD was like love at first sight. It was also the starting point of a series of discouragements and joys. A linear array CCD camera especially designed for astronomy was born. At the beginning of 1985, the family grew with a mosaic CCD camera, letting us make deep sky images all at once. On January 30, 1985, the camera expressed itself for the first time by catching the light of the Orion nebula.

That the first images were not of the highest quality did not matter. A few friends and I felt that those few square millimeters of silicon marked the beginning of a real revolution that would soon sweep up amateur astronomy. A few years later we still feel this way, and the pressing questions that I regularly receive finally convinced me to write something on the subject. Throughout I have attempted to maintain simplicity without sacrificing merit. Following is a top-line overview of the book's contents:

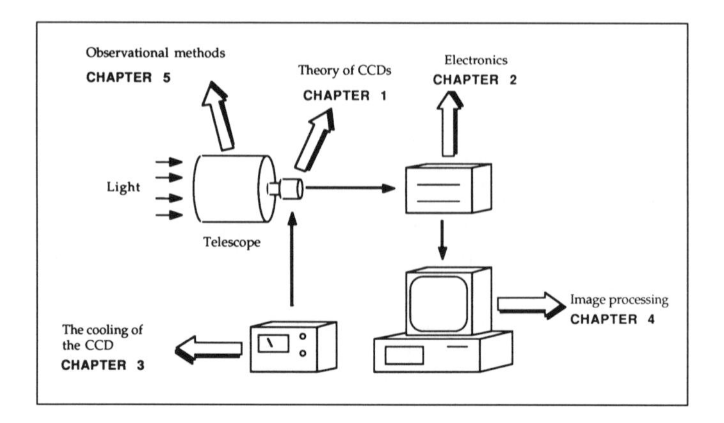

The Adventure That Awaits You

I do not want to minimize the problems that one will encounter in building and using a CCD camera—you will have to deal with computer programming, solid state electronics and even mechanics. Today CCD's have a minuscule sensitive surface compared with photography and a CCD image is less sharp than an image taken with a 2415. However, the construction of a CCD camera will take you on two exciting adventures.

First, a technological adventure. The CCD is still evolving. Manufacturers are trying hard to make a nearly perfect photonic detector and they will eventually succeed. The readout noise on some components is now only a few electrons and the size of the sensitive surfaces is larger than a 35 mm negative. Already in professional publications it is hard to distinguish, solely by appearance, a CCD image from a photograph. CCD's sensitive into the infra-red are beginning to appear, and a new field of study is on the horizon.

Each CCD has a personality—peculiar strengths and weaknesses. Just like with human beings, we have to know how to deal with them. This sometimes poses relational problems, but it's fascinating to discover that a piece of silicon has moods!

Second, an observational adventure. The CCD camera is a magic carpet to new wonders. The objects in the NGC catalog are too easy and the amateur will soon need to turn to more advanced works like *Uppsala General Catalog.* The CCD is ten times more sensitive than the best current photographic films, representing a gain of about two magnitudes. But that is not all. Unlike film, a CCD image contains data that you can readily evaluate with truly state of the art techniques. I do not risk anything by saying that the intelligent amateur, the least bit informed, with imagination, perspicacity and a CCD camera, will be able to make discoveries that have great importance because the field of study open to him is so vast and is only partly covered by professional astronomers. In a way, we are in an age like that of Galileo, when a telescope just had to be pointed anywhere in the sky to make a major discovery.

Acknowledgements

Throughout this book you will repeatedly encounter the word "we" — this is no accident. The work presented here would certainly never have come into being if I had not been surrounded by friends who either helped me in my task or encouraged me, such as Richard Szczepaniak who saw the birth of the first area array CCD camera in February, 1985 and with whom I wrote the first acquisition software in assembler on a good, old Apple II.

More recently, Guylaine Prat and Eric Thouvenot have collaborated with me on several cameras and image processing software. Further, with them and Paul Bertincourt we have built an observatory devoted exclusively to the study of the sky with a CCD. Jean-Jacques Quicot (CNES) was always helpful with questions about electronics. Serge Chevrel and Patrick Pinet stressed the importance, whatever the price, of the *quality* of the results.

Olivier Zuntini, Patrick Roth, and Alain Klotz participated in observing runs at the Pic du Midi. The research and technical teams at Observatoire Midi-Pyrénées always opened their doors to me, a most precious association for any amateur astronomer. Special mention must be made of Pierre Laques, astronomer at OMP, who let me use the prestigious one-meter telescope at the Pic du Midi at a time when my technique was less than perfect.

Finally, I wish to thank Alain Maury, Jean-Pierre Dambrine, Michel Laporte, Ronald Kaitchuck, Arne Hendon and Mark Trueblood who read the manuscript and made suggestions on how it might be improved. Whatever errors remain solely belong to me.

Dear reader, now it's your turn!

Christian Buil
Toulouse
France

Table of Contents

Chapter 1

Principles and Performance of the CCD

1.1 Introduction

The Charge Coupled Device image detector or CCD was developed in 1970 by Boyle and Smith at Bell Laboratories. The original CCD was designed to store and transfer analog information in the form of packets of electric charges within a semi-conductor structure. The charges are memorized in storage sites usually composed of MOS capacitors (Metal Oxide Semi-conductor). There can be several hundred to several thousand MOS capacitors on a single silicon chip. The sites are coupled together by transfer circuitry which makes it possible to move the charges in an orderly manner to a point where they can be measured.

CCD's have various uses: memories, delay lines, correlators and optical detectors. This last category is of most interest here. In a photosensitive detector with CCD registers, the electrical charges are created through the photoelectric effect. The storage sites are also photoelements (they are sensitive to light). They are organized either in lines (linear array CCD) or in a matrix (area array CCD).

After a period of being exposed to light, called the integration time, the photocharges are transferred, one behind the other, to an output stage. The electric signal delivered at the output stage is proportional to the incident illumination falling on the read photosite. We therefore observe at the output the variation of an electric signal, synchronized with the readout rhythm of the CCD, and corresponding to the number of charges contained in the packet. The injection of charges, their storage, their transfer and their readout are the basic functions of the CCD.

1.2 The MOS Capacitor

A MOS capacitor is composed of a doped semiconducting substrate, covered by an insulating layer (silicon oxide, SiO_2), on which a metallic electrode, also called a gate, is placed (see Figure 1.1). The insulating oxide is a thin layer of silicon several tens of micrometers thick. The metallic electrode is a deposit of aluminum or polycrystalline silicon heavily doped to become a conductor.

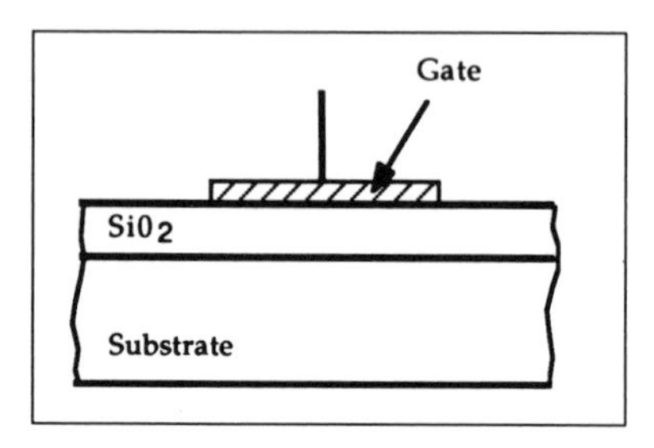

Figure 1.1 *A cutaway view of a MOS cell. The insulating silicon oxide layer is typically about a tenth of a micrometer thick.*

The layer of silicon oxide makes the MOS structure insulating and forms a capacitor. Let us study the mechanism that stores the electrical charges in an MOS capacitor. A silicon atom has its valence band occupied by 4 electrons. Each valence electron associates two by two with those of neighboring atoms to form a crystalline lattice. This is the principle of covalent linking. In the case of crystalline silicon, 8 electrons would complete the valence band. If trivalent impurities (a trivalent atom has 3 electrons on its external electronic layer) are introduced into the lattice, each atom of impurity introduces only 3 electrons, and there will be only 7 electrons on the outer layer. A "hole" is therefore created. This hole is caused by the absence of an electron and because of this deficiency we say that the semiconductor is P-type doped (for positive). On the other hand, N-type semiconductors are doped with pentavalent atoms which have 5 electrons on the valence band (there is one electron too many).

In what follows we will suppose that the substrate is of P-type doped silicon as is most frequently the case with CCD's. The holes are then called majority carriers. In this type of semiconductor there are several free electrons, resulting from the action of thermal energy which breaks the covalent bonds and brings the electrons from the valence band toward a band where they can move freely: the conduction band. These electrons are called minority carriers.

Let us positively bias the gate of a MOS capacitor with about 10 volts. At the instant that the bias is applied, the majority carriers (the holes) present in the neighborhood of the SiO_2–Si interface are pushed back into the interior of the substrate. A zone that is almost empty of majority carriers is then created in the area of the interface. A zone where charges are rare is called a depletion zone or a space-charge zone. Figure 1.2 shows the profile of the voltage V along the interface SiO_2 – Si.

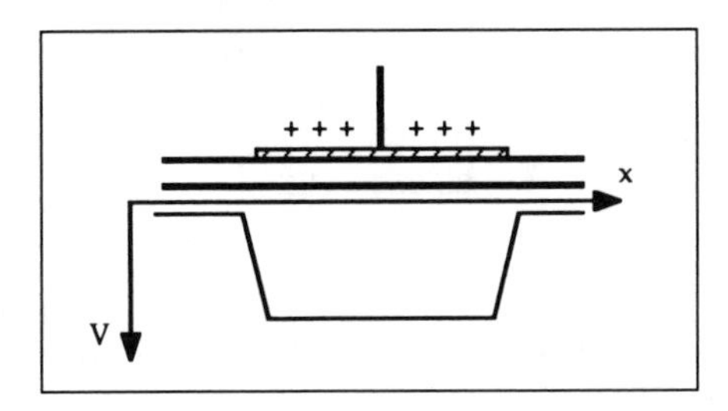

Figure 1.2 *Diagram of the surface voltages on a MOS cell when the gate is positively biased. V = bias voltage profile along the* SiO_2 – Si *interface.*

A space-charge zone is not in a state of equilibrium. Over time, electron-hole pairs are either generated from within the depletion zone or diffused at the frontier of this zone. These electron-hole pairs are separated by the electric field and the electrons accumulate in the neighborhood of the SiO_2–Si interface. This concentration of minority carriers of a type opposite to those of the substrate creates an "inversion layer" (here, a majority of electrons in a P-type substrate). The presence of minority carriers reduces the voltage of the surface from V_1 to V_2 (decrease in the thickness of the depletion zone). Figure 1.3 depicts this phenomenon as a well that fills with minority carriers. This analogy is often used—and allows us to visualize a well of surface voltage or, more simply, of a *voltage well*. The shaded zone symbolizes the stored charges (electrons in the case of a P doped substrate). Depending on the substrate technology, the voltage well typically extends to a depth of 3 to 6 μm from the silicon-oxide interface.

After a certain period of time called "thermal relaxation time," there are as many electrons on the surface as there are holes in the substrate. This phenomena is known as *dark current*. The state of equilibrium that results from this situation is reached between one and several dozen seconds after biasing the electrode. Relaxation time depends on the type of silicon, on the state of the surface between the silicon and the oxide and, of course, on the temperature.

The CCD is used when the MOS capacitor is in disequilibrium; that is when the minority carriers produced by thermal effect are negligible. The "useful" (photoelectric) e^- carriers are those that are generated by *photoelectric effect* and not those caused by dark current. "Useful" e^-

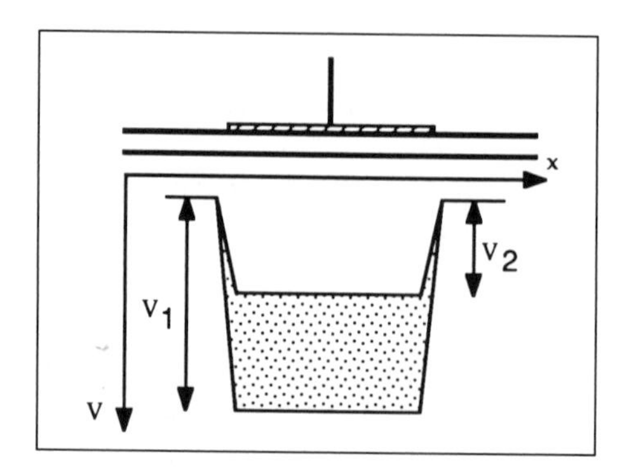

Figure 1.3 *A CCD voltage well. Here V = bias voltage profile along the* $SiO_2 - Si$ *interface; V_1 = the space charge zone; $V_1 - V_2$ = inversion layer.*

can only be accumulated for a period that is less than the relaxation time. These carriers are stored at the SiO_2–Si interface, and the inversion layer thus produced carries the information.

Notice that the operations of charge injection, transfer, and readout must take place in a very short time compared to the thermal relaxation time. This point is fundamental. In astronomy the production of charges by photoelectric effect can last several hours because of the rarity of photons. To have relaxation times of several hours the CCD must be drastically cooled.

In section 1.5 we will explore in detail how charges accumulate within a photoelement.

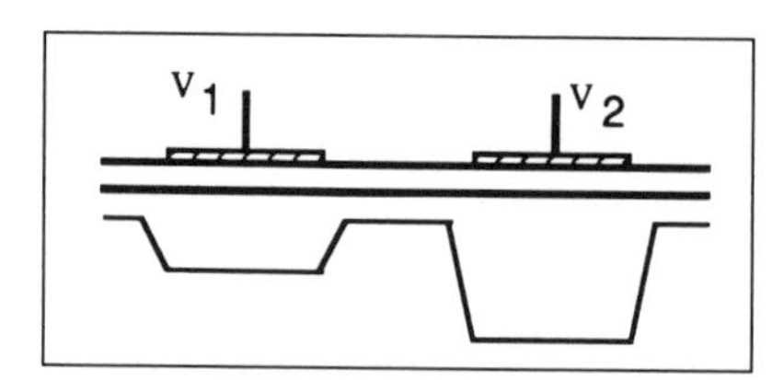

Figure 1.4 *The wells form because the gates are far apart.*

1.3 The Charge Transfer Mechanism

Let there be two distinct MOS capacitors, placed side by side, and having respectively the gate voltages V_1 and V_2. If the gates are far apart, two voltage wells will form separated by a barrier (Figure 1.4).

For an intergate distance of about a micron, the voltage barrier disappears, and the depletion zones communicate. When this situation occurs, there is capacitive coupling between the two MOS cells (Figure 1.5).

The ability to make the adjoining MOS capacitor voltage wells communicate is the basis of the *charge transfer mechanism*. Indeed, by applying a variable voltage to neighboring capacitors, it is possible to transfer charge-packets by degrees. The charges accumulate where the voltage well is the deepest. The command signals applied to the electrodes are sequential and are called clocks.

Several electrodes can be systematically arranged so that several charge-packets are transferred simultaneously along the CCD register. A group of

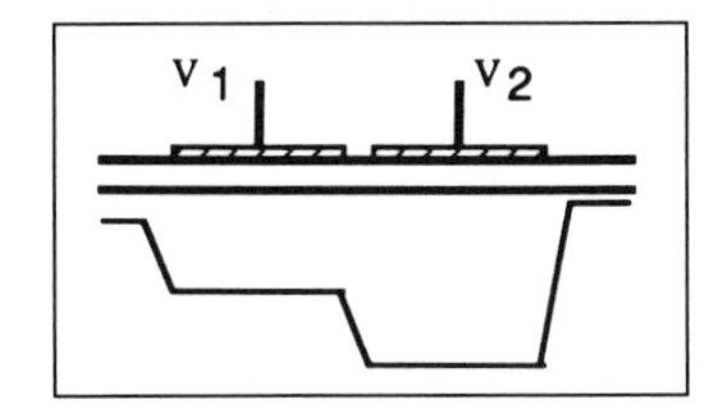

Figure 1.5 *Coupling of voltage wells.*

electrodes with a common electric link is called a *phase*. Each phase is driven by a distinct clock signal. All the clocks together form a clock timing sequence which must be carefully adjusted (to optimize elements such as the speed, the number of clock signals to be sent to the CCD, the relative phasing of the clocks, etc.) for each type of CCD. Sometimes the clock waveform recommended by the manufacturer can be modified for special applications (as is often the case in astronomy). A complete understanding of the working of the CCD is obviously essential before successfully undertaking such adaptations.

There are numerous transfer methods, specific for a given type of CCD, and they generally differ by the number of phases involved.

1.3.1 The Three-phase Transfer

Figure 1.6 illustrates the principle of the three-phase transfer. In Figure 1.6A, a deep voltage well is produced under electrode 1 by a strong bias voltage. The charges therefore accumulate under this electrode. In Figure 1.6B, electrode 2 is progressively biased (clocked "high") while at the same time electrode 1 is debiased (clocked "low"). The charges then flow from 1 to 2. The same operation is then reproduced between electrodes 2 and 3. Notice that at least one electrode separates two consecutive packets of charges, preventing any mixing of information.

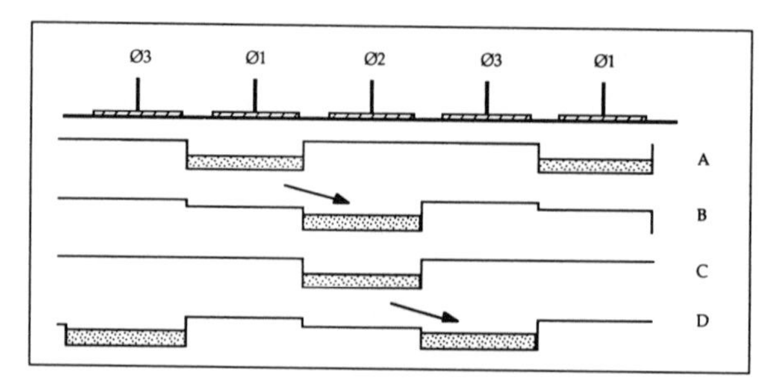

Figure 1.6 *Three-phase transfer.*

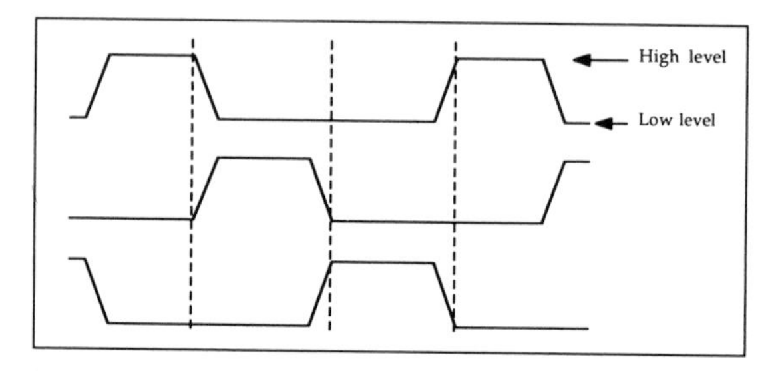

Figure 1.7 *Three-phase transfer clock timing sequence. The horizontal axis is time; the vertical axis is the amplitude of the signal applied to the phases. The high level corresponds to the maximum bias of an electrode. Note that the clocks of the different phases cross at intermediary levels to improve the flow of charges from one electrode to another.*

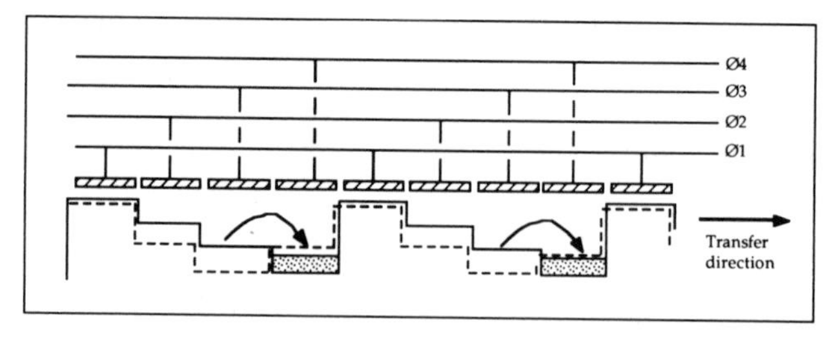

Figure 1.8 *An example of charge transfer in a four-phase CCD register. The dotted line shows the initial state of the voltage wells; the solid line shows the final state.*

1.3.2 The Four-phase Transfer

In the four-phase transfer, there is a double voltage barrier between two successive charge packets. This technique prevents any risk of charge backflow even when the clocks are not optimally phased. The four-phase transfer is usually used when high readout speeds are required—typically at clock speeds higher than 10 MHz.

Three- or four-phase registers are called non-directional because the direction of the charge packets is set by a sequential order of clock signals and not by the intrinsic structure of the registers. The direction of the transfers in the CCD can be changed by modifying the clock timing sequence.

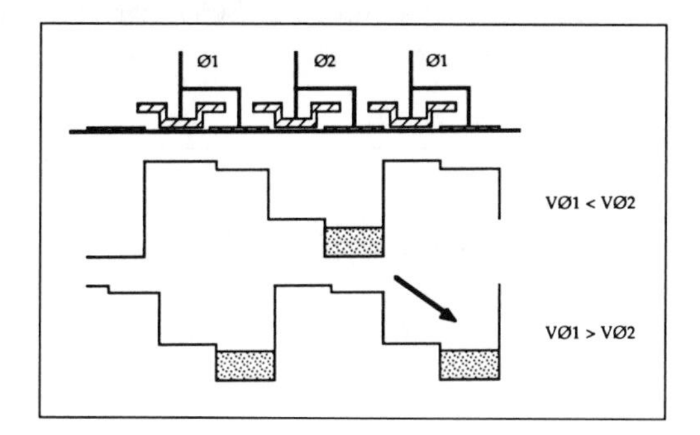

Figure 1.9 *Two-phase transfer.*

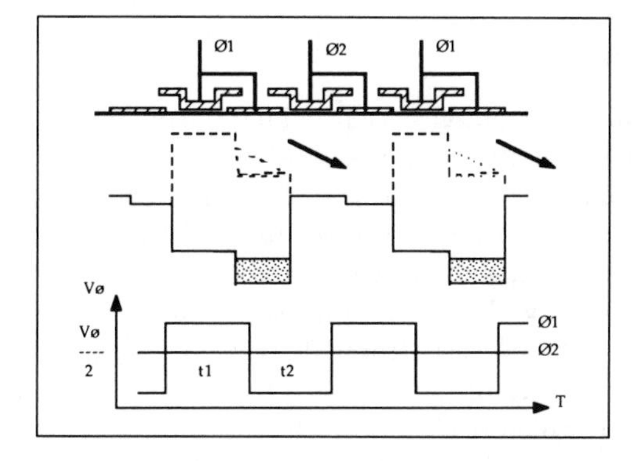

Figure 1.10 *Diagram of voltages and command signals in monophase functioning.*

1.3.3 The Two-phase Transfer

In the case of a two-phase transfer, the direction of the charge transfer is indeterminate. To solve this problem, each capacitive cell is formed by an electrode rendered asymmetric through a difference in thickness or doping. This structure induces a voltage well that is itself asymmetric (Figure 1.9). The displacement direction is set by the particular form of the electrode.

A variation of the transfer with an asymmetric electrode consists in maintaining a phase at an intermediate level, while the voltages applied to the other phase vary on both sides of this level (see Figure 1.10). We then say that the CCD is working in a monophase mode. The advantage of this method is the simplicity with which it is carried out, but the performance is average in terms of the quantity of charges transferred in a packet.

Another variation is the monophased virtual phase transfer. The asymmetry of the electric field produces results from an ionic implantation of variable thickness. A Texas Instruments CCD based on this principle equipped the camera of the Giotto probe which flew by Halley's comet in March, 1986.

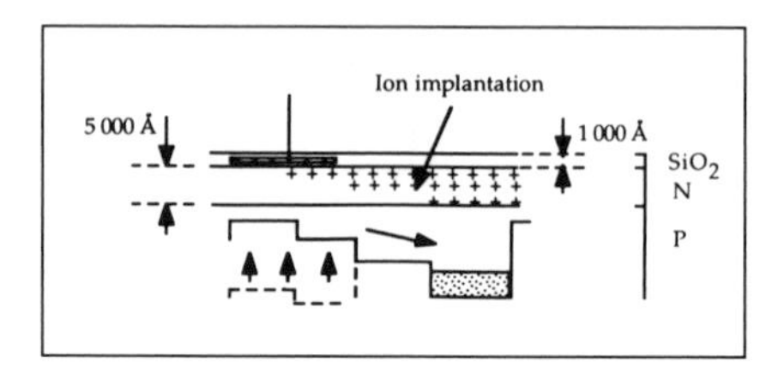

Figure 1.11 *A virtual phase CCD. Note the ionic implantation in an N doped silicon layer.*

1.3.4 Transfer Quality

A certain number of charges is left behind during phase switching. These charges are either recombined in the substrate or recovered by the following packet. This problem is known as transfer inefficiency. The recombination is greatly reduced by biasing the substrate (P-type) to a negative voltage.

The remaining principal source of transfer inefficiency is linked to the trapping of charges by chemical impurities present at the silicon dioxide-silicon interface. The carriers trapped there are freed during the passage of the following packets of charges. To reduce this phenomenon we have to use a structure which drives away the interface carriers. To do this, an N-type doped implant is introduced between the insulation (SiO_2) and the substrate P. CCD's with this type of implant are called BCCD's (Buried Channel CCD). CCD's without this implant are called SCCD's for Surface Channel CCD.

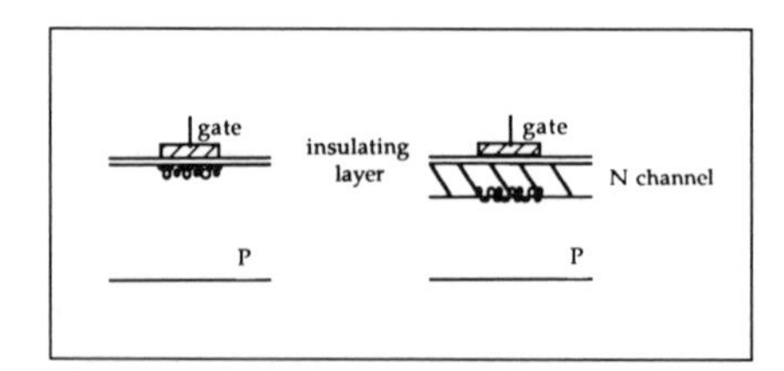

Figure 1.12 *On the left, a SCCD. On the right, a BCCD.*

Compared to SCCD's, BCCD's transfer charges more efficiently (typically a single transfer is 0.999990 efficient) and more quickly (the carriers' mobility is greater in the silicon's interior than near the SiO_2 interface). On the other hand, SCCD's have a higher charge storage capacity. Today, BCCD's are used much more than SCCD's for all applications.

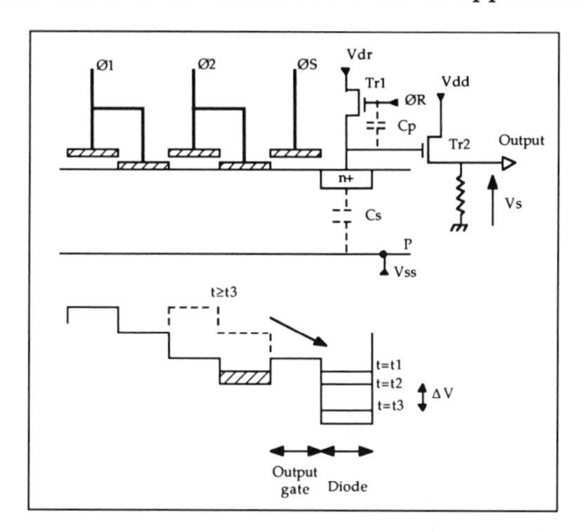

Figure 1.13 *Diagram of a floating diode output stage.* (V_{dd} *=driver drain voltage,* V_{dr} *= Reset transistor drain voltage,* V_{ss} *= substrate voltage)*

1.4 The Output Stage

The function of the output stage is to convert the charges into a voltage measurable by the user at the CCD's output pin. Figure 1.13 shows a simplified diagram of an output stage of a two-phase CCD register. This stage is said to have a *floating diode*. The principle is based on precharging a diode which acts as a capacitor at a reference level. The capacitor is then partly discharged by the packet of charges read. The difference in voltage between the final state of the diode and its precharge value is linearly proportional to the quantity of carriers contained in a packet.

Let us study in greater detail the working of this stage (see Figures 1.13 and 1.14). At instant t_1, a signal ØR (Reset signal) is switched to high level. This signal makes the transistor Tr1 a conductor and precharges the output capacitor C_s to the voltage V_{dr} (drain voltage of the Reset transistor). At instant t_2, the switch Tr1 is closed by ØR going low, which isolates the capacitor. Note that at this moment the precharge level is slightly modified following the presence of a stray capacitance (Cp) in the transistor (Reset parasitic). At time t_3, electrode Ø2 is clocked low, freeing the charges of the packet to be read in the output capacitor and causing a drop in voltage at its terminals. Electrode ØS, biased at a constant level, creates a barrier between the charge packet and the diode until instant t_3. The variation in voltage (ΔV) registered at t_3 through the follower transistor Tr2 is the desired signal. The diode is then set back to the reference voltage, and the next charge packet can be read.

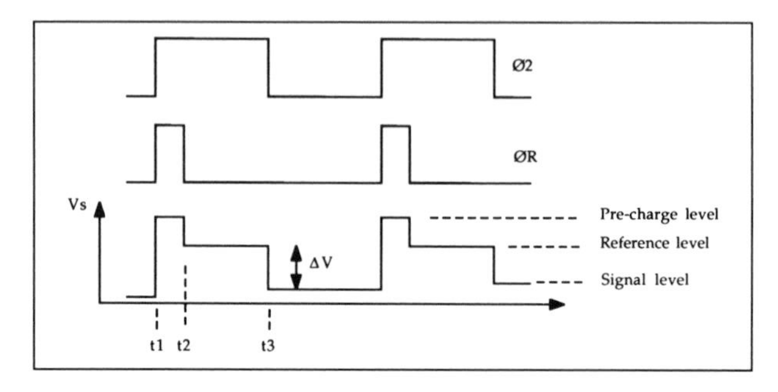

Figure 1.14 *Clock timing sequence and aspect of the video signal.*

The relation between the number of electrons contained in a packet and the corresponding variation in voltage is

$$\Delta V = qN\frac{G}{C_s}$$

where:

ΔV = the variation in voltage between the reference level and the signal level (in Volts),

q = the electron's charge (1.6×10^{-19} Coulomb),

N = the number of electrons per packet,

G = the gain in the integrated amplifier, and

C_s = the capacitance of the output diode (in Farads). The capacitance has a typical value of a tenth of a pico Farad.

The difference in voltage ($\Delta V_{\text{parasitic}}$) between the precharge level and the reference level is a function of the amplitude of the signal ØR:

$$\Delta V_{\text{parasitic}} = \varnothing R\frac{C_{\text{p}}}{C_{\text{p}} + C_{\text{s}}}.$$

The output transistor should discharge into a resistor called a charge resistor. According to the type of CCD, either this resistor is built into the chip (like the Thomson array TH7852), or the user has to place a resistor between the video output and the ground. The advantage of the latter system is that the resistance can be adjusted according to the kind of performance desired. Thus, a high resistance will lower the current discharged by the output transistor, consequently limiting the frequency response of the amplifier but also decreasing the noise of the output stage.

The integrated amplifier has a gain generally lower than unity, about 0.8. However, the exact value of the gain is often unknown and to simplify things, from now on we will take it as equal to unity.

It is not absolutely necessary to reset between two consecutive charge packets. In this case, the charges will be summed in the output diode until a ØR is sent. This technique, sometimes used in astronomy, is called *binning*. Binning increases the signal but decreases geometrical resolution. It is used when we don't really need the information of an individual packet of charges or when the quantity of charges contained in a single packet is too low to give usable information. Of course, we have to be careful not to saturate the output stage by summing too many charge packets. In practice the binning factor (number of packets added) is of the order of 2 to 5.

1.5 The Photoelements

In an image detector type CCD, charge injection is done through the photoelectric effect. When a photon penetrates into the silicon, it transmits its energy to the atoms of the crystal and this breaks the covalent bonds by producing an electron-hole pair. This pair is separated by the electric field prevailing in the neighborhood of a biased MOS capacitor, and the minority carrier (electron) is attracted toward the Si–SiO_2 interface. As the photoelectrons are produced under the influence of the light, an inversion layer is created under the gate of the MOS capacitor.

Two points should be noted:

1. The number of electrons produced is proportional to the number of incident photons. The gate's voltage must be kept constant during the exposure. This duration is analogous to photographic exposure time and is called *integration time*. The linear character of the reaction to a luminous stimulus and the integration time, which can be very long, make it possible for the CCD to record extremely weak flux.

2. If several MOS capacitors are placed side by side, the optical image is made spatially discrete. We can also say that it is sampled. The optical image is therefore transformed into an electronic image. The spatial resolution is determined by the size of the photoelements, also called *photosites* or *pixels* (from "picture elements").

A photoelement is typically 10 to 30 μm long. To avoid problems that are very difficult to overcome during reduction, especially in photometry, the CCD should have a square pixel. Most CCD chips marketed for television applications have rectangular pixels, making them less useful for astronomy.

The intensity of the photon flux, F, varies as a function of the thickness z of the silicon, following the law

$$F(z) = F(o)e^{-\alpha z}$$

where α is the coefficient of intrinsic absorption of the silicon and $F(o)$ is the incident flux on the silicon.

Typical values for α are tabulated below as a function of wavelength (λ) and of temperature.

λ(nm)	$\alpha(\mu m^{-1})$ T = 300K	$\alpha(\mu m^{-1})$ T= 77K
400	5.0	4.0
600	0.5	0.25
800	0.1	0.005
1000	0.01	0.002

We can easily verify that 90% of the radiation penetrating the structure is absorbed after a distance of $2\alpha^{-1}$. Thus, blue photons (400 nm) are stopped at less than 1 μm, while infrared photons (1000 nm) can cover more than 100 μm. However, photons with wavelengths longer than about 1.1 μm lack the energy to produce an electron-hole pair in the silicon. The following factors determine the shape of the spectral sensitivity curve:

1. Sensitivity in the blue is limited by the weak penetration of the photons which cannot interact in a zone of charge collection. To increase the efficiency in this region of the spectrum, the thickness of the silicon crossed by a photon must be small.
2. In the red, silicon's physical properties are such that it is impossible to directly detect a photon beyond a wavelength of 1.1 μm. For better sensitivity in this spectral domain, the silicon layer must be thickened to increase the probability of interaction.

Striving for high sensitivity in both the blue and red domains leads to a contradiction. That is one of the reasons why today we find two types of arrangement for the photoelements: those that are "frontside illuminated" corresponding to a thick substrate and optimized for the red part of the spectrum; and those that are "backside illuminated," in which case the substrate is very thin and optimized for the blue area of the spectrum.

1.5.1 Frontside CCD's

In a frontside illuminated CCD, the light first crosses the electrode and then the insulation before reaching the silicon. Practically, a thick CCD is manufactured from a silicon wafer 500 μm thick, which constitutes a massive, weakly resistant (P^{++}) substrate, on which a layer of doped P silicon is deposited by epitaxy[1] and if necessary an N-channel to make a BCCD. This epitaxial substrate, in which photons produce carriers liable to be caught under the gates, is from 5 to several 10's of micrometers thick. The advantage of an epitaxial substrate compared to a massive unique substrate is the possibility of controlling the thickness of the charge collecting zone and of optimizing certain parameters such as spatial resolution (see section 1.7.5). Light crossing several layers can produce interference phenomena that are revealed by a variation of the transmission as a function of the wavelength.

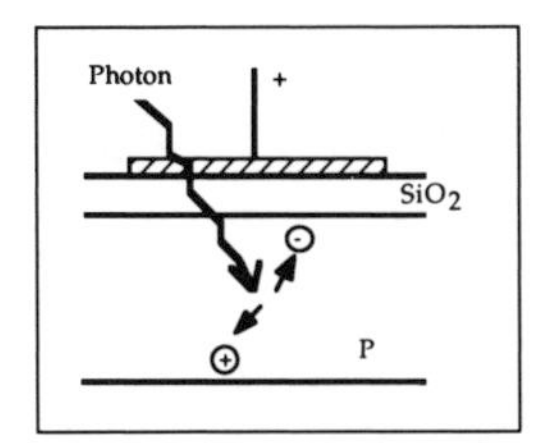

Figure 1.15 *Charge injection by photoelectric effect in a front-lit CCD. The gate and the insulation are semi-transparent, allowing the photons to interact with the silicon.*

Another problem with frontside CCDs is that the light must cross the gate and the insulation before reaching the charge collection zone. The efficiency of such a CCD generally does not exceed 50% at the peak of spectral sensitivity. Furthermore, the layers crossed become less and less transparent to radiation at wavelengths shorter than 450 nm. This kind of CCD is therefore virtually insensitive below 400 nm.

On the other hand, the red sensitivity of a thick CCD is satisfactory and allows us to approach the theoretical limit in the infrared. The thicker the substrate, the greater the sensitivity. For example, going from a 20 μm thick substrate to one 50 μm thick increases the sensitivity at 8500 Å by 50%. There are, however, limits for increasing the substrate's thickness that are linked to the generation of a growing parasitic thermal current and to the loss of spatial resolution (charge diffusion).

[1]A technique commonly used in microelectronics to deposit different layers of substances on a substrate, either in a liquid phase or in a gaseous phase.

There are ways to make a thick CCD sensitive in the blue and even in the ultraviolet. One technique uses thin layers of fluorescent matter that are deposited on the sensitive surface. When a photon penetrates this layer, fluorescence is produced at the point of impact. If the emitted photon has a wavelength detectable by the CCD, the ultraviolet radiation is recorded. The deposit of fluorescent matter thus acts as a wavelength converter from the ultraviolet to the visible.

The first use of thin layer techniques was in 1980 with coronene. This organic material covers the Space Telescope's CCD's and extends the sensitivity to 1000 Å with an average efficiency in the ultraviolet of approximately 10%. The flux absorbed by the coronene in the UV is reemitted around 5500 Å. The efficiency of this substance drops for wavelengths greater than 3500 Å. But the CCD is only sensitive beyond 4000 Å so that a treated CCD shows a marked loss of sensitivity between 3800 Å and 4200 Å.

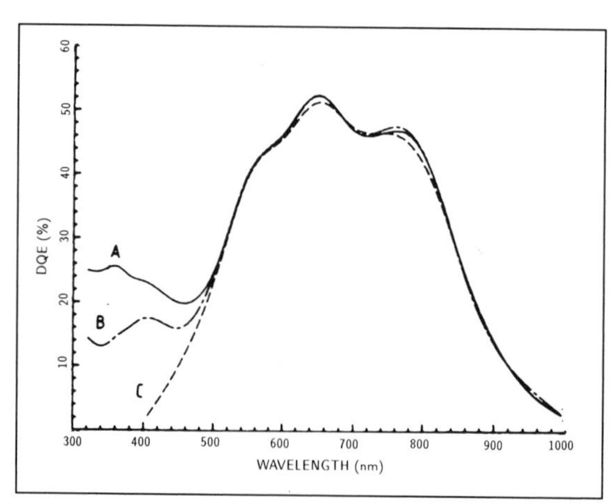

Figure 1.16 *Detective quantum efficiency (see section 1.7.2) as a function of the wavelength of a CCD recently treated with a layer of laser dye (curve A). After 6 months of exposure to prevailing light, performance has decreased slightly (curve B). Curve C represents the response of the untreated CCD. (from M. Cullin in* Optimization of the Use of CCD Detectors in Astronomy*)*

Today we more often use plastic acrylic films containing dye used in lasers to shift their emission wavelength (laser dyes). Different dyes are skillfully mixed to increase the efficiency. The first dye converts the distant UV into a UV radiation at a longer wavelength; then a second dye converts the light emitted by the first into a radiation at a still longer wavelength, and so on. In this way cascades of 5 or 6 dye layers move the UV information into a radiation bandwidth detectable by the CCD. From the blue to the red, the most common dyes are PPO, BIS-MSB, COUMARIN

481, COUMARIN 540, and HOSTASOL GG. The latter emits light around 5500 Å and the whole presents an efficiency of 20% for wavelengths greater than 3000 Å. This is obviously a substantial gain that can be exploited at high altitude observing sites (the atmosphere's cutoff wavelength is 3200 Å at sea level). Performance remains constant for several years if the CCD is not exposed to very strong light and if it is always stored in a vacuum.

The treatment we just described is not easy to carry out. To reach the sensitive surface of the CCD, one must remove the glass cover. An acrylic resin is dissolved in a solvent (toluene, methanol). This resin must of course be transparent to UV and mechanically firm, especially when the CCD is drastically cooled (−100°C). Some CCD's "tolerate" a given resin better than others (the roughness of the sensitive surface may more or less bond to the deposit). The dyes are then mixed with the plastic solution and several tens of microliters are deposited on the CCD's surface. Next, it is spun (1000 to 1500 RPM) so that the film is evenly spread over the CCD's active area. This film is 3 to 4 μm thick. If a problem with the film arises, the deposit is dissolved in the appropriate solvent, and the treatment begins again. It is important to know that some of the products mentioned above are very toxic, and these operations should therefore take place only by qualified personnel using the proper safety equipment as recommended by the manufacturer.

Lumigen is also sometimes used as a radiation converter. This product is a yellowish looking phosphorus that is found in the ink of certain fluorescent marking pens. It absorbs radiation below 480 nm and re-radiates it around 525 nm. Lumigen is deposited on the CCD by sublimation in a vacuum.

A thick CCD is relatively easy to make, a fact which decreases its cost and generally gives good reproducibility of opto-electronic characteristics.

1.5.2 Backside CCD's

With a backside CCD the light arrives from the side with the silicon substrate. The efficiency of a backside CCD should theoretically be maximum because the photon does not cross any intermediary layers. However, for the electrodes' charge collection to be effective, these charges should be generated in the depletion zone. This zone, however, is very thin, and since the rest of the silicon is very opaque, the substrate should be considerably thinned (Figure 1.17). A correctly thinned CCD is very efficient but limited by the reflection coefficient of the silicon. By applying an anti-reflection treatment to the sensitive surface, 80% of the photons can be recorded at the peak of spectral sensitivity.

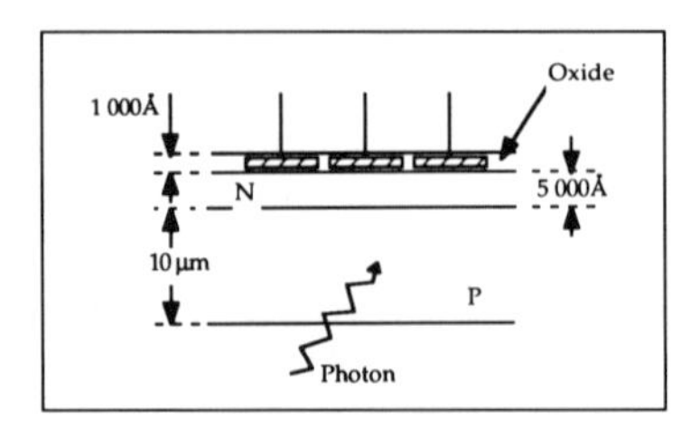

Figure 1.17 *Dimensions of a back-lit CCD. Note the depth of the substrate which is reduced to a thin membrane about 10 µm thick. Note also that the gates can be submerged in the silicon oxide.*

On the other hand, the thinned silicon layer is the cause of interference fringes of equal thickness which strongly modulate the sensitivity across the detector's surface. These fringes are especially visible when the CCD is used in the red or when it is lit by a source involving spectral emission lines (the light of the night sky has several lines of this type). They can be removed from an image (more or less) by photometric reduction methods.

A consequence of thinning the substrate is that backside CCD's are usually less sensitive in the red than frontside CCD's. On the other hand, the sensitivity in the blue is satisfactory because the thinning allows short wavelength photons to interact efficiently with the silicon. The sensitivity in the UV is strongly dependent on variations in the substrate's thickness, which is usually only controlled to within a micron. As a result local variations in sensitivity can be important. Sensitivity in the UV is also affected by the presence of a fine layer of silicon oxide originating at the surface of the substrate. This layer, 20 Å thick, is at the origin of a real voltage well that traps any photon which doesn't penetrate more than a few hundred Angstroms into the silicon—this is especially true for UV photons.

The surface trapping phenomenon is severe enough to justify research into how to decrease its effects. One of the solutions proposed consists in vigorously lighting the CCD by a UV source before the exposures. For this process we use spectral discharge lamps (mercury, zinc, cadmium) giving a strong line near 2500 Å. The oxygen in the air reacts to this radiation by producing a negative ion which settles on the CCD's surface, creating an accumulation of negative charges which can remain and repel the photoelectrons toward the front during image acquisition. This technique, *UV flood*, is efficiently used in many observatories. It has the advantage of a gain in sensitivity in the blue without modifying the technology of the CCD. The oxygen in the air plays an essential role in this treatment which must therefore be done in the prevailing atmosphere. After exposure to the UV flux (about 20 minutes), the CCD is immediately placed in a vacuum. If the temperature is then reduced to below $-100°$C, the charges produced on the surface are immobilized and the field is permanent for several months.

However, the results are sometimes variable and dependent on the CCD, length of exposure to the UV flux, wavelength of this flux, the presence of water vapor, etc. In many ways this technique makes the astronomer look like the sorcerer's apprentice who hasn't mastered all the elements of his experiment. Furthermore, gaseous treatments (nitric oxide—very dangerous for the user!) or treatments by corona discharge (very dangerous for the CCD!) have been used in the past, but are no longer used today because they are difficult to implement and the performance is unpredictable.

Sensitivity in the UV can be increased in a completely different way: a very fine metallic layer is deposited on the back side of the CCD to produce a static charge which repels the photoelectrons toward the voltage wells at the front side. This is the "flash gate" technique, developed by Jet Propulsion Laboratory. A quantum efficiency of 30% can be reached between 1000 Å and 4000 Å. The film, made of gold or platinum, is about 5 Å thick, and thus nearly transparent. A variation that gives a more stable and reproducible performance consists in placing a fine electrode of about 30 Å on the back side. It is then negatively biased (Biased Flash Gate technique—BFG). A simpler method with similar results consists of placing a P^+ implantation on the back side (this technique is used in France by Thomson for its thinned CCD's).

Surface treatments of the CCD are sometimes used to increase the CCD's efficiency in the visible domain. Some anti-reflection treatments (TiO_2 and Al_2O_3 for example) permit quantum efficiency in the neighborhood of 90% around 6000 Å.

A thinned CCD, which resembles a very fine membrane about ten μm thick, can be installed as it is on a rigid support. This assemblage is fragile and as the support expands and contracts, the membrane can stretch and give an uneven sensitive surface. With very fast optics ($\approx$f/3) the defects in evenness can result in focusing that is not identical at all points of the detector.

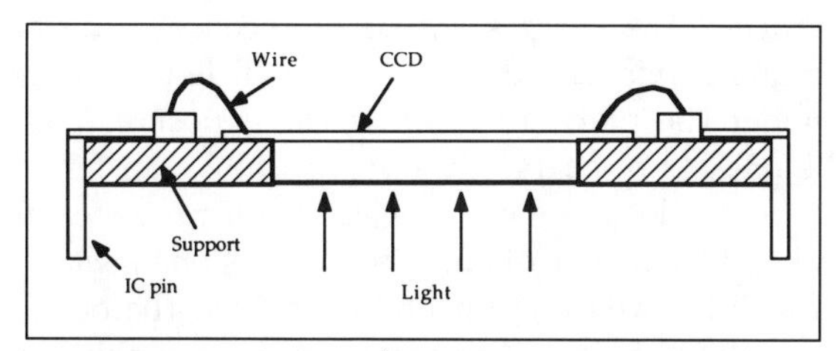

Figure 1.18 *Mounting of a thinned CCD.*

Sometimes the membrane is glued to a glass support, and the light crosses the glass before reaching the silicon. Besides the rigidity of this mounting, the variation of the refraction index between the silicon and its environment is less abrupt (silicon's index 4, glass's index 1.5), and loss by reflection is decreased.[2] In other cases, the membrane is glued onto a ceramic support on the side with the electrode (Tektronix CCD, for example). In all cases, the CCD's support should have an expansion coefficient identical to that of the CCD to avoid any separation.

Thinning a CCD is a very delicate operation. The main problem is obtaining a uniformly thick membrane so that the detector's sensitivity is homogeneous. Thinning is generally done by chemical attack with strong acids (hydrofluoric acid, for example). Mechanical polishing is rarely used because the resulting surface is rougher and the classical problems encountered by telescope mirror makers arise (for example, a turned down edge).

Manufacturing difficulties still make backside CCD's very expensive because they are only produced in small quantities. This situation is all the more unfortunate because their performance makes them almost perfect detectors when used properly.

1.6 General Organization of a CCD

1.6.1 Linear CCD's

A linear CCD is composed of a photosensitive row and one or more transfer registers. The photosensitive row can have from a hundred to several thousand pixels. Currently, the largest arrays are 60 mm long with pixels about 10 μm wide.

Figure 1.19 shows the typical structure of an array. The photosensitive row is framed by transfer electrodes. This arrangement of two lines of registers optimizes the CCD's size and its functioning (for example, capacity for storing a larger number of charges in the registers or a decrease in the number of transfers). One of the registers receives information from the even pixels; the other, from the odd pixels.

After integration, the charges are simultaneously transferred into the two transfer registers via the transfer gate which makes sure that they are confined under the photosites during the exposure (command signal ØP). Then the packets of charges are sequentially transferred to registers (Register Transport Clock). These registers are protected from light by aluminum masks. The output diode only receives one packet of charges at a time, once from the even register and once from the odd register. The information from the photosensitive row is restored at this level. Following

---

[2] The reflection coefficient of silicon in the air is approximately 35% between 1 μm and 0.6 μm; then it increases as it goes toward the blue: 40% at 0.5 μm, 50% at 0.4 μm, 65% at 0.3 μm.

the diode, there is an amplifier and if necessary a sample-hold circuit that, at a precise moment, holds the video signal at a constant value so that it can be easily processed.

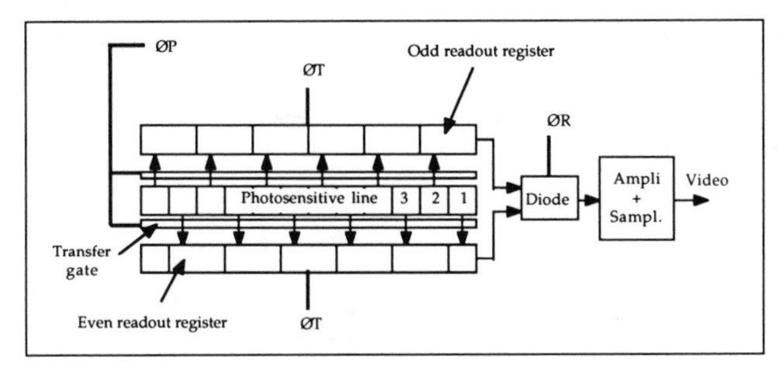

Figure 1.19 *Diagram of a linear array CCD.*

During the transfer of the charge packets, a new "image" is registered in the photosensitive row. In its turn it will be transferred into the shift registers and so on. In this scenario the integration time corresponds to the array's readout time. It is also the minimum integration time (1 millisecond for a 1000 element array read at a frequency of 1 MHz). On the other hand, nothing prevents a longer exposure; one need only hold the clocks at the proper level.

In Appendix A you will find a very simple example of setting up a linear CCD.

1.6.2 Area Array CCD's

To obtain an image with a linear CCD, the observed scene or the array itself must be moving in a direction perpendicular to the photosensitive row. The array is read out regularly during the motion. The lines thus acquired restore the scene once they are placed side by side.

This principle works perfectly when, for example, a document is to be digitized. Linear CCD's are used intensively for such applications (fax machines, computer scanners). In astronomy, however, the luminosity of objects is generally very weak, thus requiring an extremely slow scanning rate to integrate a sufficient number of charges. Long acquisition times, and the extremely regular scanning that must be maintained during that time prevent any use of arrays for deep sky observation.

The answer is area array CCD's: these are assemblages of photoelements arranged in lines and columns. This arrangement registers a two-dimensional image at one time. There are essentially two organizations of CCD arrays, differing by their transfer techniques: interline and frame transfer.

Interline Transfer

Interline organization is shown schematically in Figure 1.20. Each photosensitive column is associated with an adjacent shift register. Since the latter has the same optoelectronic properties as the photosites, it must be optically desensitized by an opaque aluminum mask placed on its surface. The information carried in the vertical registers is thus not disturbed by the incident luminous flux.

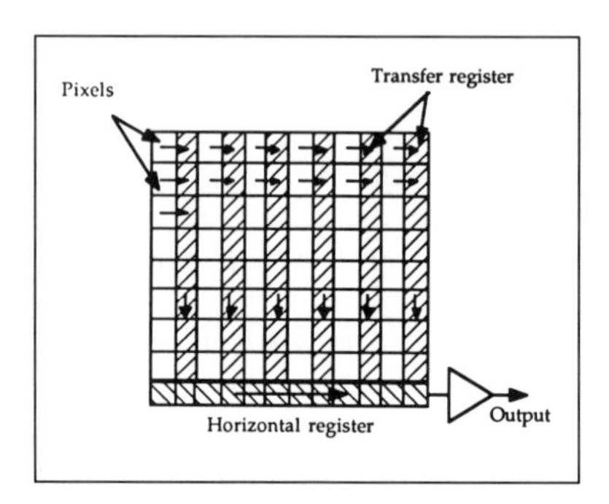

Figure 1.20 *Interline transfer array.*

The interline transfer array works in the following way. At the end of the integration time, the charges generated in the photosites are simultaneously transferred to the adjacent vertical readout registers. This transfer is brief, about a microsecond. While the photoelements collect a new image, the packets of charges that are present in the vertical registers are transferred row by row into a horizontal register. This register is read in such a way that the information arrives sequentially at the output stage. After the complete readout of the horizontal register, it receives the contents of a new image row, and so on. Of course the presence of transfer registers between the lines of detectors decreases the sensitive surface (typically 50%). Furthermore, this structure causes an aliasing phenomenon in the final image (Figure 1.21).

Frame Transfer

In this organization there are two distinct arrays on the same chip. One is light sensitive and constitutes the image zone; the other, identical in structure but covered with a mask to prevent light from coming in, is the memory zone (Figure 1.22).

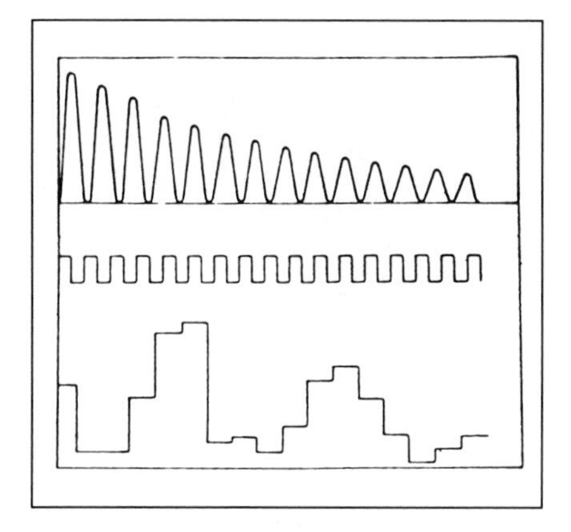

Figure 1.21 *"Beating" is produced when the spatial frequency of the recorded signal is of the same order of magnitude as the sampling step. At the top is the input signal; in the middle is the sensitivity profile of the CCD (succession of active zones and insensitive zones corresponding to the photosites and the transfer registers); at the bottom is the resulting signal, characterized by the presence of a very low frequency modulation. This is the phenomenon of aliasing.*

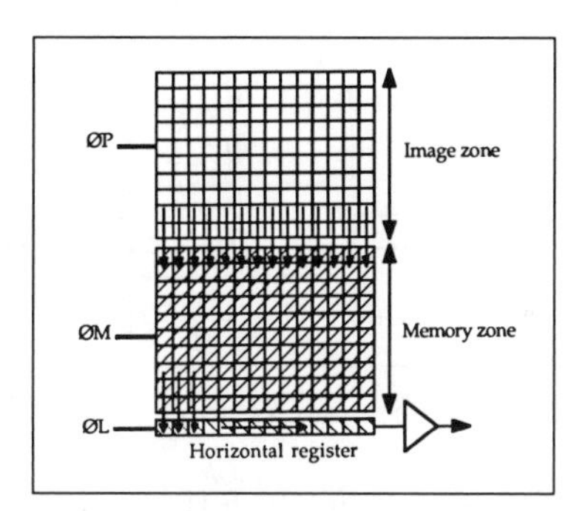

Figure 1.22 *The organization of screen transfer.*

After integration in the image zone, the charges are brought as fast as possible into the memory zone, away from light. For this process, the photosites themselves are arranged in vertical transfer registers. During the passage of the charges from the image zone to the memory zone, the clocks that order the transfer in these two sections of the CCD are identical (ØP = ØM). At each stroke of clocks ØP and ØM, the contents of one row are simultaneously transferred into its neighbor. The information contained in the memory is then transferred row by row toward a horizontal register, as in the interline organization.

Figure 1.23 *A screen transfer CCD array. The photosensitive zone is the black area. The memory zone, covered with an aluminum mask, is the light area. The 1 Franc coin gives the scale. Model TH7863 of Thomson-CSF.*

In the vertical axis, the minority carriers are confined under the photosites by the presence of the differential electrostatic field between the successive electrodes. In the horizontal axis, the lines are separated by *channel stops* which prevent the charges from spreading in this direction. Channel stops are almost transparent to light.

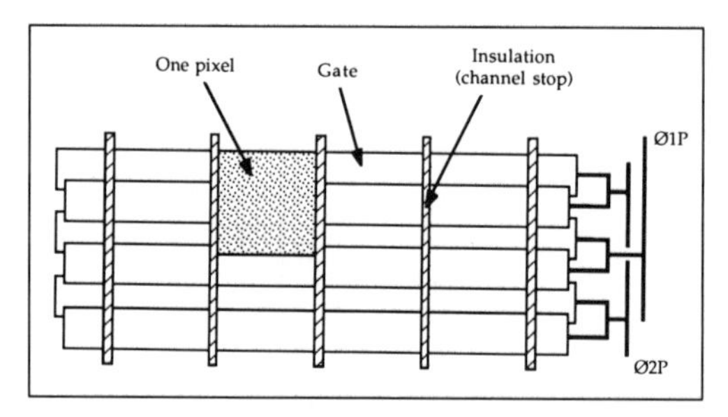

Figure 1.24 *Pixel separation in a screen transfer array with a two-phased register.*

Figure 1.25 shows the manner in which the charges are transferred from the memory zone to the horizontal register.

The transfer time from the image zone to the memory zone should be very short compared with the exposure time to avoid any *smearing* in the image. This is almost always the case in astronomy because this transfer time is usually several hundred micro-seconds, whereas integration time can be over an hour. The complete readout time of the array is usually short compared to exposure time (from about a 100 milliseconds to several dozen seconds according to the technique used and the number of pixels). The memory zone is therefore not always necessary. For this reason CCD manufacturers furnish arrays where the memory zone is converted into an image zone, thus doubling the sensitive surface. Still the careful astronomer always adds a mechanical shutter in front of the CCD to prevent a particularly bright object from disturbing readout.

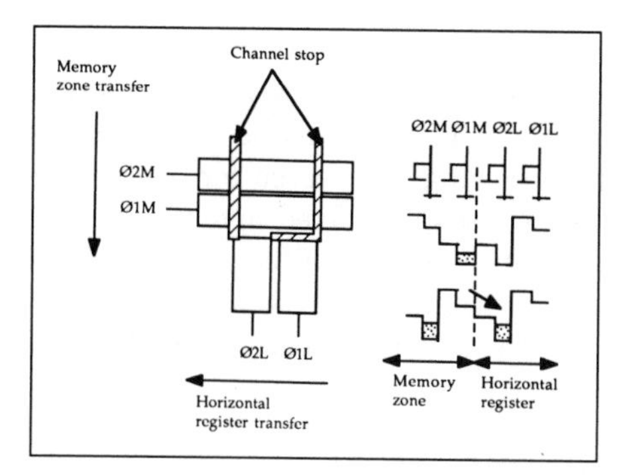

Figure 1.25 *Arrangement of electrodes insuring the passage of the charges from the memory zone to the horizontal register. The registers are the two-phased type. The diagram of the associated voltages shows that the passage of charges takes place at the leading edge of Ø1M while Ø1L is at low level.*

The entire image zone is photosensitive in a frame transfer organization; consequently, aliasing phenomena are almost nonexistent. Because this is a decisive advantage over interline transfer, only frame transfer technology is used in astronomy.

An imaging technique specific to CCD's—TDI (Time Delay Integration)—is possible with an array working in frame transfer mode. This technique moves the optical image on the sensitive surface at the same time that the image zone readout takes place. The motions of the image and the charges are synchronized; that is, the displacement of an image point happens at the same speed as the charge transfer from one photoelement to the next. The integration time of an image point is equal to the time it takes to cover the length of the CCD's photosensitive zone or, equivalently, to the duration of the transfer from the image zone to the memory zone. The smearing effect caused by the TDI function has an amplitude of a pixel, which is not detrimental to the image resolution in most applications.

There are two main advantages to the TDI mode:

1. Since one point of the image has been integrated on all the photosites of a column, the radiometric quality is excellent. Let us imagine a speck of dust placed on a photoelement. In the framework of a normal readout of the CCD, we would have a final image with a black spot corresponding to the speck of dust. In the TDI mode the dust's effect is averaged with the contribution from all the pixels belonging to the column (that is, more than 100 pixels). The dust therefore becomes almost invisible on the final image.

2. It is possible to acquire bands of images by the "kilometer". One need only let the observed scene pass by the detector at the same time that the image lines are being read out regularly and continually.

The dimension of the final image then depends only on the duration of the passage (but also secondarily on the storage capacity of the acquisition system). The TDI mode is used in certain programs of automatic sky surveying—to search for asteroids and comets. With this method the telescope is fixed and the diurnal motion produces the displacement of the image on the sensitive surface. The frequency of the CCD readout must be perfectly synchronized with the speed of the scene's motion in the focal plane (fine tuning of the clock's frequencies might be necessary).

1.6.3 Color CCD's

CCD's are used in consumer video cameras where there is a strong demand for color images and devices have been developed that obtain color images directly. For this process, photoelements are alternately covered with tiny pixel-sized colored filters, generally red, green, and blue. The information coming from the three groups of pixels is separated by appropriate processing of the video signal. Three distinct images acquired through each of the three filters can thus be obtained and electronically combined to produce a color image.

To be frank, using color CCD's is not recommended for astronomical imaging. The physical separation of the pixels required to produce the three color components significantly lowers the spatial resolution by a factor of three when compared with a "monochrome" CCD. The fact that "blue" pixels are not localized at the level of the "green" pixels, themselves not coincident with the "red" pixels, poses difficult problems for photometric processing. This problem can be complicated even more by the fact that with some CCD's the pixels corresponding to the different colored components are not all the same size. In addition, the spectral transmission of the filters used in color CCD's is very different from the transmissions used by astronomers for their astrophysical work, a disparity which is highly detrimental to any comparative study.

1.6.4 The Dimensions of CCD's

One of the few complaints that can be made about CCD's is the small size of the sensitive surface, especially when it is compared to a photographic plate. A typical CCD has 512×512 pixels, each about 20 μm square, providing a sensitive surface of about 1 cm^2. A 35 mm photographic negative presents a surface that is 8.5 times larger and a 9×12 plate covers a field about 100 times greater at the focus of the same observing instrument!

Of course, CCD manufacturers are trying to remedy this drawback. That is why it is already possible to find arrays more than 50 mm long. However, making such components with reproducible quality remains a difficulty. Such a large CCD occupies the whole surface of the silicon wafer from which several dozen small ones are usually made. A defect, even very localized on the wafer, means that the whole large sized CCD must be thrown away, whereas with small detectors only one or two chips out of several would have been discarded. Therefore large sized CCD's can be produced only at a very slow rate, using the latest technology which is always expensive.

A cheaper solution for increasing the field size is to use several CCD's side by side. This method is technically called "butting." The CCD's are usually glued close together on a common support. There is a dead zone, several tens of micrometers wide, where the chips meet, but this is still negligible compared to the whole area covered by assembling the CCD's.

Another alternative for increasing the field while avoiding butting is separating the observed field into several parts by an optical procedure, the image of each field being sent to a distinct CCD. Generally the CCD's are placed opposite the faces of an optical pyramid, with as many CCD's as there are facets on the pyramid. The archetype of this kind of mounting is the "4 shooter", a camera equipping the 5 meter telescope at Mount Palomar. This camera is made up of a 4 facet pyramid which distributes the field toward four 800 × 800 CCD's.

With a large CCD, there are so many photoelements that transfer losses can become important. In addition, readout times become prohibitively long. In very large arrays (over 1000 × 1000) there are always at least two, and sometimes four, output stages. In this case, according to the zone of the detector, the charges are transferred toward separate output stages. The outputs work at the same time, thus proportionally decreasing the readout time (see Figure 1.26). Sometimes the outputs are not identical; they are selected according to the readout speed desired (outputs optimized for rapid readout are also noisiest and therefore rarely used in astronomy where very weak signals are routinely sought).

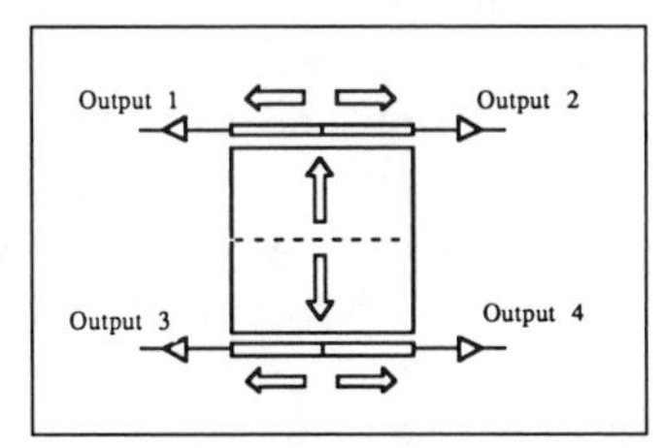

Figure 1.26 *Example of a array with two horizontal registers and four outputs.*

1.7 Performance

1.7.1 Ruggedness

The silicon chip is usually contained in a Dual-in-Line ceramic case. The light crosses a glass window to reach the sensitive surface. The latter is therefore not directly accessible and cannot be accidentally contaminated. This arrangement confers mechanical stability to the CCD (an advantage in astrometry, for example).

The reliability is very good as long as the electrical conditions specified by the manufacturer are respected. Like all MOS circuits, a CCD can be destroyed if subjected to strong electrostatic charges. It is therefore necessary to store it with the pins stuck into conducting foam. During handling, one should wear a grounded wrist strap (with a 1 meg ohm resistor acting as a current limiter) to dissipate static electricity.

Unlike photomultiplier tubes, the CCD's photosensitive retina can be subjected to strong lighting without any damage.

1.7.2 Quantum Efficiency

Quantum efficiency (QE) is defined by the following relation:

$$QE = \frac{\text{average number of detected photons/pixel/second}}{\text{average number of incident photons/pixel/second}}.$$

Because the quantum efficiency is always less than 1, all the incident photons do not cause the appearance of an excited electron. In the case of the CCD, we have seen that the efficiency ranges from 40% to 80%. This, however, is not a bad ratio compared to a photographic emulsion whose efficiency is presently 2% to 4%.

If a photon produces a photoelectron, the photoelectron will not necessarily be measured at the detector's output. There are several reasons for this situation. For example, the photoelectron could be lost during transfer, or drowned in the signal fluctuations generated at the level of the output stage. These phenomena, whose details we will give below, produce noise that can blur the information and thus prevent its detection. For this reason, quantum efficiency is not necessarily a good criterion of quality for a sensor. It can often be replaced by "equivalent quantum efficiency" or "detective quantum efficiency" (DQE). Let us examine the definition of this quality.

The signal-to-noise ratio of a perfect sensor is linked only to the statistical fluctuation of the number of photons received per unit of time. In these conditions, the expression of the square of the signal-to-noise ratio is written

$$\left(\frac{\text{signal}}{\text{noise}}\right)^2_{\text{ph}} = \frac{N^2}{\sigma^2},$$

where N is the number of photons received during the measurement and σ^2 is the variance of the measured signal. It can be shown that the number of photons per unit time follows a Poisson law when N is large, allowing us to say that the variance is equal to the average number of incident photons. With the hypothesis of large N, the number of incident photons during the measurement can, in a first approximation, be considered equal to the average number $\langle N \rangle$ that we would find if we took many measurements. The signal-to-noise ratio of the perfect detector then becomes

$$\left(\frac{\text{signal}}{\text{noise}}\right)^2_{\text{ph}} = N.$$

The perfect detector is said to be limited by photon noise. In fact, we have seen that all detectors are affected by intrinsic noises so that the signal-to-noise ratio is always less favorable. The DQE (a real criterion of a detector's quality, i.e. affected by its own noises) is defined by the relation

$$DQE = \left(\frac{S}{N}\right)^2_{\text{measured}} \Big/ \left(\frac{S}{N}\right)^2_{\text{ph}}.$$

It immediately follows that

$$\left(\frac{S}{N}\right)_{\text{measured}} = \sqrt{DQE \cdot N}.$$

In other words, everything happens as though the detector had seen only $DQE \times N$ photons. The DQE is always less than or equal to QE.

The DQE can also be written as the ratio of the variance of the photon noise on the variance of the signal coming out of the detector:

$$DQE = \frac{\sigma^2_{\text{ph}}}{\sigma^2_{\text{measured}}} = \frac{\sigma^2_{\text{ph}}}{\sigma^2_{\text{ph}} + \sigma^2_{\text{intrinsic}}}.$$

We see that the real noise of the detector is equal to the quadratic sum of the photon noise and the intrinsic noise. With a CCD, the high quantum efficiency allows many photons to be recorded, often making the second term negligible compared to the first. This is an important asset.

It should be noted that the "efficiency" can be higher than unity when the CCD is lit by very energetic radiation. In fact, the energy carried by an X or γ photon is sufficient to pull out the valence electrons of several atoms of the crystalline lattice. As an example, a thinned CCD receiving a Lyman α photon (1216 Å) produces an average of 3 electron-hole pairs. To take this phenomenon into account, a new quality criterion was established to measure CCD performance when it is lit by energetic radiation (extreme UV or X-rays). This is called charge collection efficiency—CCE. CCE translates the ability of the CCD to collect all the charges produced by a unique photon in a single pixel. The CCE depends on the rate of recombining charges and the manner in which the charges are distributed onto the neighboring pixels.

Good uniformity of quantum efficiency over the whole sensitive surface is a highly desired quality in a CCD. Any variation in efficiency appears as a local change in gain, causing noise. Despite manufacturers' efforts, the sensitivity is never sufficiently uniform for most astronomical work, especially since the effects of dust and eventual optical vignetting must be added to the intrinsic variations of the CCD. We will see how to correct this non-uniformity in Chapter 5.

Local variation in quantum efficiency can have a more subtle origin: the QE sometimes depends on past lighting of the detector. This is the hysteresis phenomenon. It is found by lighting the CCD relatively strongly with a contrasted scene and then taking a long exposure of a weak object. The latter then shows in filigree those parts of the bright scene that were previously observed. The "ghost" image is positive, indicating that the quantum efficiency is higher in the areas that were previously strongly exposed. This memory effect can last several hours, even several days if the CCD is maintained at low temperature. It has been observed on thick CCD's as well as on thinned ones.

A way of decreasing hysteresis is to light the CCD uniformly with a 7000 Å source, then taking many rapid readouts before beginning the long exposure. Of course, a basic precaution is to avoid violently lighting the sensitive surface between each acquisition. In any case, the appearance of hysteresis is always very slight and it can even pass unnoticed on most amateur cameras.

1.7.3 Spectral Sensitivity

In the preceding pages we saw some methods for extending the spectral sensitivity of a CCD. However, without any special "tinkering", the response is close to that of silicon, i.e. the CCD can record, without too many problems, radiation from 0.4 μm to 1 μm. The possibility of observing radiation beyond 0.7 μm is a new and important point (the eye and most photographic plates are insensitive beyond that limit).

The spectral sensitivity can change slightly when we pass from one pixel to another. This change must be taken into account when precise radiometric corrections of images are to be made. We will see in Chapter 5 that it is relatively easy to do so.

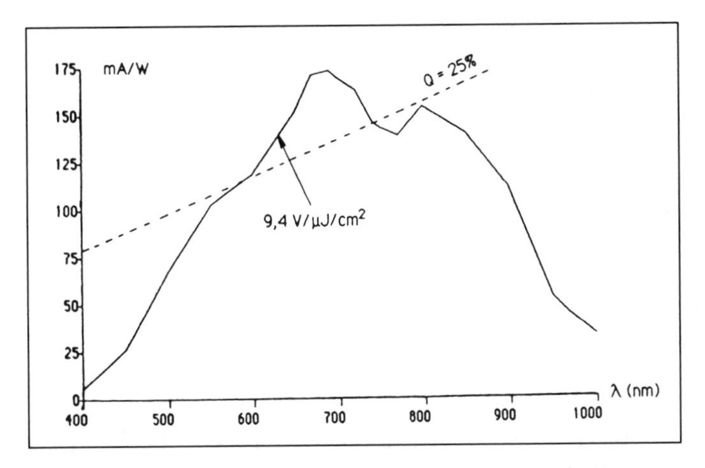

Figure 1.27 *Spectral response of a Thomson TH7852 array, expressed here in a current at the entry of the output diode as a function of the luminous flux in Watts penetrating into a pixel. The slanted line plots the positions where there is a quantum efficiency of 25%. We see that the response is relatively uneven, a situation which is characteristic of CCD arrays.*

1.7.4 Transfer Efficiency

During the transfer from one stage of the register to another, a certain number of charges is left behind. We will now quantify this phenomenon which we mentioned earlier in section 1.3.4. If N_o is the number of charges under a gate and if N_t is the number of charges under the following gate after the transfer, the Charge Transfer Efficiency (CTE) is written

$$\mathrm{CTE} = 1 - \frac{N_o - N_t}{N_o}.$$

Only a slight transfer inefficiency can be tolerated in a CCD. For example, if 1% of the charges is left behind at each transfer, a packet initially containing 100 electrons, will contain after 100 transfers

$$100e^- \times (0.99)^{100} = 37 \text{ electrons.}$$

In today's large arrays a packet must be moved an average of 500 times! This is why CCD manufacturers are trying to reach a nearly perfect rate of transfer—small losses at each transfer can really add up at the output. Today CCD registers have a typical transfer efficiency of 0.999990. Transfer efficiency depends on temperature. The phenomenon is important below −100°C where the carriers' mobility decreases and blocks the operation of the CCD.

A fraction of the charges can be left behind if the phase switching is too rapid to allow the charge to flow from one well to another. In astronomy it is always possible to read the CCD slowly, thus making this kind of problem negligible.

Transfer efficiency decreases when the packet contains a small number of charges (less than 1000 electrons). The problem is caused by a charge trapping mechanism that results from the presence of chemical impurities in the silicon or because of the poor condition of the surface at the interface level. These "threshold phenomena" limit the CCD's linearity under weak lighting.

One method for eliminating charge trapping is to illuminate the CCD weakly and uniformly just before the exposure (a brief flash by an LED is enough). This operation saturates the traps ("fat zero") and facilitates the photocharge transfer when the exposure is finished. This "preflashing" is often used in astronomy—for both CCD's and photography. Its disadvantage is that it adds supplementary noise to the image (photon noise from the flash).

The threshold phenomenon can only appear if the charge level is very slight. Thus with a moderately cooled CCD, a thermal current sufficient to saturate the traps appears in a few seconds. There is noise again, but this time it is of thermal origin.

1.7.5 Spatial Resolution

A CCD's spatial resolution for a given set of optics focusing an image on it is a function of the following factors:

1. The dimension of the pixel. The bigger the size, the worse the resolution for a given plate scale since the fine details are spatially averaged. Photoelements are usually 10 to 30 μm long. A CCD thus resolves much less for a given plate scale than a good photographic emulsion whose grain can be smaller than 5 μm. However, the final resolution is also a function of the size of the image that is formed on its surface. The magnification of the instrument should therefore be greater with a CCD than with a photographic plate to obtain an equivalent result.
2. Transfer inefficiency that slightly "dilutes" the charges of a pixel into the following pixels (on a time point of view) as the transfer is carried

out. The resolution is then a function of the object's position on the photosensitive retina.

3. The diffusion of the charges in the substrate during integration. This charge diffusion is particularly acute for radiation at long wavelengths, which is absorbed deep into the substrate. This region is nearly neutral since it is not under the influence of the voltage well. The electrons produced in this zone are either recombined or displaced by diffusion, and they can be caught by the voltage wells of neighboring photosites. The uncertainty about the photoelectron's destination is called *charge diffusion*, and it produces a "fuzziness" in an image taken in the infrared spectral range. To diminish charge diffusion we have to decrease the thickness of the P doped layer so that many of the charges produced by the long wavelength photons form in the supporting substrate where they rapidly recombine and thus do not participate in the electronic image. However, this technique causes a decrease in the red sensitivity. We should note that diffusion can also create "cross talk" between photosites and the adjoining transfer registers.

Figure 1.28 shows charge diffusion in a Thomson TH7852 array.

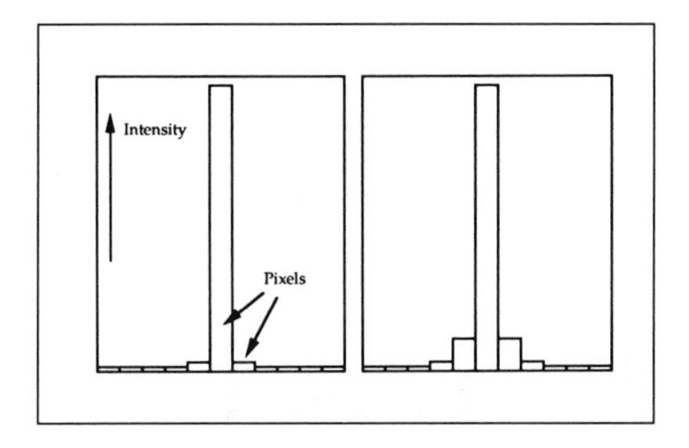

Figure 1.28 *Charge diffusion phenomenon.*

To carry out the test in Figure 1.28 we formed the image of an 8 μm wide slit on the photosensitive surface with a high quality lens. The slit is oriented along the line direction (i.e., perpendicular to the direction of charge transfer in the image zone). On the left of the figure, we find a photometric cross-section along the row direction of the slit in the center of the array for a wavelength of 0.51 μm. On the right, the same slit is lit at 0.89 μm. Since the slit is narrower than the pixel (30 μm), we measure the percussional response of the CCD. Because of charge diffusion, the signal is detectable on the pixels preceding and following the lit pixel, especially in the infrared image. Charge diffusion is, however, very limited in this particular array, which has a specially designed epitaxial layer.

Spatial resolution is often expressed in terms of rate of modulation. Let us form on the CCD the image of a test pattern containing a succession of black and white bands with a regular interval. The modulation rate is then a function of the observed contrast of the test pattern. If the test pattern has a rectangular profile, it is called contrast transfer function (CTF). If it has a sinusoidal profile, we measure the modulation transfer function (MTF).

Let $S_{\max}$ be the signal picked up in the bright part of the test pattern and $S_{\min}$, the signal in the dark part. The MTF is given thusly:

$$\mathrm{MTF} = \frac{S_{\max} - S_{\min}}{S_{\max} + S_{\min}}.$$

An MTF of 100% corresponds to a maximum resolution in the considered spatial frequency. The MTF is recorded for different spatial frequencies, usually expressed in pairs of lines/mm, which give the curves of Figure 1.29.

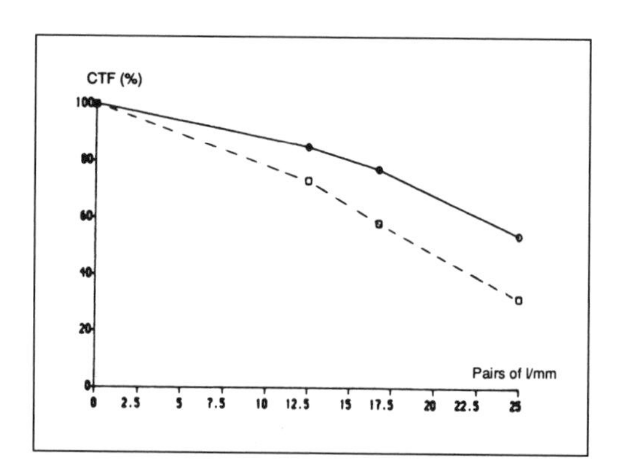

Figure 1.29 *Rate of modulation of a TH7852 array. The bars of the test pattern, with a rectangular profile, are turned in the direction of the transfer "image zone—memory zone". Shown as a solid line, the wavelength is 0.51 μm; as a dotted line, the wavelength is 0.89 μm. The drop of CTF in the infrared is caused by "cross-talk."*

The MTF of a CCD is the product of the elementary MTF's caused by the sampling effect (pixel geometry), by the transfer inefficiency, and by the charge diffusion:

$$\text{Total MTF} = \text{geometric MTF} \times \text{transfer MTF} \times \text{diffusion MTF}.$$

The geometric MTF is given by the formula

$$\mathrm{MTF}_{\mathrm{geom}} = \sin\left(\frac{f}{f_{\mathrm{sam}}} \cdot \frac{\pi\,\mathrm{dx}}{p}\right) \Big/ \left(\frac{f}{f_{\mathrm{sam}}} \cdot \frac{\pi\,\mathrm{dx}}{p}\right),$$

with

p = interpixel distance,

dx = pixel size,

f_{sam} = spatial frequency of the image sampling (usually $f_{\mathrm{sam}} = 1/p$),

and

f = spatial frequency of the input signal.

For example, if $p = 30 \times 10^{-3}$ mm, $\mathrm{dx} = 30 \times 10^{-3}$ mm, $f_{\mathrm{sam}} = 1/p = 33.33\ \mathrm{mm}^{-1}$, at the frequency of 12 pairs of lines/mm, we find an MTF of 0.80.

The transfer inefficiency MTF is given by the formula

$$\mathrm{MTF}_{\mathrm{trans}} = \exp\left[-N\epsilon\left(1 - \cos\frac{2\pi f}{f_{\mathrm{sam}}}\right)\right],$$

where

N = number of transfers undergone by a chargepacket,

ϵ = transfer inefficiency (1 – transfer efficiency),

f_{sam} = spatial frequency of image sampling,

and

f = spatial frequency of the input signal.

Using the data of the preceding example and supposing an average of 200 transfers for a packet and an inefficiency of 1×10^{-5} we find, for 12 pairs of lines per millimeter, an MTF of 0.996. For more precise modeling, we should also consider the fact that transfer efficiency is different in the horizontal and vertical registers. Precise measurement of MTF variation at different points of the sensitive surface is a means of determining transfer inefficiency. It is, however, useful to see that transfer inefficiency contributes very little to the loss of resolution in an medium-sized CCD.

The diffusion MTF is more difficult to calculate since it depends on the physical characteristics of the silicon. In the case of a frontside CCD, its expression is

$$\mathrm{MTF}_{\mathrm{diff}} = \left(1 - \frac{e^{-\alpha d}}{1 + \alpha L}\right) \Big/ \left(1 - \frac{e^{-\alpha d}}{1 + \alpha L_o}\right),$$

with

d = depth of photon interaction

α = absorption coefficient of the silicon

L_o = diffusion length of the minority carriers in the silicon

and

$$L^{-2} = L_o^{-2} + (2\pi f)^2.$$

Typical values for d go from 1 μm to 10 μm and correspond to the depth of the voltage well. We will suppose that the wavelength is 7000 Å. In this case $\alpha = 0.2\mu\text{m}^{-1}$ (see the values tabulated in section 1.5), and $d = 3\mu$m. The diffusion length is typically 50μm. However in this example we will suppose that we are dealing with a TH7852 array in which the epitaxial zone is thinned to about 20 μm, thus $L_o = 20$ μm. For $f = 12\text{mm}^{-1}$ we find an MTF of 0.93.

The total MTF of our CCD is then

$$\text{MTF}_{\text{total}} = 0.80 \times 0.997 \times 0.93 = 0.74.$$

In practice, it is quite difficult to build test patterns with a sinusoidal profile, and most of the time we calculate a CTF. An approximate relation between the CTF and the MTF is written

$$\text{CTF} = \frac{4}{\pi}\text{MTF}.$$

In the case studied, the CTF would be 94% at 12 pairs of lines/mm.

1.7.6 Linearity

Let E be the illumination received by the detector and let S be its response. The detector is said to be linear if, whatever the illumination, we verify the relation

$$S = k_1 \cdot E + k_2$$

where k_1 and k_2 are constants.

In practice, linearity is limited to a working range of intensity. For low illumination, linearity is no longer possible because of threshold phenomena such as charge trapping; at high light levels, the detector becomes saturated (the voltage well of a CCD with a thick substrate then contains between 10^5 and 10^6 electrons). What then happens is that a "space charge" occurs in which the e^- repel each other and leaks over to the next well in the column, causing blooming along the column.

Linearity is generally measured by the difference between the real characteristic and the ideal straight-line calculated by linear regression between the limits E_1 and E_2 (Figure 1.30). This high degree of linearity permits

CCD's to be used for *area* photometry with better than 1% (0.01 magnitude) accuracy.

$$\text{Linearity} = \sqrt{\frac{1}{N}\sum_{n=1}^{N}\left[1 - \frac{V_{\text{real}}(n)}{V_{\text{ideal}}(n)}\right]^2}$$

with N the number of measures carried out between E_1 and E_2.

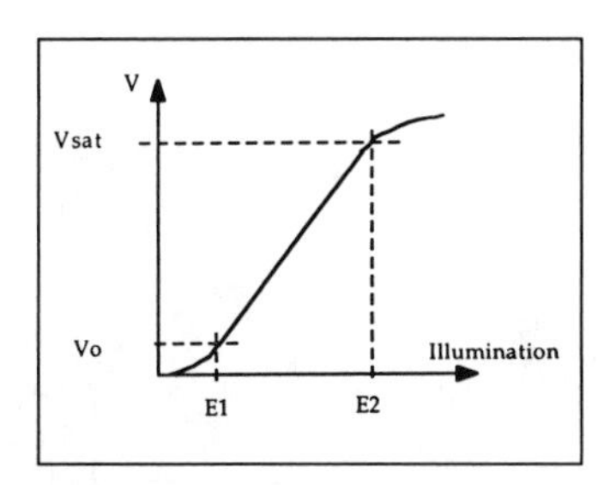

Figure 1.30 *Response curve to the illumination of a CCD detector. To make the point easier to see, the non-linear zones have been greatly exaggerated. The illumination E_1 is located at a very low level, and illumination E_2 produces an abrupt malfunction of the CCD.*

The linearity of a CCD is remarkably good ($< 10^{-3}$), to the point that it is sometimes difficult to make this defect conspicuous when working with illuminations between E_1 and E_2 (see Figure 1.30).

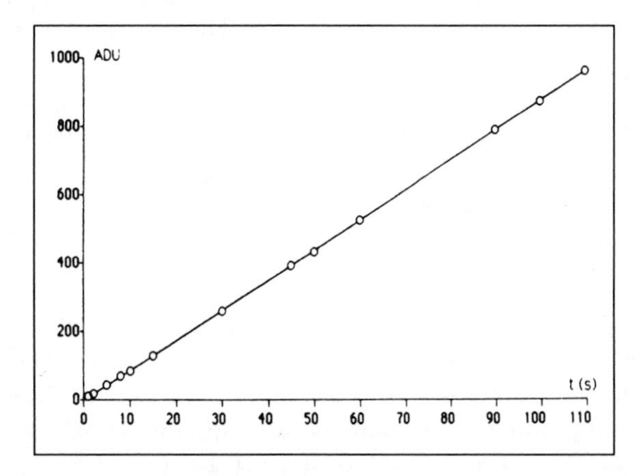

Figure 1.31 *Measurement of the linearity of the TH7852 array and the associated video electronics. The vertical scale is in "Analog Digital Units" (see definition in section 1.7.12).*

The linearity can be measured by illuminating the CCD with a very stable luminous source and by tracing the detector's response as a function of integration time.

Another measurement method consists in taking perfectly calibrated photometric images of stellar fields. We then check to see whether a linear relation exists between the signal measured on various stars and the brightness of these stars. Of course, the measurements are carried out in precise spectral bands to take into account the color of the stars (indices) and of atmospheric transmission. This technique measures the detector's characteristic under real working conditions.

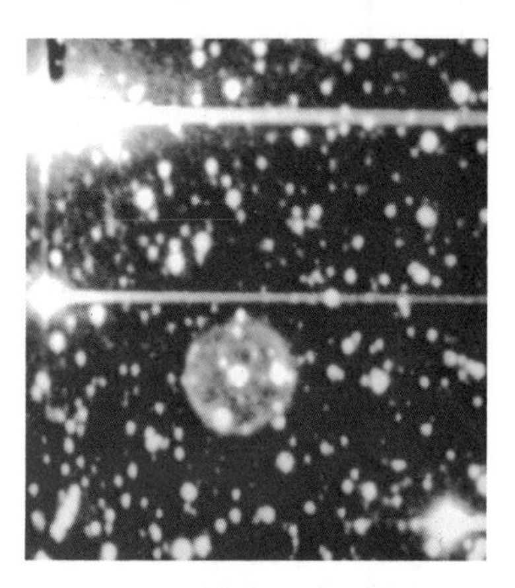

Figure 1.32 *The planetary nebula PK36+17.1 is surrounded by bright stars. These stars heavily saturate the sensor (here, a TH7863 array) causing the horizontal artifacts. They are due to a strong transfer inefficiency caused by a number of charges far above the saturation threshold and to the pollution of the points passing under the bright stars during the transfer from the image zone to the memory zone. (The transfer direction is from right to left in this example.) The strong diffusion is only partly caused by the CCD. Dust on the telescope mirror is also responsible for this phenomenon. The image was taken by integrating 20 minutes on the T60 telescope at the Pic du Midi Observatory.*

1.7.7 Blooming

When a photosite is subjected to excessively strong illumination, the accumulated charges can become so numerous that they spill onto adjacent photoelements. A saturated pixel produces a characteristic diffusion similar to the halo surrounding bright stars on a photographic plate. In addition, the number of charges accumulated in a saturated well can be such that its contents cannot be emptied in one or more transfers. A trail starting at the saturated point then appears in the direction of the transfer of rows. This effect, called *blooming*, is often the signature of a CCD image.

A strongly illuminated CCD retains parasitic charges that are progressively restored during successive readouts. Consequently, it is important not to saturate the CCD between each stellar exposure.

Anti-blooming devices implanted on certain CCD's allow us to limit the effects of saturation. They are electrodes placed on the edges of the photosites to drain the overflow of a well (see Figure 1.33). The flux can then reach up to 100 times the saturation value of a pixel without the appearance of blooming. On the other hand, the diffusion onto neighboring pixels, although limited, does not disappear completely, mostly because of the charge diffusion already described (section 1.7.5).

An anti-blooming device is undeniably useful for the observation, for example, of a weak object close to one that is much brighter. However, the anti-blooming drain produces an area that is photosensitively "dead" in the middle of the image zone, decreasing by at least 10% the sensitivity of a array equipped with one.

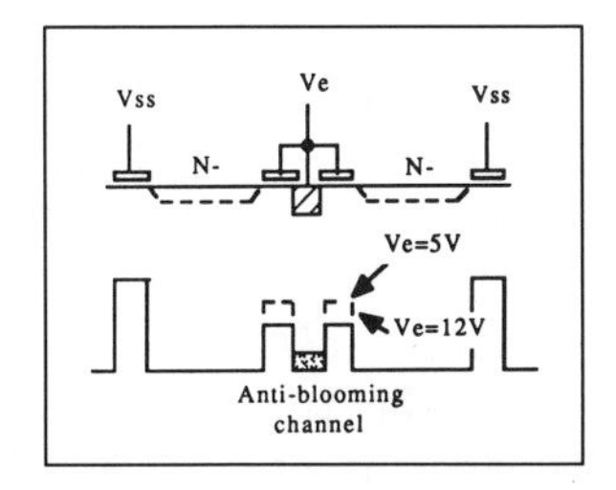

Figure 1.33 *Structure of the anti-blooming device of the TH7852 array. An insulating channel on two of the columns is made of a transparent poly-silicon gate (biased* V_{ss}*). The anti-blooming device is inserted between these insulating cases and it is common to both columns. At the bottom of the figure, we see the diagram of the voltage corresponding to this arrangement. When a well is filled with charges, the excess ones flow into the drain. The system's efficiency can be set by adjusting the voltage* V_e.

1.7.8 The Dark Current

The main enemy of the CCD user is the dark current. Even when the device is placed in complete darkness, a signal is observable at the output of the detector. This signal is the result of charges created in the silicon because of thermal agitation of the crystalline lattice.

Charges of thermal origin are created in the photoelements as well as in the transfer registers (which share the same technology as the CCD). Their number, like that of the photocharges, is obviously tied to the integration time. The production rate is such that at room temperature a standard CCD in darkness is saturated after only a few seconds of integration. On the other hand, the dark current becomes nearly negligible at a temperature of about $-100°$C. In a first approximation, the thermal signal decreases by a factor of two for a 7°C drop in operating temperature.

The theoretical expression of the dark current is

$$S = AT^{3/2}e^{-V_g \cdot q/(2 \cdot k \cdot T)},$$

where A is a constant depending on the units used for S, T is the temperature in Kelvin, V_g is the gap voltage (physical characteristic of the semiconductor), q is the electron's charge (1.6×10^{-19}C) and k is Boltzmann's constant ($1.38 \times 10^{-23} J/°K$).

Figure 1.34 shows an example of experimental determination of the dark current in a Thomson array TH7852. From this study we conclude that V_g = 1.36V and $A = 3.36 \times 10^6$ when expressing the output signal in volts. For the tested array, the dark current as a function of integration time t (in seconds) and of temperature T (in ° Kelvin) is written

$$U = (3.36 \times 10^6 \times T^{3/2} \times e^{-7884/T}) \times t + 0.002(V).$$

The constant term at the end of this equation (V) represents an electric offset independent of temperature and integration time (difference between the reset level and the dark current level). We will see this in more detail in section 2.3.3.

For example, with a temperature of 0°C an integration of about a minute can take place, and at $-30°$C this time can be increased to 10 minutes. However, in both cases, the dark current is not sufficient to create significant noise in the image. The variation in the generation of thermal charges from one pixel to another is the main limiting factor. It is possible to eliminate this lack of uniformity by subtracting from the original image an exposure obtained under identical conditions (same temperature, same exposure time) but in darkness. We will call this a "thermal map" or "dark map".

The subtraction of the thermal map from the associated stellar image is an operation that must be carried out systematically when the CCD's temperature is not low enough to make the production of thermal charges negligible during the exposure. Unfortunately, with this method it is impossible to get rid of the thermal noise, which is proportional to the square root of the number of thermal charges.

When the CCD is cooled to a very low temperature, the silicon's characteristics change and can provoke malfunctions. Thus the charge transfer efficiency can drop. In practice, a CCD is never cooled below $-120°$C. At low temperature the spectral sensitivity curve changes. There is also a shift toward the blue on the infrared side of the response curve. This shift is typically 2.5Å/°C.

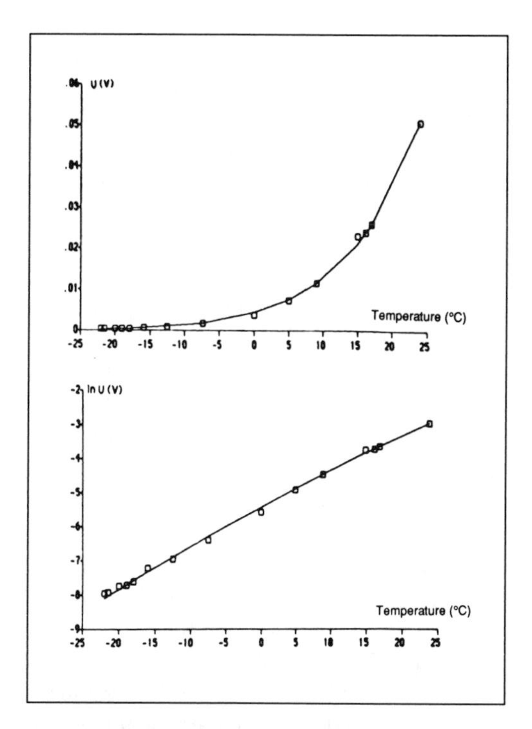

Figure 1.34 *Value of the dark voltage of a TH7852 array as a function of the temperature for a 1 second integration time. At the top the scale of the axis of the voltages is linear; at the bottom it is logarithmic.*

Particularly optimized CCD's, manufactured on special substrates, have remarkable dark current performances. The Tektronix TK512 CCD has a thermal charge production rate of $3500e^-/s/pixel$ at room temperature. This rate means that we reach only 10% of saturation (10^5electrons) in 30 seconds. The same CCD used at 173°K produces $0.01e^-/s/pixel$.

Technology recently developed proves to be even more promising. Researchers started from the fact that most of the thermal signal generated came from the Si–SiO_2 interface (in a more marginal way, charges are created in the depletion zone or diffuse in the neutral volume of the silicon). The thermal signal coming from the Si–SiO_2 interface is higher when the density of impurities in the interface is significant but also when the density of free carriers (electrons or holes) in the interface is low. When it functions normally, a buried channel CCD produces a strong depletion at the interface level and thus repels the carriers from it. It is necessary, therefore, to fill the interface zone with carriers. We can manage to do so by inverting the biasing of the gates, i.e. by applying a negative voltage to them if the underlying channel at the Si–SiO_2 interface is N-type (in this case the holes migrate toward the interface). This bias should be such that the voltage of the Si–SiO_2 interface is less than that of the substrate (V_{ss}).

If we bias negatively and globally all the gates of a CCD, the voltage wells disappear and there is no more accumulation of photocharges nor possible transfers. Today, however, it is possible to build CCD's where there are implants at the level of each pixel, inducing a well while having a phase inversion. This technique is called the multi-pinned phase (MPP). The performance of these CCD's is exceptional. There are components with such a weak dark current that it takes 10 minutes for the detector to saturate with only the thermal charges! The decrease in the thermal signal can be of two orders of magnitude between an MPP CCD and a standard one. This rate of decrease, of course, affects the design of the detector's cooling system which can be much less efficient and therefore less costly in an MPP CCD than in the case of a standard CCD (see Chapter 3). On the other hand, the charge storage capacity of an MPP CCD versus a standard CCD is reduced by a factor of 2 to 5.

Theoretically, the MPP technique can be partially applied to some standard CCD's (in other words, Everyman's CCD). Suppose that we are working with a four-phase CCD. During charge integration it is perfectly possible to invert two phases out of four (typical voltages of -2 to -5V), the remaining phases being normally biased to a positive value to create a voltage well. The performance expected from this arrangement is a decrease in the dark current by half, which is not negligible.

However, most standard CCD's have diodes to protect against static electricity at the level of the clock inputs. These diodes prevent any biasing less than -0.6 Volt, which means that the MPP technique cannot be used.

The MPP technique, now in its infancy, might someday become accessible to the amateur thanks to the possibility of partial inversion.[3] This is a subject for experimentation during cloudy nights.

1.7.9 Sensitivity to Cosmic Rays

Cosmic rays are very high energy particles that can interact with silicon to produce a large quantity of electric charges. Usually only the pixel closest to the impact is affected, the impact showing on the image as a white spot limited to one pixel. Sometimes a cosmic ray does not arrive alone and up to 5 or 6 impacts can be counted on one image. The histogram of the energies of cosmic rays presents a pronounced lobe so that the cosmic ray impacts

[3] *Inversion* is where minority carriers become majority carries—*partial inversion* is where some of the phases are given negative (instead of zero) charge to reduce the dark current, by the process described above, which in some phases converts minority to majority carriers.

produce spots having approximately the same amplitude. Sometimes the ray arrives sideways on the CCD; then its signature on the image is a characteristic streak. Thick CCD's are more affected by the bombardment of cosmic rays because the possibility of interaction is greater in a thick substrate.

The probability of impact increases at high altitudes because the atmosphere is thinner and does not attenuate cosmic rays. At 2800 meters (Pic du Midi Observatory) we have observed up to 100 events/hour/cm^2 with a TH7852 array. At sea level this rate is 2 to 5 times lower.

It should be noted that even the CCD's surroundings can generate radiation. Some glass used for the window in some cameras may be slightly radioactive.

Because the tracks left by cosmic rays look like stars, it is extremely difficult to detect them. Sometimes the very punctual character and the large amplitude of the events help us to distinguish them, but the best method of discriminating them is to take two exposures of the same field and compare them.

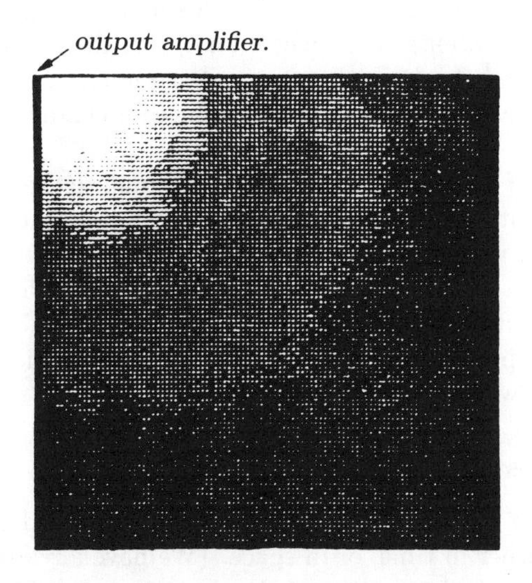

Figure 1.35 *Example of electroluminescence caused by the output amplifier.*

1.7.10 Electroluminescence

The output amplifier of certain CCD's can emit a weak light by electroluminescence. This occurrence is reported fairly often. It is very troublesome because the CCD can be saturated by its own luminous emission in just a few minutes of integration.

The complete solution to the problem is turning off or sufficiently lowering the voltage supplying the amplifier circuit during the exposure. The amplifier is then reactivated only a few seconds before the readout of the target. In this case, a small relay or a transistor (setup as a switch) controls the power supply of the amplifier.

We have never had this problem with a array like the TH7852 because of the technology used (epitaxial substrate) and the image zone and the output stage are completely separated by the memory zone.

1.7.11 Cosmetic Defects

We call a cosmetic defect any defect locally affecting the quality of the sensitive surface. It is common to find isolated pixels with greater dark current than their neighbors (hot spot), a completely insensitive pixel (dark spot), or sometimes a completely defective column. A problem that occasionally comes up is the trapping of charges at pixel level. If the defective site is located in the image zone or in the memory zone, a part of the contents of a column is affected (a trap like this in the horizontal register is a disaster since the contents of the whole image are disturbed). The trapped charges are restored slowly in the points of the image passing through this site during the transfer. This process produces a characteristic comet tail effect coming from the defective pixel.

The number of defects is obviously proportional to the area of the array. Even today, despite progress in production techniques, it is uncommon to find a CCD with no defects. Usually manufacturers offer several levels of quality for a given type of CCD. Since none are perfect, 5 or 6 black points uniformly distributed throughout the image and a defective column can be tolerated. Image processing techniques (cosmetic corrections) can diminish their effects afterwards.

1.7.12 Sources of Noise

Noise appears as random variation of the registered signal about an average value, either in time, or in space. (We have already seen an example with the inter-pixel variation of the dark current.) These fluctuations obviously fundamentally limit detectivity. It is therefore important to understand their origin to be able to decrease their effects.

When the CCD is used at room temperature and if the integration time is longer than about a hundred milliseconds, the preponderant noise is caused by the temporally random generation of the dark current. It is important to understand the difference between a thermal *signal* and thermal *noise*. For simplicity we will restrict this discussion to a single pixel. A long exposure with a moderately cooled CCD will give a certain level of dark current. By holding the CCD's temperature constant and taking a large number of identical exposures it is possible to determine a *mean thermal level.* Then, this mean thermal level, or *signal*, is an offset which is simply added to the useful signal. This offset is not a problem if its amplitude is small in comparison to the detector's dynamic range. An extreme case would be when the detector is so saturated by dark current that it is impossible to acquire a useful signal. The thermal signal may be removed during processing by a simple subtraction. However a warning is in order, the thermal signal's value is spatially dependent—the dark current may be different from one pixel to another.

The thermal noise is far more critical. Again we restrict the discussion to a single pixel. If we make a single exposure in darkness and if we compare the value of this exposure to a mean value determined from a great number of identical exposures we will see a slight difference. This difference is thermal *noise* and is a random component of the signal. This uncertainty about the true level of the signal in a single exposure reduces the detectivity of the sensor.

Statistical laws enable us to quantify the thermal noise. It can be demonstrated that thermal noise depends on the mean level of dark current, i.e., the temperature and the integration time. If N_t is the number of thermal charges created, the standard deviation in number of electrons in the signal registered—also called RMS (root mean square) noise—is written

$$\sigma_o = \sqrt{N_t}.$$

With a dark current of 10^5 electrons, (10% of the saturation), the thermal noise reaches 300 electrons.

When a charge packet moves along the CCD, we have seen that a certain fraction of this packet stays behind at each transfer. This transfer inefficiency varies in a random way as a function of the quantity of charges carried (N_s), of the number of transfers carried out (n), and of the transfer inefficiency (ϵ). If we suppose each transfer to be independent, the noise is written

$$\sigma_e = \sqrt{2\epsilon n N_s}.$$

We should keep in mind that N_s represents the sum of charges produced by photoelectric effect and by thermal effect. In the case of a transfer inefficiency of 10^{-5}, of 300 transfers, and of a quantity of charges carried per packet of 10^5, the noise is 25 electrons.

At the the output stage of the CCD register, the floating readout diode is brought to a reference voltage before the readout of a new packet. A noise (called "reset" noise) is associated with this charge of the output capacitor.

Its value is given by the relation

$$\sigma_r = \frac{1}{q}\sqrt{kTC},$$

where

$q =$ the electron's charge, in Coulombs

$k =$ Boltzmann's constant,

$T =$ the temperature in Kelvins, and

$C =$ the capacitance of the output diode in Farads.

For a temperature of 20°C and expressing C in pico Farads, the relation becomes

$$\sigma_r = 400\sqrt{C}$$

The typical value of C is 0.1 pF. We deduce a precharge noise of the output diode of the order of 126 electrons. In Chapter 2 we will see a method of processing the video signal that suppresses the reset noise (double correlated sampling).

A noise that is extrinsic to the CCD is very important and is called "photon noise". It originates in the corpuscular nature of the light (the arrival of the photons follows a Poisson distribution). The value of this noise is equal to the square root of the average number of signal electrons that collected during integration time. We saw in section 1.7.2 that the number of photocharges depended on the quantum efficiency (Q) of the detector. The signal noise is then given

$$\sigma_s = \sqrt{QN_p} = \sqrt{N_s},$$

where N_p is the number of incident photons, and N_s is the number of photocharges registered.

The CCD is used with additional electronics whose role is to amplify the video signal and then to digitize it for computer processing. These circuits themselves introduce noise. We will consider here that the amplifier noise is negligible. The only noise source taken into consideration and easily calculated results from quantification error during signal digitizing. This operation consists in cutting the analog signal coming from the detector into a finite number of slices numbered from 1 to j. For a given amplitude of the signal, there is a corresponding number of a slice which will be stored in the computer. Thus for an infinity of possible values of the analog signal, digitizing substitutes a finite amount of information. The length of a step is called a quantification step, coding step, or analog digital unit (ADU).

It is handy to have digital coding using a base which is a power of 2. In this case if U_m is the maximum value of the signal that we want to convert, the quantification step p will be given

$$p = \frac{U_m}{2^n},$$

where n is the number of binary elements (bits) used to code the information. In practice, the number of bits is between 8 and 16. The quantification is carried out by an electronic circuit called an analog/digital converter. The input signal is usually a voltage. The relation between this voltage U and the number of quantification steps N observed at the converter output is obviously

$$U \approx p \cdot N.$$

The $\approx$ sign is due to an error which is introduced during digitizing since the exact value of the signal is replaced by an approximate value. In other words, there is an uncertainty about the position of the analog signal inside the quantification step corresponding to this signal. This uncertainty becomes weaker as the quantification step is narrower or, equivalently, is digitized by a larger number of bits.

The quantification error is called quantification noise since it has the character of a random noise. Statistical laws show that this noise is of the form

$$\sigma_q = \frac{p}{\sqrt{12}}.$$

If p is expressed as the number of electrons at the CCD's output, the noise itself will be expressed in number of electrons. To make the contribution of the quantification noise negligible, we have seen that we need a coding step corresponding to a small number of electrons. In professional cameras this means a quantification step representing only several dozen electrons at the CCD's output. Let's take an example. For example, if the maximum value of the signal is $N_{\text{max}} = 10^6$ electrons and if the quantification noise is less than 10 electrons, the minimum number of bits necessary to digitize the signal is computed as follows: the value of the quantification step is

$$p = \frac{N_{\text{max}}}{2^n},$$

the quantification noise is

$$\frac{p}{\sqrt{12}} = \frac{N_{\text{max}}}{2^n\sqrt{12}} \leq 10 \text{ electrons},$$

and if $N_{\max} = 10^6$:

$$2^n \geq \frac{10^6}{10\sqrt{12}},$$
$$2^n \geq 28870,$$
$$n \geq 14.8.$$

To take advantage of the full dynamic range of a low noise CCD requires a 15 bit converter. This corresponds to the performance of a professional camera. In the case of amateur cameras, 12 bit digitizing is the norm because of the cost of the converter and the lengthy duration of the conversion.

To make things clear, we will determine the noise of a CCD poorly cooled to $-20°$C. For this example, we suppose that the integration time is 300 seconds and that during this period of time we will register 10^5 electrons. The quantification step represents 128 electrons (12 bit digitizing in the case of the camera described in this book).

The dark voltage is 0.12V after the 300 seconds of integration. If we suppose that the value of the CCD's output capacitance is 0.1×10^{-12} Farads, this voltage corresponds to a number of electrons equal to (see section 1.4)

$$\frac{0.12}{0.1 \times 10^{-12} \times 1.6 \times 10^{-19}} = 75000e^-.$$

Thermal noise $(\sigma_o) = \sqrt{75000} = 274$ electrons;

transfer noise $(\sigma_e) = \sqrt{2 \times 10^{-5} \times 300 \times (75000 + 10^5)} = 32$ electrons;

reset noise $(\sigma_r) = 0$ (double sampling);

signal noise $(\sigma_s) = \sqrt{10^5} = 316$ electrons; and

quantification noise $(\sigma_q) = 128\ /\sqrt{12} = 36$ electrons.

The total noise is obtained by taking the quadratic sum of all the elementary noises:

$$\sigma_t = \sqrt{274^2 + 32^2 + 316^2 + 36^2} = 421 \text{ electrons}.$$

In the absence of a luminous signal, the preponderant noise is caused by the dark current.

We call *readout noise* the noise measured when the thermal signal is almost nil and when the CCD is in the dark. In addition, we will always consider that the readout noise includes not only the CCD's intrinsic noise but also that of the associated electronics.

When the CCD is cooled to about $-100°$C, the dark current becomes negligible and the total noise in darkness is not more than a few dozen electrons. It is relatively common to find a CCD with a readout noise of the order of $10e^-$ (besides reset noise) at the focus of a professional telescope. On the basis of what is being done in laboratories, CCD's at $2e^-$ are being forecast. This remarkable characteristic, associated with a quantum efficiency greater than 50%, makes the CCD a nearly perfect detector. And things probably won't remain there! Nondestructive readout of the floating diode may eventually reach a noise of less than 1 electron. The principle is to give a back and forth motion to the charge packet that is positioned between two electrodes during its reading. One of the electrodes is none other than the floating diode. It is thus possible to read the same packet successively several times. For example, with this technique the noise can be reduced by a factor 10 by averaging 100 measurements of the packet.

1.7.13 Experimental Determination of Noise

It is important to determine camera noise experimentally to detect any anomalies and to optimize its operation by proper adjustments. Thermal noise can only be decreased by cooling the CCD. This is an important problem that we will discuss in Chapter 3.

The CCD's readout noise depends on many factors such as the values of the DC biasing voltages, the form of the clock signals, the stability of the amplifier's gain, parasitic coupling in the connections, etc. There are many parameters that have to optimized.

Let us first recall some definitions. The average value of the signal provided by an image pixel is calculated

$$\bar{y} = \frac{1}{n}\sum_{i=1}^{n} y_i$$

where y is an element of a set of n measured values.

The noise power is given by the calculation of the variance

$$V = \frac{1}{n-1}\sum_{i=1}^{n}(y_i - \bar{y})^2.$$

The RMS noise is equal to the standard deviation

$$\sigma = \sqrt{V}.$$

The standard deviation expresses the degree of dispersion of the sample values compared to their average value. Thus the interval having the half-width of a standard deviation relative to the average value contains 68%

of the samples. We find 95% and 99.7% of the samples respectively in the intervals of more or less 2 and 3 standard deviations.

Let's return to the CCD. If N is the number of electrons contained in a packet of charges, and q is the electron's elementary charge, the packet's total charge will be written

$$Q = N \cdot q$$

with $q = 1.6 \times 10^{-19}$ Coulomb.

The total charge can also be expressed as a function of the capacitance C of the CCD's output diode and of the observed voltage U

$$Q = C \cdot U$$

from which

$$N = \frac{C \cdot U}{q}.$$

The RMS signal noise expressed in number of electrons is written

$$\sigma_s = \sqrt{N} = \sqrt{\frac{C \cdot U}{q}}.$$

We will call the gain, the voltage equivalent of one electron

$$g = \frac{q}{C}.$$

This relation allows us to convert the noise expressed in number of electrons into a noise expressed in volts. In these units the power of the noise becomes

$$\sigma^2_{s(\text{volt})} = \left(\frac{C \cdot U}{q}\right)\left(\frac{q^2}{C^2}\right) = \frac{q}{C}U = g \cdot U.$$

In practice, different operational noises are added to the signal noise. We will put these additional noises (reset noise, thermal noise, etc.) under the generic name of readout noise σ_l, the formula for which is

$$\sigma^2_{s(\text{volt})} = g \cdot U + \sigma^2_{e(\text{volt})}.$$

For different values of the illumination of the CCD, we observe U and the variance σ_s^2; then we plot the curve σ_s^2 = f(U), called *photon transfer curve*. The ordinate at the origin of this curve gives us the readout noise of the CCD (we assume the dark current to be negligible, either because the detector is sufficiently cooled or because the integration and readout times are short). This noise can be obtained more simply by measuring the fluctuation of the signal coming from a given pixel on a large number of successive images taken in complete darkness.

The slope of the curve $\sigma_s^2 = f(U)$ is merely the gain g. From it we deduce the value of the CCD's output capacitance and from there the relation between the voltage measured and the number of electrons contained in the corresponding packet of charges.

$$C = \frac{q}{g}.$$

The transfer equation can also be written in analog digital units. Let N be the signal in number of electrons and let g' be the gain of the whole amplifier + analog/digital converter expressed in analog digital units per electron. The signal measured in ADU at the output of the converter will be

$$S_{(\mathrm{ADU})} = g' \cdot N.$$

The signal noise expressed in analog digital units is written (we suppose that the electronics don't add any noise)

$$\sigma_{s(\mathrm{ADU})} = g' \cdot \sqrt{N}.$$

The variance is then

$$\sigma_{s(\mathrm{ADU})}^2 = g'^2 \cdot N.$$

Let us take the ratio of the variance to the measured signal

$$\frac{\sigma_{s(\mathrm{ADU})}^2}{S_{\mathrm{ADU}}} = \frac{g'^2 \cdot N}{g' \cdot N} = g'.$$

We find an expression similar to that already defined. Thus

$$\sigma_{s(\mathrm{ADU})}^2 = g' \cdot S_{\mathrm{ADU}}.$$

As before, in this equation we have to include noises that don't depend on the signal. Thus

$$\sigma_{\mathrm{ADU}}^2 = g' \cdot S_{\mathrm{ADU}} + \sigma_{1(\mathrm{ADU})}^2.$$

This form of the transfer function is especially interesting since the inverse of the slope $(1/g')$ directly gives the number of electrons corresponding to an analog digital unit.

The transfer curve has to be obtained with a very stable luminous source. Professionals use tritium radioactive sources that emit weak light with a stability of 10^{-6}/hour. The amateur will have to be satisfied with illuminating the CCD with a LED connected to a well regulated power supply. For correct results at least 100 measurements have to be made on a pixel and the variance calculated on these 100 values. The precision is increased if the variance of several pixels is measured simultaneously. Typically, the resolution of the analog/digital converter should be sufficient to detect a tenth of a millivolt at the output of the CCD but this value depends upon the gain of the amplifier.

We should finally note that the normally straight-line of the photon transfer function can curve by taking a null slope for very weak illumination (only if it is possible to obtain a very low readout noise). The phenomenon can be attributed to non-linear effects from charge trapping (see section 1.7.4).

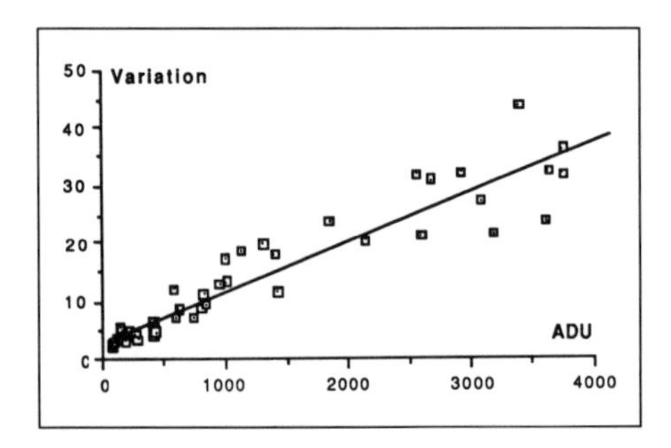

Figure 1.36 *Transfer curve of a camera equipped with a TH7852 array.*

Figure 1.36 shows an example of experimental determination of the transfer straight-line for a Thomson TH7852 array and a 12 bit converter. The detector's temperature was about $-20°$C. The distribution of points about the best fit straight-line, calculated by linear regression, shows the difficulty of obtaining the transfer curve.

The slope of the straight-line is 0.009 from which we get the number of electrons corresponding to an ADU

$$\frac{1}{0.009} = 111 \text{ electrons.}$$

The full scale of the 12 bit analog-digital converter is 10 V. With an amplifier gain of 22, the ADU taken at the output of the CCD has the value

$$\frac{10}{4095 \times 22} = 1.1 \times 10^{-4}\text{V}.$$

From this we can calculate the capacitance of the output diode by the formula

$$C = \frac{qN}{U} = \frac{1.6 \times 10^{-19} \times 111}{1.1 \times 10^{-4}} = 0.16 \times 10^{-12} F = 0.16\text{pF}.$$

This is the typical value forecast by Thomson for its arrays.

The variance at the origin (CCD in darkness) is 2.0 ADU, a standard deviation of 1.4 ADU. Since we know the number of e^- corresponding to an ADU ($111e^-$) we can deduce that the readout RMS noise expressed in number of electrons is $1.4 \times 111 = 155$. With a working temperature of -45°C, the noise is brought to about $100e^-$RMS.

1.7.14 Some Useful Equations

The first equation links the CCD sensitivities expressed in Joules (energy) and in Watts (power). $\mathrm{T_i}$ is integration time.

$$\text{Response}\,(\mathrm{V}/\mu\mathrm{J}/\mathrm{cm}^2) = \frac{\text{Response}\,(\mathrm{V}/\mu\mathrm{W}/\mathrm{cm}^2)}{\mathrm{T_i}(s)}.$$

The second equation establishes a relation between the response expressed in $V/\mu W/cm^2$ and the response expressed in A/W.

$$\text{Response}\,(\mathrm{V}/\mu\mathrm{W}/\mathrm{cm}^2) = \frac{10^6 \times \text{Response}\,(\mathrm{A/W}) \times \text{Pixel Surface}\,(\mathrm{cm}^2)}{\text{Diode Readout Capacitance (pF)}}.$$

The next equation calculates the ouput voltage in V as a function of the value of the output diode capacitance and of the number of electrons contained in a packet.

$$\text{Output Voltage (V)} = q \times \frac{\text{Number of Electrons}}{\text{Diode Readout Capacitance (F)}}$$

with q the electron's charge (1.60×10^{-19} Coulombs).

The last relation lets us compute the quantum efficiency at a certain wavelength as a function of the detector's response in A/W at this wavelength.

$$\text{Quantum Efficiency} = \text{Response(A/W)} \cdot \frac{h \cdot c}{q \cdot \lambda}$$

with

h = Planck's constant (6.63×10^{-34} J.s)

c = the speed of light (3×10^8 m/s)

The term $(h \cdot c)/\lambda$ represents the energy of the photon in Joules (wavelength expressed in meters).

By replacing the numerical values and expressing the wavelength in micrometers and the efficiency in percentage, we find

$$\text{QuantumEfficiency}\ (\%) = 124.0 \times \frac{\text{Response (A/W)}}{\text{Wavelength}\ (\mu\text{m})}.$$

Figure 1.37 *The barred spiral galaxy NGC 7741. The exposure was 600 seconds using a 280 mm f/6 telescope. The CCD is Thomson-CSF's model TH7863. Note the blooming streak originating at the brilliant star (the CCD is read from left to right).*

Figure 1.38 *A good example of sky pollution. During the observation of the galaxy pair NGC 7443-4, an airliner left traces of its signal lights on the CCD even though the field covered by this image is only about 13 arc minutes. Observation was made with a 280 mm telescope and a TH7863 CCD. The integration time is 600 seconds, but the duration of the passage of the airliner was only a fraction of a second.*

Figure 1.39 *ARP 77 (NGC 1097) is a spectacular galaxy in Fornax. Note the distortion of the spiral arm due to a small companion galaxy. 5 minute integration time using a TH7863 CCD with the 24-inch T60 telescope at Pic du Midi Observatory.*

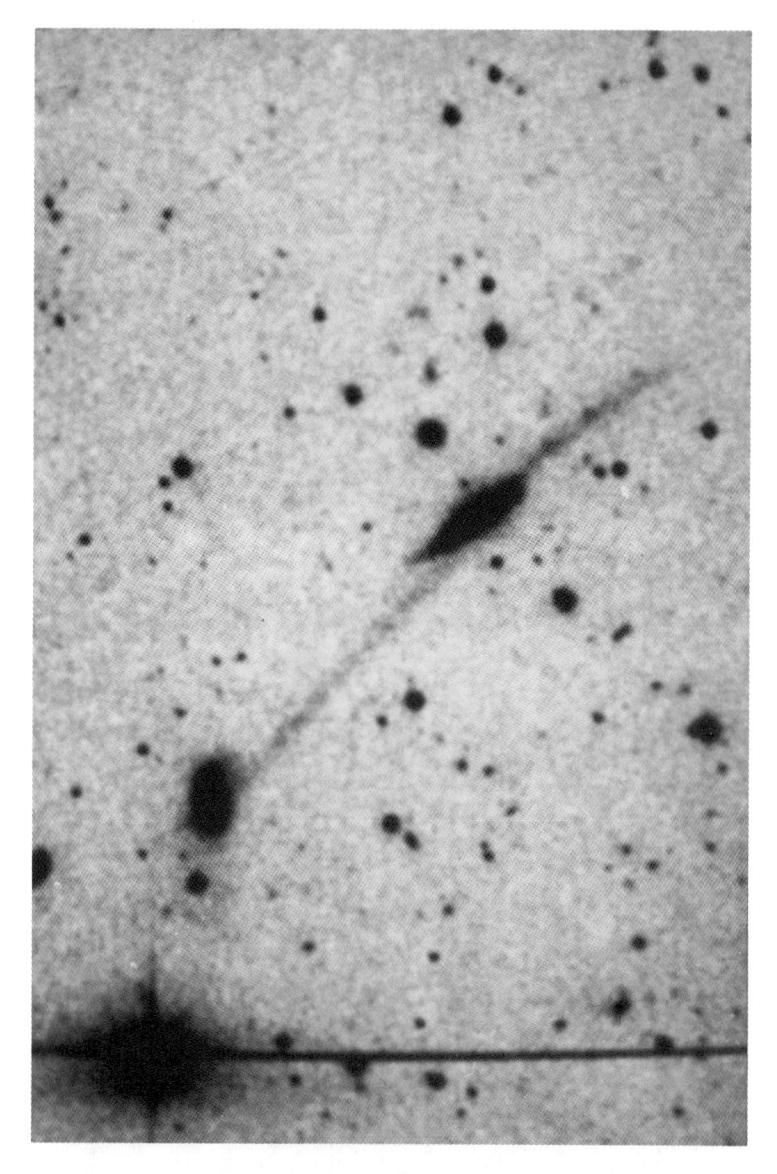

Figure 1.40 *ARP 295 is an impressive interacting pair of galaxies in Aquarius. 10 minute integration time using a TH7863 CCD with the 24-inch T60 telescope at Pic du Midi Observatory.*

Chapter 2

The Electronics of a CCD Camera

2.1 Introduction

All astronomical CCD cameras share the following characteristics:

1. a CCD image detector,

2. electronic circuitry to transfer the CCD image to a computer,

3. a computer with software to display (visualize) and process the images,

4. a cooling device to reduce the CCD's dark current during long exposures of faint objects.

The CCD and the computer complement one another. While a video signal can be easily digitized, the resulting files are large and complex. A computer can manipulate these large digital image files with an ease unheard of in photography.

The designer of a CCD camera should recognize that it is an interactive tool. With it, one should be able to visualize the image and process it very soon after the exposure. This capability is a major asset because if a problem should be found "in real time", the acquisition can take place again immediately. The notion of interactivity also includes the ease of image manipulation using mass storage devices (floppy disk, hard disk, magnetic tape unit, etc.).

The need to cool the detector is certainly one of the major difficulties in the construction of a CCD camera. Cooling techniques are generally poorly understood, and if a design has not been well thought out, it is easy to end up with a huge, complicated system. Chapter 3 is, therefore, devoted to cooling the detector.

Figure 2.1: *The observatory of the ALCYONE Association from which many of the images illustrating this book were taken. This observatory, run by Paul Bertincourt, Guylaine Prat, Eric Thouvenot and the author, is located not far from Toulouse. The CCD camera (TH7863) is mounted at the focus of a 280 mm Schmidt-Cassegrain telescope. The long white tube is the guiding telescope. The acquisition computer is on the floor below.*

2.2 The Main Choices

2.2.1 CCD Selection

Worldwide there are about ten CCD manufacturers. Some of the more familiar are: Thomson, Tektronix, GEC, Ford Aerospace, Reticon, Kodak, Texas Instruments, and Fairchild. Although this number is relatively small, it is big enough to pose problems when it comes selecting a CCD—every manufacturer offers a wide range of different products.

The first consideration in the CCD selection process is cost. A camera that will perform at professional levels requires a sizable investment. A Tektronix 2048 × 2048 array currently (1991) costs about $5,000. Professional observatories often use CCD's that were specifically developed for astronomy and often they represent the cutting edge in technological achievement.

Since the astronomy market is very small and manufacturers must recoup their costs, selling prices extend into thousands of dollars.

On the other hand mass produced CCD's for industrial or consumer use are available at very low prices, well below $100. CCD's manufactured in Japan by Sony and others fall within this category. However, do not expect exceptional performance from these components. The opto-electronic characteristics needed for astronomy can be only partly found on these inexpensive products. Their limitations—cosmetic defects that can be corrected by supplementary circuits in the camera, reduced dynamics, low sensitivity, limited spectral range, too small pixels, shape of the pixel not square, etc.—are all problems when the CCD is applied to astronomy.

Fortunately for the amateur astronomer, there is a middle ground between the manufacturers of custom-made, world-class CCD's and the mass manufacturer who judges output by the pound. Thomson in France and EEV in England are two companies that specialize in custom-made CCD's but have production runs large enough that the prices are reasonable for our purposes. They manufacture components primarily for imaging in medicine, space, nuclear physics, and astronomy. Their overall volume allows them to sell excellent quality devices like the TH7852 with 144 $\times$ 208 pixels for about $350. A Thomson array of 288 $\times$ 384 pixels (TH7863) costs about $1,000. These prices depend on the quality desired and the purchase quantity. Thomson, like GEC, also custom manufactures components, some specifically for astronomy applications—thinned CCD by Thomson, UV coating by GEC—but the price for these devices are usually prohibitive for amateurs. For a given type of CCD, manufacturers offer different classes of quality corresponding to the number of defects found on the sensitive surface. Among the more common problems are insensitive pixels or lines, hot spots, and uneven sensitivity along the surface.

Besides the cost, the size of the array is another factor to be considered. Initially, we might seek the largest sensitive surface possible. But largeness is not necessarily an advantage that we can effectively exploit: the quantity of information to be processed grows geometrically as the CCD array increases in size. A digital image of 144 $\times$ 208 pixels coded on two bytes contains about 60KB (1KB = 1024 bytes), a 288 $\times$ 384 array represents 220KB, and the biggest array currently in production, a Tektronix device with 2048 $\times$ 2048 pixels needs a storage capacity of more than 8 MB for a single image! Ford Aerospace has announced an even bigger array—4096 $\times$ 4096.

Having cast a dark shadow on bigness let us note here that there are clever ways to deal with a big array and a small computer. For example it is always possible by electronic or computer methods to process only a part of a large array (windowing). Using the entire surface of the detector can then be put off until a sufficiently powerful computer system has been acquired.

Depending on the type of usage intended, either matrices or linear arrays will be used. The latter are preferred in spectroscopy, the spectrum being spread along the photosensitive line. Thomson, Reticon, and others make special components for this application.

2.2.2 The Electronic Architecture

The electronics associated with a CCD can be arranged into several subgroups:

1. generation and shaping of the clocks,
2. generation of the biasing voltages, for the CCD as well as for some of the electronic parts,
3. amplification of the CCD's low amplitude video signal,
4. analog/digital conversion (ADC) of the video signal to create digital images that can be handled by a computer, and
5. interface with the computer.

There are systems designed to directly digitize a video signal generated by a standard TV camera. It therefore might seem possible to use a camcorder with a computer. However, without special electronics, it is still impossible to take long exposures. A camcorder can, because of its frame rate, record only down to third magnitude with a 200mm telescope. Furthermore, the digitization frequency imposed by the TV standard is such that it is difficult to work with resolution higher than 8 bits in intensity (256 levels) which is inadequate for most astronomical work. Rapid digitizers working with a large number of bits are beginning to be sold, but they are expensive and their digitizing precision, which is limited by floating reference levels, is always lower than what we should expect from an astronomical CCD camera. A video cassette recorder is also not an alternative primarily because dynamics, noise, and nonlinearity prevent any serious processing of images.

Based upon the discussion above one can conclude that the astronomer has two choices: purchase a CCD camera from a specialized manufacturer or assemble one from available parts. Knowing the budget of most amateur astronomers, we will explore the second alternative, presently the most economical.

Without claiming to be exhaustive, we will briefly review types of architecture for CCD cameras that can be used in astronomy.

Architecture 1

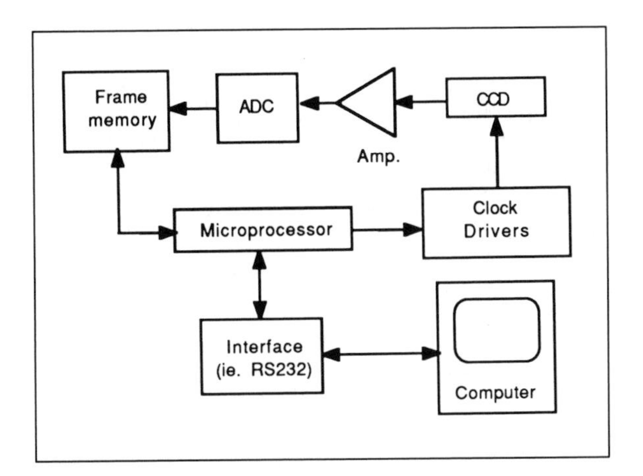

Figure 2.2 *Overview of a camera based on the Architecture 1. Here, the microprocessor has a central role. It provides the pathway between the image memory and the computer and between the image memory and the ADC. The computer also creates the clock timing sequence by software.*

In Architecture 1 (see Figure 2.2), the basic element is a memory plane, or image memory, into which the digitized CCD signal is sent. This memory is then read by the computer via a standard serial (RS232) or parallel (Centronics, IEEE488) interface. The interface as well as the image memory is managed by a microprocessor. All of this can be managed by a single chip microcomputer containing dynamic memory, interfaces and even a language. An advantage of this architecture is that the camera can be made to work with any computer having a standard interface, i.e., almost all computers presently available. Except for the IEEE488, which is difficult to set up, these interfaces are very slow. It will take at least 3 minutes to transfer a 200KB image by 9600 baud RS232 interface (1 baud = 1 bit/s in a single serial line interface). Obviously this limits interactivity and is a great handicap.

The main advantage of an image memory independent of the computer is that we can design a system in which the image is examined by the computer while the next one is being exposed.

Architecture 2

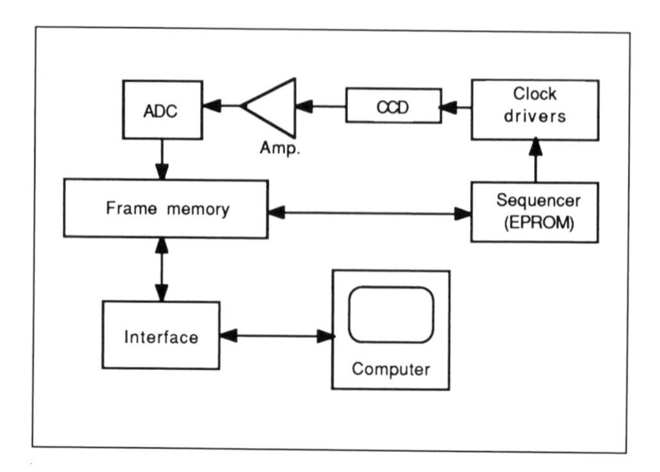

Figure 2.3 *Overview of a camera based on Architecture 2. Note the link between the image memory and the sequencer that synchronizes these two elements.*

In Architecture 2 (see Figure 2.3) we retain the image memory found in Architecture 1, but this time it is read by the computer via an optimized non-standard link. Should we change computer systems, we merely have to modify the interface, which is not usually a problem. The interface can be very fast, eliminating the main disadvantage of the Architecture 1.

If we require functions such as variable integration time or several clock timing sequences, the electronic diagram becomes complex with Architecture 2. An excellent way to simplify the electronics is to use an EPROM to hold the sequence of commands to be sent to the CCD. By incrementing the address bus of the EPROM we can create clock signals on the EPROM's data bus pins. Since the EPROM's addressing can be cyclical, we can digitize several successive frames. To change the type of clock timing sequence, we modify the programming of the EPROM. This solution requires an EPROM programmer, i.e., a device that is fairly specialized but not very expensive and simple enough to be a made at home (diagrams of EPROM programmers are frequently published in electronics magazines).

To improve the flexibility of Architecture 2 we can replace the EPROM with RAM memory whose contents are loaded from the computer. In the same way, we can use *timers* that can be programmed in cascades (e.g. Intel timer IC model number 8254). With the right programming, these circuits can produce a relatively complex clock timing sequence inexpensively.

Architecture 3

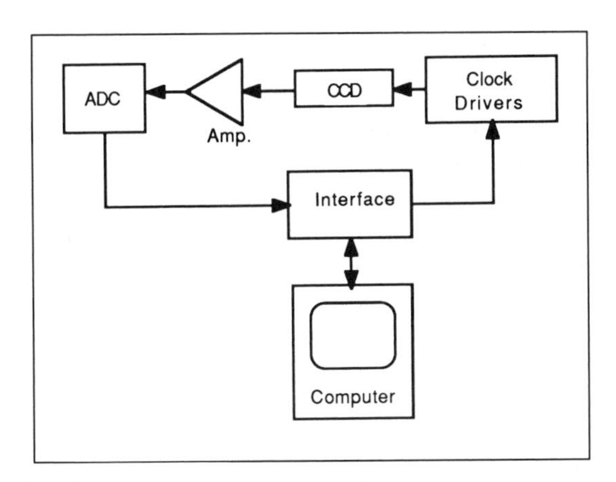

Figure 2.4 *Overview of a camera based on Architecture 3. This is the simplest configuration. The computer simultaneously acts as a sequencer and as image memory.*

With Architecture 3 (see Figure 2.4) the digitized image is transferred directly into the computer's central memory, either by a DMA (direct memory access) channel or through a parallel input/output circuit (PIO = peripheral input/output). This architecture is very dependent on the type of computer used. On the other hand, transfer problems are simplified because an intermediary element, the image memory, is eliminated. Getting rid of the image memory is beneficial because producing it is always difficult and expensive.

Architecture 3 is the easiest to put into practice. It also offers great flexibility since the computer itself generates the clocks through software. The CCD's operation can be adapted to very different situations by changing a few lines of the program, which is quicker and less expensive than wiring circuits or modifying the contents of an EPROM. On the other hand, the readout speed is dependent on the speed of machine used. For example, it is less than 1MHz for an AT PC unless the DMA mode is used but at the price of greatly complicated electronics. Still, given the long integration times commonly used in astronomy, readout time is not very critical (except in solar and sometimes planetary observations). Other factors which will limit our speed are certain types of noise which decrease with low digitization speed. We will spend considerable time later on ways to reduce noise.

For our cameras we adopted the third architecture, which will be explained in detail.

2.2.3 Choosing a Computer

Several elements must be considered when choosing a computer system: performance, interfacing ease, abundance of documentation and software, durability, and price.

To carry out most processing algorithms the computer's dynamic memory must be large enough to hold two or three images simultaneously. For instance an image generated with a Thomson TH7852 array requires nearly 200KB of memory. An 8 bit micro-computer like the 6800 or the 8080 cannot handle this. With larger arrays even larger amounts of memory are required which cannot be handled efficiently with the memory addressing overlay technique used in the MS-DOS systems. Further, with MS-DOS one cannot directly manage more than 640KB. But even this amount of memory is insufficient to comfortably work with images larger than 150KB. Fortunately, EMS (Expanded Memory Specifications) techniques can go beyond the MS-DOS 640KB limit—up to several megabytes. Using an EMS memory is unwieldy but relatively simple (there is much information available on the subject today). The OS/2 system lets one break the 640KB limit more elegantly, but there is a high price to pay because an 80386 microprocessor and 3 MB of memory are necessary. Other more recent machines do not have the same addressing problems—for example machines (like Macintosh II or NeXT) that are equipped with the 68000 series 32 bit microprocessor manufactured by Motorola.

Considerable data storage capacity is required when working with CCD images—one can easily acquire more than 10 MB of preliminary data in a night of observation. A hard disk is indispensable and one with a capacity of 40 MB lets the astronomer work comfortably. Archiving will be a big problem for the active observer—expect to store your data on a lot of floppy disks! While reasonably reliable, many floppy disks become a logistic nightmare—a cartridge tape drive may be a desirable convenience. A 60 MB tape drive now sells for about $400 with the price declining and capacities increasing steadily. But the most appropriate solution to the storage problem may be coming in new storage devices—digital optical disks, either WORM type (Write Once Read Many) or rewritable. The cost of these devices is still high, but since a single disk can contain about 500 MB of data, the price per byte can be competitive. Even more interesting are DAT tape units capable of holding nearly a gigabyte on a tiny cartridge costing about $15.00 *and* with access to any file on the tape in 20 to 30 seconds!

Processing speed is linked not only to the type of microprocessor (8, 16 or 32 bits) but also to its clock frequency. Therefore, XT type PC compatibles, that run at 4.77MHz, are impractical for image processing because of their slowness. A 80286 PC AT with a 8MHz clock frequency is about the minimum for processing images of average size (256 × 256 pixels). For larger images a PC equipped with a 16MHz 80386 is not a luxury. While all the digitized images can be stored as integers, many processing methods require the data to be handled as real (or floating-point) numbers—this almost dictates the use of a math coprocessor. A math coprocessor can speed up processing by a factor of 10 when working with real numbers. Most micro-computers have a coprocessor socket mounted on the mother board. Installing a math coprocessor therefore requires only the insertion of an LSI-sized-chip.

Graphic monitors are an absolute necessity for CCD image processing. The number of gray levels is as important as the monitor's overall resolution. With less than 16 gray levels, the gaps between each level are noticeable and produce contours that harm the realism of the image (it might not, however limit its interpretation). At about 64 gray levels, the eye has difficulty distinguishing between two successive levels. High quality graphics boards do not code gray levels beyond 8 bits (i.e., 256 levels).

On the other hand, determining the optimum number of colors is more difficult. If images are to be displayed in pseudo-colors or false colors, a technique often used for a more quantitative representation of the image than by gray levels, 16 simultaneously displayable colors is sufficient. However, to visualize an image in true colors, for example from 3 exposures made in the 3 basic constituents (red, green, blue), one needs a much wider palette. Ideally, each of the three colors should be coded on 8 bits (256 levels of blue, 256 of green, and 256 of red), and the three image-planes that result should be displayed simultaneously on the screen. On some micro-computers we are beginning to see graphics boards with these capabilities (24-bit resolution).

The number of displayable points on the monitor determines the image sharpness. Sometimes the number and arrangement of these screen points is insufficient to properly display the image. For example, the vertical and horizontal resolution may not be the same—the screen pixel may not be square. Under these circumstances a distorted image is displayed—the shape differs according to the axis under examination—this effect is quite noticeable when what should be a circle appears as an oval. There are two remedies for this problem: to produce scaling in the software to compensate for the distortion introduced by the screen, or to correct the expansion of the image by adjusting the height and width settings of the monitor. Prior to purchasing a monitor one should be sure that these "fixes" can be accomplished satisfactorily.

A monitor resolution of 320 × 200 points represents the minimum for acceptable image visualization. The VGA standard (IBM video graphics array) resolution is 640 × 480 and is quite satisfactory. VGA graphics boards can be purchased for less than $250. A VGA board can have a palette of 16 simultaneously displayable tones selectable from among 262,000 which are displayed with a resolution of 640 × 480. VGA boards also have a 320 × 200 mode which can display 64 levels of gray or 256 colors simultaneously which in turn can be selected from a palette of 262,000 colors. Today there is a wide choice of "super VGA" boards capable of displaying 256 colors or 64 levels of gray simultaneously with a resolution of 640 × 480 or even 800 × 600. These boards have spectacular capabilities, but unfortunately such VGA extensions are not standardized and consequently each requires unique programming.

For Apple's Macintosh II, there is an assortment of graphics boards with exceptional performance suitable even for professional observatories. These boards are expensive but in line with the cost of the Macintosh. Some computers have perfectly good graphics boards as standard equipment. Good examples are the Atari ST and especially Commodore's Amiga.

To exploit the capabilities of a VGA graphics board an analog monitor is a necessity. The best choice is a multi-frequency monitor which automatically adapts to the resolution of the graphics board. A wide frequency bandwidth is also a necessity when working with resolutions of 800 × 600 or 1024 × 768.

The computer should be capable of accepting a wide variety of interfacing methods. Standard interfaces such as the RS232 are too slow and inflexible to be usable with a CCD. We therefore need access to the system's buses. When working at the bus level, address decoding and CPU conflicts can arise. To deal with these potential problems good documentation about the machine is necessary. Probably the best bus in this regard is the IBM-PC. This machine has a dedicated interface bus for peripheral boards—installing an interface board in these machines is reasonably simple and reliable—provided you set the hardware and software switches properly. From a practical point of view this bus has become the industry standard and is fully documented in readily available literature from a wide variety of sources.

Today, a major trend in the computer industry is toward "intelligent" buses. In these machines, when a new board is added, the central processor dialogs with the new board to configure it to operate within the system. This has advantages in that it allows multiprocessing (and many other features) but it requires a much more complex interface board. This can be particularly troublesome when designing and debugging prototype boards. Some chip manufacturers offer sets of integrated circuits that simplify the construction of such boards. This type of bus is found in machines such as the Macintosh II (Nubus) or IBM's PS/2 (MCA bus). Other interface possibilities exist via industrial buses such as the VME (Motorola), but for the time being the cost of boards running in such environments is prohibitive.

The documentation for the computer must be complete and of good quality. As noted above the PC architecture is well documented and data are available from a variety of sources—there are hundreds of good books describing this machine from all points of view. This is true for both hardware and software. Apple's Macintosh is also well documented, especially the software (the *Inside Macintosh* series edited by Apple is a gold mine of information for programmers).

Sometimes detailed documentation of a computer is provided only to those who have the status of a developer for the company. This status is not given to everyone, and it should be considered before choosing a machine.

The quality and quantity of programming languages is an important consideration when choosing a computer. It is almost impossible to acquire and process images with a program written in an interpreted language. A compiler is therefore needed for a satisfactory execution time. Among compilers, whatever the language used, execution speeds are very close. At the moment C language is very popular because the machines' capabilities can be fully utilized—in certain cases its speed is close to assembler language. However C is not a panacea. While its flexibility is appealing to system programmers, it can be bewilderingly complex to all but an expert programmer. Until recently BASIC was shunned by professionals, but remarkable compilers such as Borland's Turbo Basic, Microsoft Basic 7.0 for PC and Macintosh, and Atari's GFA Basic are changing professional attitudes. These BASIC's have numerous extensions of the fundamental language, borrowing quite heavily from Pascal and C, and thus seem rather hybrid, but they are extremely efficient yet easy to learn and to use. Pascal has been reinvigorated with the introduction of Turbo Pascal by Borland and now has quite a few users.

I personally use Microsoft BASIC Professional Development System 7.0 which has many commendable features. Its language is extensive and has no known "bugs" (errors). It accommodates tables bigger than 64KB, and there is an excellent error diagnostic program. Moreover you can call routines written in other languages: Assembler, C, FORTRAN, and Pascal.

Today's compilers are so efficient that it is seldom worth resorting to writing instructions in Assembler. For example, routines that create the clock timing sequence in one of my cameras are easily carried out directly in MS Basic 7.0 on a 8MHz PC. The added effort required to write in Assembler may be justified only in very time sensitive applications such as image display.

The astronomer should avoid computers that do not enjoy a large user base. A machine that has few users or is no longer sold gets no new software and can therefore no longer evolve. For this reason, and despite their age, even machines based on IBM-PC 8086 architecture are still a good investment because the large number in service is the guarantee of their durability and software support. In fact, because of this huge user base most software written for the 8086 will run on an MS-DOS 286 or 386 based machine.

The more machines there are of the same type, the easier it is to exchange data with other users. This is particularly important if one wants to make an image bank. It is also desirable that CCD camera users work with the same image processing software to facilitate comparisons. For these reasons the AT compatible computer will be adopted as this book's "standard" computer.

2.3 The Electronic Layout

In this section we will describe the complete electronic circuit built around a Thomson TH7852 CCD. It is fairly easy to adapt this circuitry to more complex CCD's. As an example, at the end of this section we will give some information about setting up a larger CCD—a Thomson TH7863.

All the electronics are contained on five boards:

1. A CCD board. It supports the CCD and is therefore placed in the optical head of the camera;
2. A board to amplify the video signal, located immediately next to the optical head;
3. An analog-digital conversion board;
4. A clock board;
5. A computer interface.

The links between these different boards are shown on Figure 2.5.

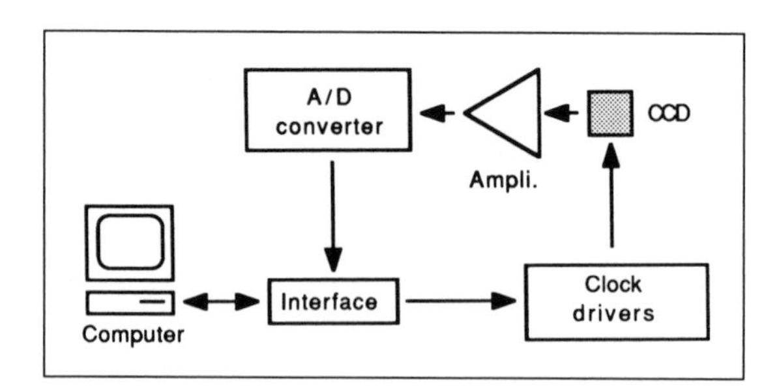

Figure 2.5 *Overview of the electronic layout for the CCD camera described in this book.*

The kind of CCD used obviously determines the electronic diagram; that is the reason why we will first look at the main characteristics of the TH7852.

2.3.1 The Thomson TH7852 CCD

The Thomson TH7852 will be the CCD around which our camera is built.[1] This component delivers high performance at low cost (about $350). In addition, the transfer mode is two-phased, which greatly simplifies the circuit design, especially when compared to 3 or 4 phase CCD's. However, the photosensitive retina is small, 6.2 × 4.3 mm.

The TH7852 is a front side, buried channel frame transfer type CCD (see Figure 2.6). It also has an anti-blooming device. The data sheets specify a usable array of 288 lines and 208 columns. Normally, this array is used to generate two interlaced TV frames. For our purposes we will sequentially select first one of these frames and then the other. At each integration the high levels of the phases Ø1P and Ø2P will toggle the image zone to enable us to "artificially" double the vertical resolution of the CCD.

The TV interlacing mode cannot be used in astronomy because one single frame forms the final image (single shot functioning). Consequently, during integration, the same phase is always set at high level, either Ø1P or Ø2P, according to the wishes of the user. This has the effect of reducing the number of useful lines to 144. Note that the pixel is formed by a pair of electrodes (Ø1P and Ø2P) each of which in turn is composed of two zones. Further, only the zone collecting a maximum of charges is shifted by half a pixel.

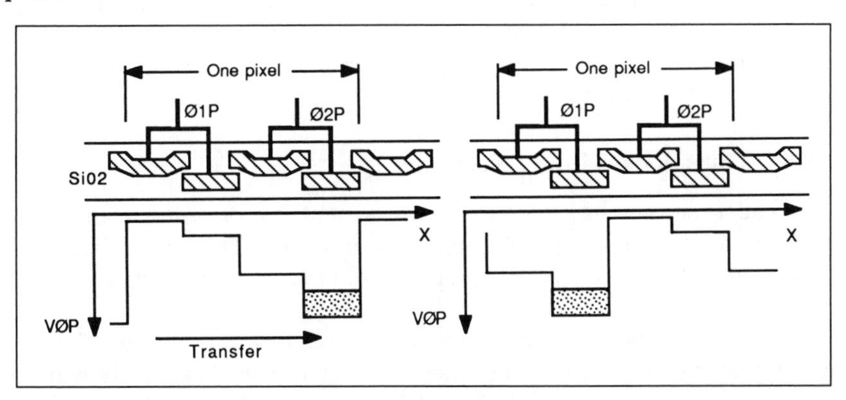

Figure 2.6 *Detail of a photosite of the TH7852 array. The pixel area has a two-phase structure. The collection zone shifts by a half-pixel when alternately biased by Ø1P and Ø2P during the exposure.*

[1]See Appendix B—Sources of Components for address.

Figure 2.7 shows a pixel seen from above. The inter-pixel distance is 30×28 μm, but because of the presence of the anti-blooming drain, the useful surface of the pixel is about 30×19 μm.

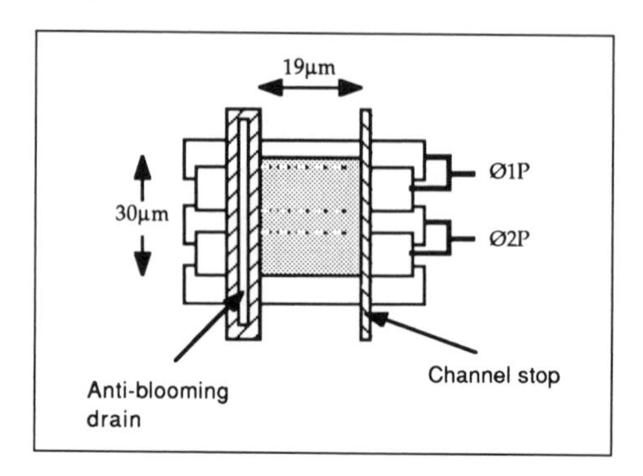

Figure 2.7 *A pixel seen from above. The photosensitive part is shown by the shaded area.*

The TH7852 has lines and columns that are either not electrically connected or concealed from light. Some of these are used to insulate the photosensitive part of the CCD from zones outside the chip, while others are dark references. Lines dedicated to these purposes have the effect of reducing the TH7852 CCD array to 145×218 light sensitive pixels.

Mechanically, the detector is packaged in a dual-in-line 24 pin ceramic case. Its dimensions are 30mm long, 15mm wide, 3mm thick. The light enters the device through a glass window. The space between the window and the array contains an inert gas. The CCD's pin spacing ($^1/_{10}$ inch) and the distance between the two rows of 12 pins (0.6 inch) match standard integrated circuit mounts. Not all CCD's have this feature and some need special hard-to-find mounting hardware.

2.3.2 The Clock Timing Sequence

Recall that the clock waveforms are a sequence of orders that direct the CCD to execute integration and readout stages. This sequence can be graphically represented by a series of high and low clock levels as a function of time.

We have to know how to interpret the clock timing sequence found in the manufacturer's data sheets. The clock timing sequence is usually defined for a television mode, CCIR, which requires a specific readout frequency. Manufacturers usually do not provide a specific clock timing sequence for astronomical applications. This is not a serious problem if one understands how the device operates.

To make the TH7852 work, the following clocks have to be created:

1. Clocks Ø1P and Ø2P. These phases control the image zone. During the integration period, one is set low and the other is set high. In the TV interlaced mode these two clocks are transposed at each integration. Between a readout of the odd and even shift register, the center of gravity of the charge collecting zone shifts by half a pixel. This capability is not used in astronomy because the phase is held high during the entire exposure. After integration and during the transfer from the image zone to the memory zone, phases Ø1P and Ø2P are toggled high and low 145 times. The transfer procedure is done as fast as possible to avoid smearing.
2. Clocks Ø1M and Ø2M. These phases control the memory zone. During the line transfers from the image zone to the memory zone, phases Ø1P and Ø1M are identical, as are Ø2P and Ø2M. For TV a new integration time starts when the entire image has been transferred into the memory zone. This integration time is used to move the contents of the memory zone, line-by-line, into the horizontal shift register. At a ØM order, the charges contained in the last line of the memory zone are sent all at once into the horizontal register. At the same time, all the other lines of the memory zone move down by one position. In the TH7852, the memory zone is protected from outside light by an aluminum mask, which allows a relatively slow readout from this part of the array.
3. Clocks Ø1L and Ø2L. These clocks sequentially transfer the charge packets through the horizontal register to the output stage. Between two ØM clock cycles, at least 218 ØL clock cycles are needed to read the entire register. Extra cycles will not harm the CCD. In fact, they are beneficial. Since the supplementary read points contain no useful information, the register is cleared better before receiving the following line if extra ØL clock cycles are applied.
4. ØR clock. This clock, which is synchronized with ØL clock, controls the precharge of the output diode. To make things simpler, the manufacturer suggests making ØR = Ø2L. However, in astronomy where it is important to get the lowest signal noise possible, we must produce these two clocks separately. The positive pulse of the ØR phase is produced at the leading edge of Ø2L. The duration of a ØR high state is always less than the duration of the Ø2L high state.

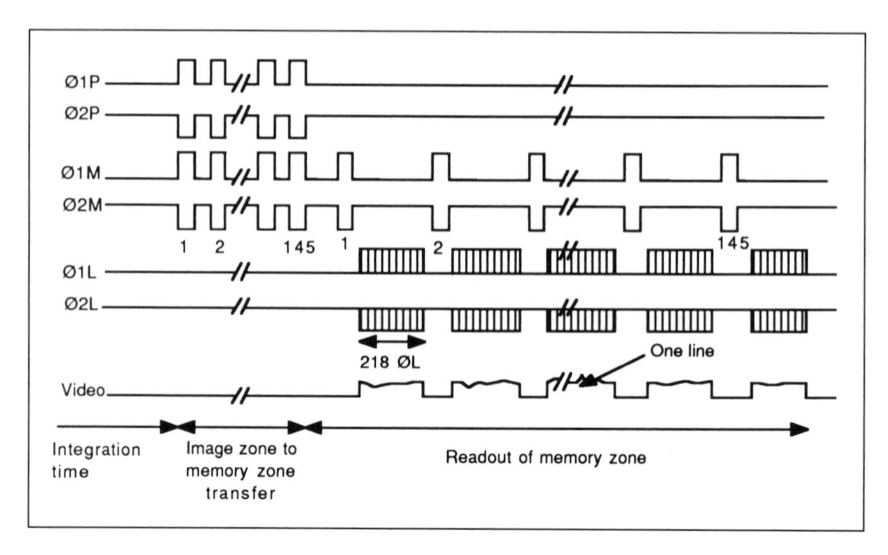

Figure 2.8 *The clock timing sequence of the TH7852 array.*

Figure 2.8 shows that the clocks are in symmetric pairs. This is a characteristic of two-phased transfer.

In the astronomical single shot readout mode, much time can elapse between two successive exposures. The detector receives light and accumulates dark charges during this period. Therefore, it is important to clear the CCD thoroughly before beginning an exposure. To do this, the entire array is rapidly read several times just before integrating the image. The last rapid readout is important because its duration sets the smallest possible integration time.

The clock timing sequence can be refined in several ways. Just after integration and just before the transfer from the image zone to the memory zone, the latter can be read rapidly to remove the thermal charges that accumulated during the exposure.

The array must be read completely. However, it is possible to process only a part of the image (windowing). The useful part is read slowly to carry out the analog/digital conversion under optimum conditions but the contents of the rest of the array are cleared by a fast sequence.

2.3.3 The Video Signal

The first 8 points of a line are columns masked from light by a deposit of aluminum. These are the dark reference points, which are never used as such in astronomy. We next find 208 points containing the real image information. Finally the two last points again correspond to two masked columns. As the first line of a frame contains no useful information, we have 144 effective lines.

Figure 2.9 shows a video signal. Note the relative phasing of the clocks and the video signal levels. The video signal is referenced to a DC voltage of about 9V.

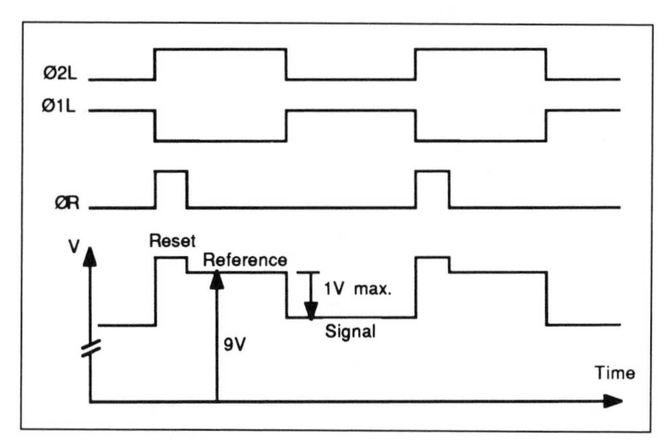

Figure 2.9 *Detail of the video signal at the output of the CCD.*

Each image point includes the following:

1. A reset level corresponding to the precharge of the output diode. The precharge is controlled by the leading edge of ØR and is almost instantaneous (several tens of nanoseconds is enough).
2. A reference level corresponding to the output capacitor's maximum charge. The difference in amplitude between the reset level and the reference level is caused by a voltage spike created when the transistor that isolates the diode from its charge source is triggered by the falling edge of ØR.
3. A signal level (or video level) equal to the charge packet arriving at the output capacitor. The capacitor's rate of discharge is in proportion to the number of electrons contained in the packet. The amplitude of the signal level is therefore always less than or equal to the reference level. The difference between these two levels becomes greater as the illumination increases. The injection of charges into the output diode is carried out at the leading edge of Ø1L.

2.3.4 The Generation of Clocks

The camera described in this book has a non-traditional clock generation method. Clock signals are usually created by a large number of logic circuits; however we will use a completely different approach. The clock timing sequence will be produced by the execution of a computer program. This method creates the clock sequences with looped instructions via an input/output (I/O) circuit. This procedure considerably reduces the number of logic circuits needed to run the camera and it has the added benefits of improved focusing and versatility.

On the other hand, the parameters of the clock timing sequence are closely tied to the machine's performance. The CCD's readout frequency is limited by the computer's clock frequency and more specifically by the number of clock cycles needed to write to the I/O circuit. On a slow machine, the readout time of a medium-sized array can be more than 10 seconds. In this case the software absolutely has to be optimized by programming in Assembler. With computers based on 16 bit microprocessors, the write speed to an I/O port is typically 100KHz, even when the program is written in a high level language. In our case, the clocks can be produced by a program written in MS Basic 7.0 and run on an 8MHz PC AT. The readout and storage time of an image from a TH7852 array is about 2 seconds with this configuration, which is fast enough for most applications.

Another possible disadvantage with a programmed clock timing sequence is a lack of precision in the output signals. Let us suppose that we are trying to generate a clock signal having a form factor of 1, i.e., in which the duration of the high level is identical to the duration of the low level. With a computer having a synchronous bus, such as the APPLE II, there is no problem—the delays can be adjusted to obtain a symmetrical shape that is stable over time. With an asynchronous bus, as is the case with PC compatibles, if the form factor we want can be approached, we see that it constantly is in a state of change but remains within a certain range. This is technically called *jitter*. For a PC AT, the jitter has an amplitude of the order of a micro-second. Setting aside the esthetic aspect of such a signal, this amount of jitter has no discernable effect for the frequency range in which we will operate our CCD.

Before starting the description of a program to generate clocks, let us study the peripheral input/output circuit—PIO.

2.3.5 The Interface Board

The 8255 PIO

The preferred input/output LSI device for PC compatibles is the PIO 8255. This device has several programming modes. The one that interests us has 24 I/O lines that can be programmed as either inputs or outputs in groups of 3 ports of 8 bits each. These 3 ports, called A, B, and C, correspond to 3 user addressable registers. Port C can be separated into two groups of 4 bits, one group capable of an access direction differing from the other.

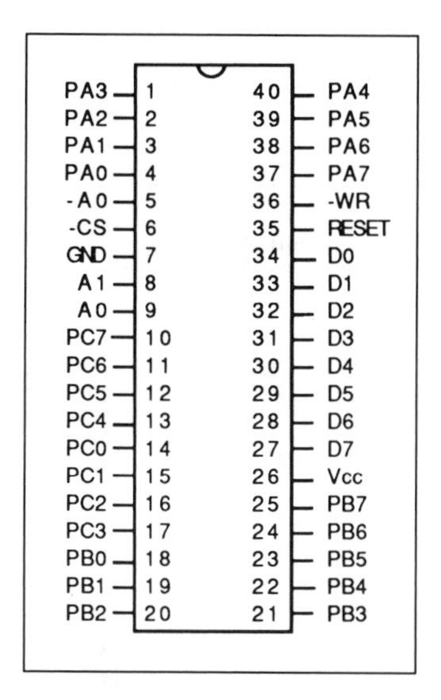

Figure 2.10 *Pin connections for the Intel PIO 8255.*

If a port is set up in output mode and if an 8-bit binary number is written to its address, we will find this binary configuration on the corresponding output pins of the chip. In the readout mode, we can determine the state of the signals applied to the pins of the port by examining its contents. The bits of a port are numbered from 0 to 7, with 0 representing the least significant bit and 7 the most significant bit.

A fourth 8-bit register in the 8255 defines, among other things, the direction in which the ports work—as inputs or outputs. This is the control register. Table 2.1 shows how the ports function as input and output according to the binary configuration of the control register (left column).

There are many sources for PC PIO boards equipped with one or more 8255 PIO's. These boards are suitable for our application and are recommended for those who want a minimum of bother and can afford them—they sell for less than $350. For those interested in building their own or those who just want to know what makes it work, here is some information about building a PIO board.

Let us see how the 8255 accesses the PC's memory. First we should consider the computer's bus (see Figure 2.11). The bus is made up of 62 lines in 2 rows (the additional 16 bit extension bus found on the PC AT is not used). All the signals are TTL compatible.

The signals that concern us are:

1. D0–D7: bi-directional data lines to permit the exchange of data between the microprocessor and the I/O circuits;
2. A0–A19: address lines, of which only lines A0 to A9 are used for decoding I/O circuits;

Table 2.1
Input/Output Addressing for an 8255 PIA

Control Register Address	Port A	Port B	Port C
10000000	PA0–PA7=O	PB0–PB7=O	PC0–PC3=O PC4–PC7=O
10000001	PA0–PA7=O	PB0–PB7=O	PC0–PC3=I PC4–PC7=O
10000010	PA0–PA7=O	PB0–PB7=I	PC0–PC3=O PC4–PC7=O
10000011	PA0–PA7=O	PB0–PB7=I	PC0–PC3=I PC4–PC7=O
10001000	PA0–PA7=O	PB0–PB7=O	PC0–PC3=O PC4–PC7=I
10001001	PA0–PA7=O	PB0–PB7=O	PC0–PC3=I PC4–PC7=I
10001010	PA0–PA7=O	PB0–PB7=I	PC0–PC3=O PC4–PC7=I
10001011	PA0–PA7=O	PB0–PB7=I	PC0–PC3=I PC4–PC7=I
10010000	PA0–PA7=I	PB0–PB7=O	PC0–PC3=O PC4–PC7=O
10010001	PA0–PA7=I	PB0–PB7=O	PC0–PC3=I PC4–PC7=O
10010010	PA0–PA7=I	PB0–PB7=I	PC0–PC3=O PC4–PC7=O
10010011	PA0–PA7=I	PB0–PB7=I	PC0–PC3=I PC4–PC7=O
10011000	PA0–PA7=I	PB0–PB7=O	PC0–PC3=O PC4–PC7=I
10011001	PA0–PA7=I	PB0–PB7=O	PC0–PC3=I PC4–PC7=I
10011010	PA0–PA7=I	PB0–PB7=I	PC0–PC3=O PC4–PC7=I
10011011	PA0–PA7=I	PB0–PB7=I	PC0–PC3=I PC4–PC7=I

3. IOR: this signal (I/O read) goes low to signal the addressed circuit to present its data on the bus;
4. IOW: this signal (I/O write) goes low to signal to the addressed circuit that data are available on the bus;
5. RESET DRV: this line sets everything to zero when the computer is turned on;
6. AEN: this address enable signal notifies the circuits that a DMA cycle is in process;
7. GND: the ground—found on pins B1, B10 and B31;
8. +5V: found on pins B3 and B29.

Figure 2.12 shows the diagram of an interface board. The two least significant bits on the bus are connected to the 8255 via a 74LS244 bus driver. The state of these bits determines the selection of one of the 4 registers in the PIO. The following table specifies the selected register as a function of the binary configuration of A0 and A1.

A0	A1	
0	0	Port A
1	0	Port B
0	1	Port C
1	1	Control

The address bits A2 to A9 of the PC bus are used to decode the 8255. The latter is addressed when its CS (Chip Select) pin is low (0). Since there are 10 lines on the PC bus, the addressable space is 1024 bytes ($2^{10} = 1024$). These addresses, to which input/output operations are allotted, are not all usable. For example the addresses from 0 to $0F (a $ indicates a hexadecimal base) are already used by the DMA controller. The addresses $20 and $21 are assigned to the interrupt controller and so on. In this memory space, addresses $300 to $31F are allocated to prototype boards such as the one we are describing. Normally, there should be no address conflict between these prototype boards and the commercial boards installed in the slots of the PC.

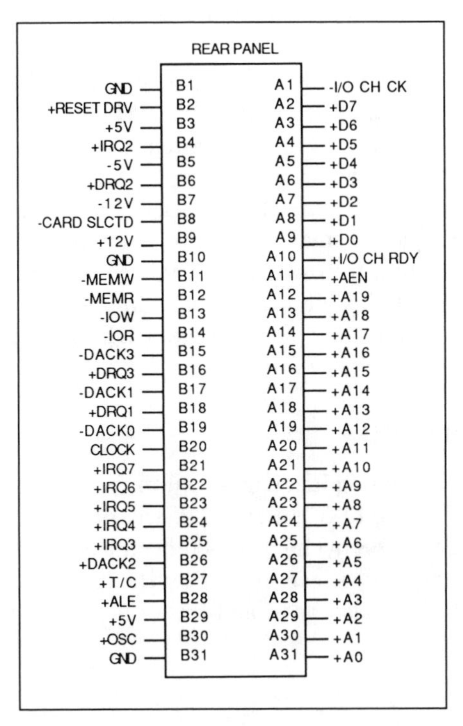

Figure 2.11 *The PC compatible I/O bus. A plus sign in front of the name of the line means that it is active at high level. A minus sign means that it is active at low level.*

Inverters (IC7 and IC8) set CS low when the address bus carries the binary configuration 1100000000 or $300 in hexadecimal. This is the basic address of the PIO. The AEN signal blocks the access to the PIO to prevent conflicts during a DMA operation. Following is the address of the registers:

Port A	$300 768 decimal
Port B	$301 769 decimal
Port C	$302 770 decimal
Control	$303 771 decimal

Let us now examine operation of the IC2, a 74LS245 tri-state device. It allows bidirectional transfer—into or out of the PIO. Obviously, at a given moment, the transfer direction is one way. The PC bus IOR line controls the 74LS245 through pin 1. The 74LS245 has an internal bus which can be placed at a level of very high impedance (a third-state). This third state is used to isolate the 8255 when it is not addressed. The signal applied to pin 19 (EN = ENable) triggers this high impedance mode. The low impedance state is obtained when EN is low and the addresses carried on the bus are valid for driving the 8255 and when the IOR or IOW signal is present (functions carried out by circuits IC3 and IC5).

This PIO circuit works with most PC compatibles (for example, it preforms flawlessly on a 80386 running at 20 MHz). It can be simplified further—the 74LS244 bus drivers can be removed because the 8255 places a negligible load on the bus.

The circuit shown in Figure 2.12 is easily wired on a printed circuit board (13 × 9cm) which is small enough to be installed in a PC short slot. This circuit can also be wire wrapped. Obviously an edge connector has to be provided to connect this circuit to the computer's bus. The port signal (including system ground) outputs are via a DB25 connector at the back of the card.

Programming the PIO

All the PC programming languages contain instructions for reading and writing an I/O address. In MS Basic 7.0, writing the number 255 to the register located at the 768 decimal or 300 hexadecimal address (translated by `&H300` in this language) is carried out by the instructions: `OUT 768,255` or `OUT &H300,255`. It is also possible to use the following variables:

```
address% = 768
value% = 255
OUT address%, value%
```

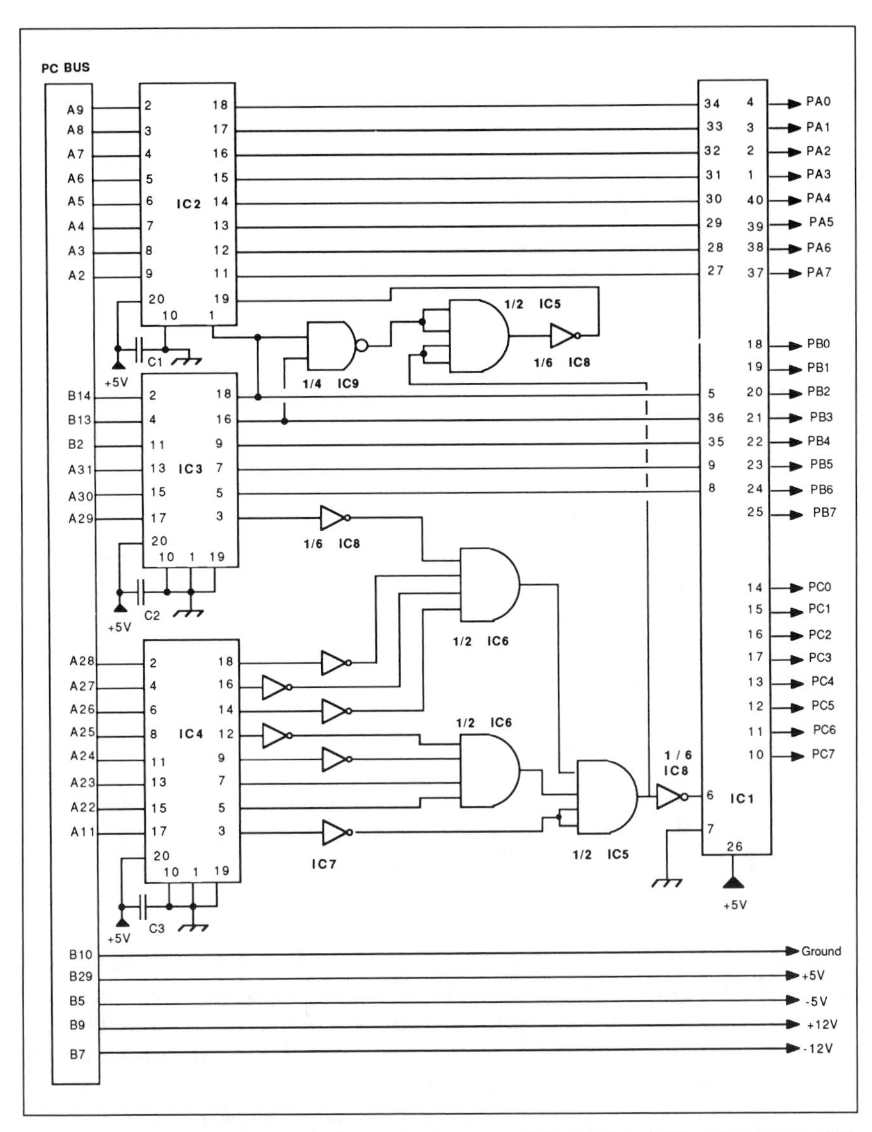

Figure 2.12 *Interface board circuit diagram. IC1 8255A; IC3,4 74LS244; IC2 74LS245; IC5,6 74LS21; IC7,8 74LS04; IC9 74LS00; C1,2 22μF and C3 22μF*

The symbol % following the name of a variable means that the variable is an integer.

The register's contents will have these read syntaxes:

```
    value% = INP(&H300)
or: value% = INP(768)
```

or: `value% = INP(address%)`

The following BASIC code generates 100 clock signals on bit 2 of port A.

```
OUT 771,128          :REM initializing
FOR I% = 1 to 100
  OUT 768,0          :REM bit 2 at low state
  X% = 0 : X% = 0
  OUT 768,4          :REM bit 2 at high state
NEXT
```

The first line initializes the PIO. The ports are all set to output mode by writing the binary word "10000000" or 128 dec. to the control register. The loop writes twice to register A: first to set bit 2 to low, second to set bit 2 high (4 decimal = 00000100 binary).

Notice the presence of two instructions (`X% = 0`) which apparently serve no purpose. Their role is to create a small wait state between two successive `OUT`s. This is a precaution to prevent a malfunction should the clock run faster than the bus jitter. On a 4.77MHz PC XT, the `X%=0` operation lasts about 8 microseconds using MS Basic 7.0 code. With an 8MHz 80286 PC AT, this wait state is 1.7 microseconds, while a 16MHz 80386 PC AT executes it in 0.55 microseconds.

2.3.6 The Clock Board

The clock board is relatively bare because we are using software to generate most of the clocks. Its role is limited to forming the clock signals that drive the CCD. Since there is room to spare we have also installed power supplies to produce the various DC voltages for the CCD. See Figures 2.13, 2.14, 2.15 and 2.16.

Referring to Figure 2.13 let us follow the path of the clock signal ØP produced on a pin of 1 of the PIO ports (we will see which one later).

The signal ends up at a series of NOR ports (circuit IC21). This IC produces two symmetrical clocks, Ø1P and Ø2P. These are two phases of the CCD register. When Ø1P is high, Ø2P is low, and vice versa. *A priori*, it would have been possible to use a simple inverter to carry out this function; however, the delay produced when the clock signal passed through the inverter would cause an edge asymmetry. To insure good charge transfer the clocks should cross precisely. Thomson recommends a cross over with an amplitude between 70 and 100% of the maximum level. Figure 2.17 shows the oscillograph of the two clocks created from the three NOR ports of our circuit. The ports are fast which enables precise location of

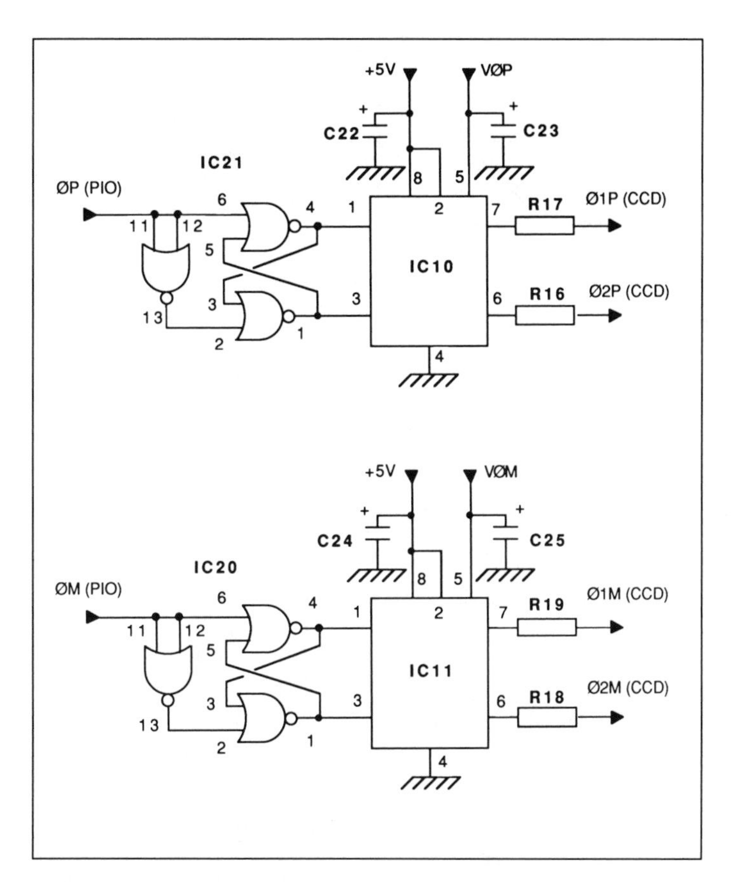

Figure 2.13 *Schematic for clocks Ø1P, Ø2P, Ø1M and Ø2M. For signals VØP and VØM see Figure 2.15.*

Clock Board Component List

T1	2N2907	R16,17,18,19,20,21,24	47Ω
D1	1N914	R5,7,11,13,15,23	470Ω
IC1	7805	R2,3,4,6,10,12,14,22	1kΩ
IC2	7815	R1	2.2kΩ
IC3,4,5	LM117	R9,25	10kΩ
IC6	7912	R8	18kΩ
IC7,8,9	LM117	P2,3,4,5,6,7,8,9	4.7kΩ
IC10,11,12	SN75361	P1	22kΩ
IC13	1Mhz Oscil.	P10	100Ωk
IC14	74LS73	C32	220pF
IC15	LM117	C4,7,10,15	22nF
IC16	SN75361	C5,8,11,12,16,18,20,28,33	0.1μF
IC17	74LS123	C17,19,21,29,34	4.7μF
IC18	74LS00	C1,13	10μF
IC19,20,21	74LS02	C3,6,9,14,22,23,24,25,26,27,30,31	22μF
		C2	47μF

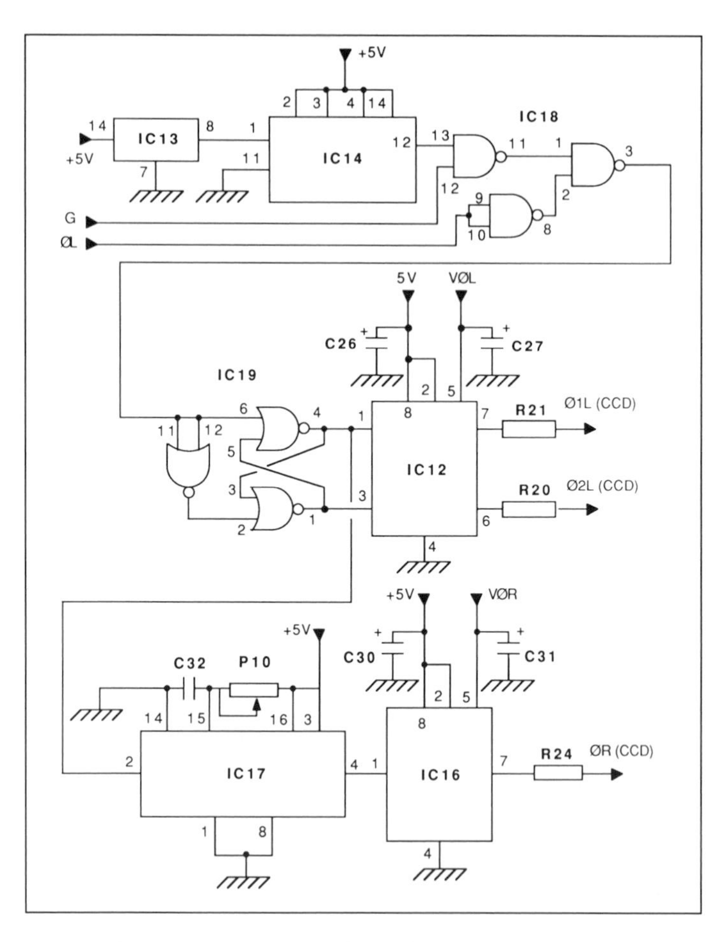

Figure 2.14 *Schematic for clocks Ø1L, Ø2L, and ØR.. For signals VØL and VØR see Figure 2.15*

Clock Board Component List

T1	2N2907	R16,17,18,19,20,21,24	47Ω
D1	1N914	R5,7,11,13,15,23	470Ω
IC1	7805	R2,3,4,6,10,12,14,22	1kΩ
IC2	7815	R1	2.2kΩ
IC3,4,5	LM117	R9,25	10kΩ
IC6	7912	R8	18kΩ
IC7,8,9	LM117	P2,3,4,5,6,7,8,9	4.7kΩ
IC10,11,12	SN75361	P1	22kΩ
IC13	1Mhz Oscil.	P10	100Ωk
IC14	74LS73	C32	220pF
IC15	LM117	C4,7,10,15	22nF
IC16	SN75361	C5,8,11,12,16,18,20,28,33	0.1μF
IC17	74LS123	C17,19,21,29,34	4.7μF
IC18	74LS00	C1,13	10μF
IC19,20,21	74LS02	C3,6,9,14,22,23,24,25,26,27,30,31	22μF
		C2	47μF

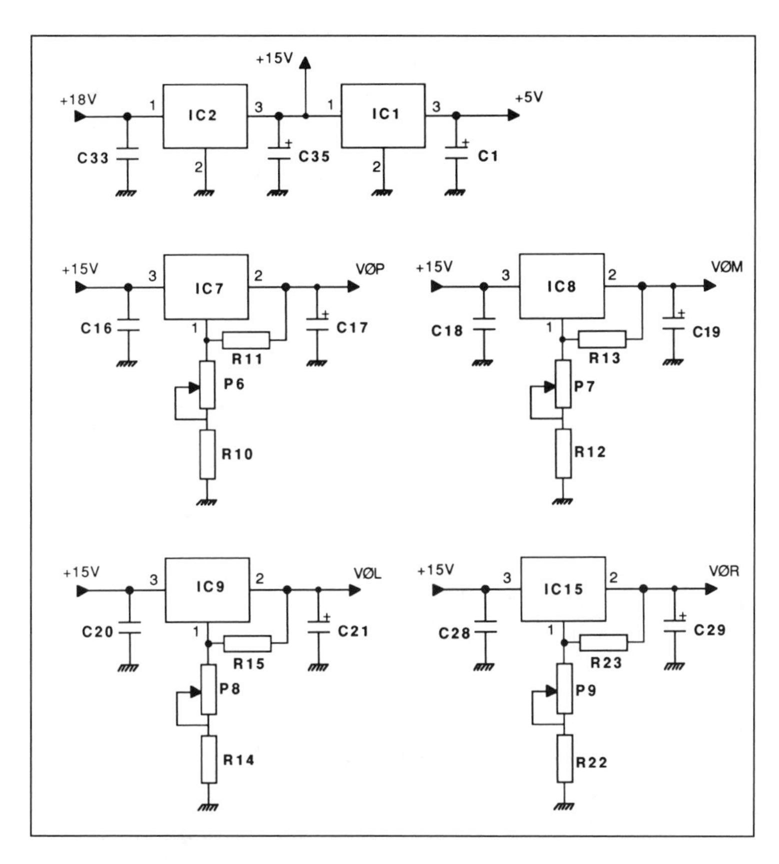

Figure 2.15 *SN75361 translator power supply.*

Clock Board Component List

T1	2N2907	R16,17,18,19,20,21,24	47Ω
D1	1N914	R5,7,11,13,15,23	470Ω
IC1	7805	R2,3,4,6,10,12,14,22	1kΩ
IC2	7815	R1	2.2kΩ
IC3,4,5	LM117	R9,25	10kΩ
IC6	7912	R8	18kΩ
IC7,8,9	LM117	P2,3,4,5,6,7,8,9	4.7kΩ
IC10,11,12	SN75361	P1	22kΩ
IC13	1Mhz Oscil.	P10	100Ωk
IC14	74LS73	C32	220pF
IC15	LM117	C4,7,10,15	22nF
IC16	SN75361	C5,8,11,12,16,18,20,28,33	0.1μF
IC17	74LS123	C17,19,21,29,34	4.7μF
IC18	74LS00	C1,13	10μF
IC19,20,21	74LS02	C3,6,9,14,22,23,24,25,26,27,30,31	22μF
		C2	47μF

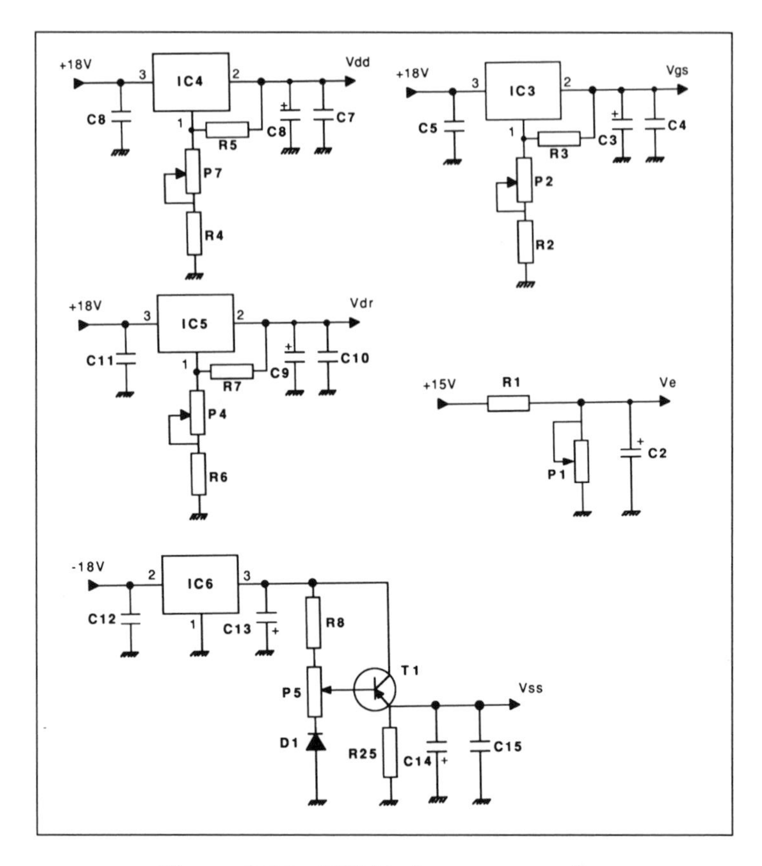

Figure 2.16 *CCD biasing power supplies.*

Clock Board Component List

T1	2N2907	R16,17,18,19,20,21,24	47Ω
D1	1N914	R5,7,11,13,15,23	470Ω
IC1	7805	R2,3,4,6,10,12,14,22	1kΩ
IC2	7815	R1	2.2kΩ
IC3,4,5	LM117	R9,25	10kΩ
IC6	7912	R8	18kΩ
IC7,8,9	LM117	P2,3,4,5,6,7,8,9	4.7kΩ
IC10,11,12	SN75361	P1	22kΩ
IC13	1Mhz Oscil.	P10	100Ωk
IC14	74LS73	C32	220pF
IC15	LM117	C4,7,10,15	22nF
IC16	SN75361	C5,8,11,12,16,18,20,28,33	0.1μF
IC17	74LS123	C17,19,21,29,34	4.7μF
IC18	74LS00	C1,13	10μF
IC19,20,21	74LS02	C3,6,9,14,22,23,24,25,26,27,30,31	22μF
		C2	47μF

the cross over point. Optimizing this parameter is difficult since one has to be able to precisely measure the transfer efficiency. With our detection level and a TM 7852 array we have found that an approximate selection of the cross over point has no visible consequences.

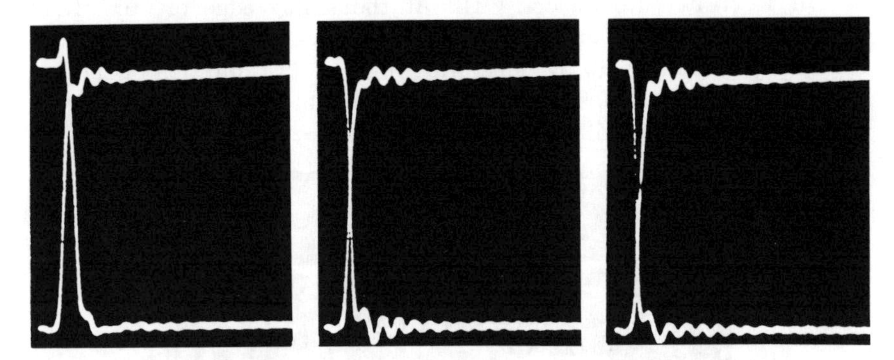

Figure 2.17 *IC's 74L02, 7402 and 74LLS02 (from left to right) double the clock ØP, and provide a means to adjust the clock cross over point.*

The CCD is a MOS device and the clocks that drive it must meet certain criteria—the voltage levels are especially critical. The TTL level signals from the IC21 are converted by IC10 (SN75361) to MOS level signals. The SN75361 can process two clock channels at a time and it acts as an inverter (a high level input signal becomes a low level output). The clock amplitude can be varied by adjusting the VØP voltage on pin 5. This voltage is about 10V for a TH7852 array. Resistors R16 and R17 insure correct coupling to the CCD. Depending on the type and length of the connecting cable between the SN75361 and the CCD, these resistances may have to be adjusted to obtain correct functioning of the CCD. The cable *absolutely* must not be too long—20 to 30cm is a maximum. This means that the clock board must be placed close to the optical head. If the cable is too long the clock signals are distorted and the CCD will malfunction.

Clock ØM is treated identically to ØP, but there is a difference for ØL (Figure 2.14) The frequency of this clock has a critical impact on the CCD's readout speed since the horizontal register is a real bottleneck. We have seen that it is important in the single shot mode to apply rapid clearing cycles to the detector before integration. To accelerate charge transfer during this procedure, clock ØL is produced by an oscillator (IC13) rather than by the software, which is not fast enough. The oscillator's frequency is 1MHz, which is reduced to 500KHz by the divider, IC14. For the digitizing operation, the computer takes over again to generate ØL. The choice of a ØL generated by the board's oscillator or by the program depends on the state of the G (Gate) signal produced by the computer. If G is high and if the bit producing ØL from the computer is low, a 500KHz clock is produced by the board. If G is low, the clock is created by the program. Clocks Ø1L and Ø2L are created like clocks Ø1P and Ø2P, Ø1M and Ø2M (circuits IC12 and IC19).

ØR is created from clock Ø1L. At the falling edge of Ø1L, IC17 (74LS123), a monostable vibrator, creates a negative pulse whose width can be regulated by P10, a multiturn trimmer. The duration of the low state is set at about 0.3μs. IC16 adjusts this signal so that its leading edge coincides with that of Ø2L.

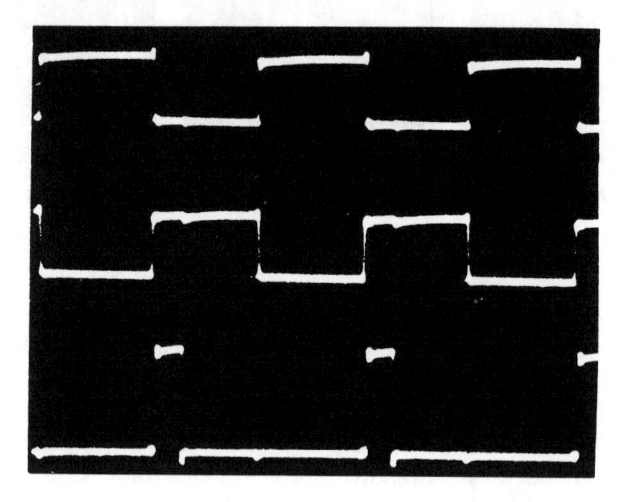

Figure 2.18 *Appearance of the clocks at the translators' output. At the top, clock Ø1L; in the middle, clock Ø2L; at the bottom, clock ØR.*

The board is powered by ±18V. A 7815 regulator (IC2) creates a +15V intermediate voltage. The +5V powering the logic circuits is supplied by a 7805 circuit (IC1). These two regulators should be mounted on small heat sinks.

The DC voltages available on most computer buses are created by switching power supplies which operate at approximately 100KHz. Switchers superimpose a considerable amount of high frequency noise on the output voltages. This noise has no effect on the computer but it is devastating to high gain CCD circuits. Therefore, a separate power supply is necessary and it must be well filtered.

The voltage for VØP, VØM, VØL, VØR are regulated by an LM117. The voltage level is adjusted by a 4.7kΩ multiturn trimmer (see Figure 2.15).

All the DC voltages powering the CCD are also regulated by LM117's except V_e, the anti-blooming voltage, which need not be very stable, and

V_{ss}, the substrate voltage, which is negative (see Figure 2.16).

Here are typical voltages for the TH7852 CCD:

$$V_{dd} = 14V \qquad V_e = 12V$$
$$V_{dr} = 12V \qquad V_{ss} = -1.5V.$$
$$V_{gs} = \ 8V$$

The best way to evaluate the operation of the clock board is to generate a clock timing sequence with the computer and then observe the signals with an oscilloscope, if possible a multichannel one. Then connect the CCD and study its video signal.

In our camera, port B of the PIO is used to generate the clocks. The bits used in this port have the following functions:

- bit 0: clock ØP
- bit 1: clock ØM
- bit 2: clock ØL
- bit 3: gate signal
- bit 4: start-convert signal

In the section on setting up the electronics, we will study a simple example of a clock generating program. In the data sheet for the TH7852, Thomson advises lowering the high level of the phase ØP to 8V during integration time, possibly as a way to decrease the dark current. If there is indeed a relation between the dark current and the level of this phase, the variation found—only a few percent in the range specified for the clock level—does not seem significant. For this reason the high level of ØP is identical during the transfer and the exposure. If the reader ever wants to analyze the influence of the level of ØP, the diagram in Figure 2.19 can be used.

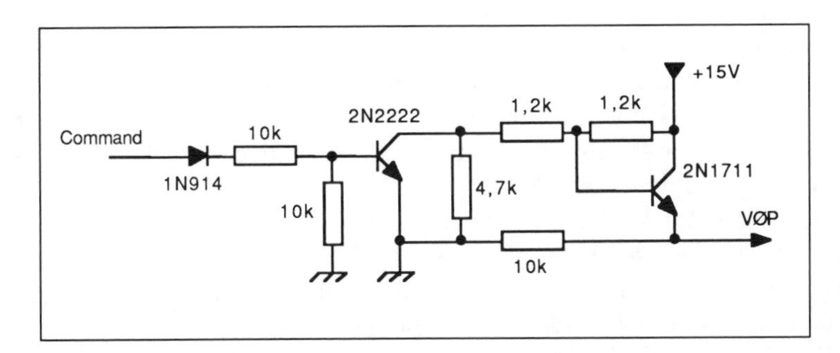

Figure 2.19 *Diagram of an optional circuit for controlling the voltage VØP. The high level of ØP is modulated as a function of a logic command signal produced by the computer (a spare bit from port B of the PIO).*

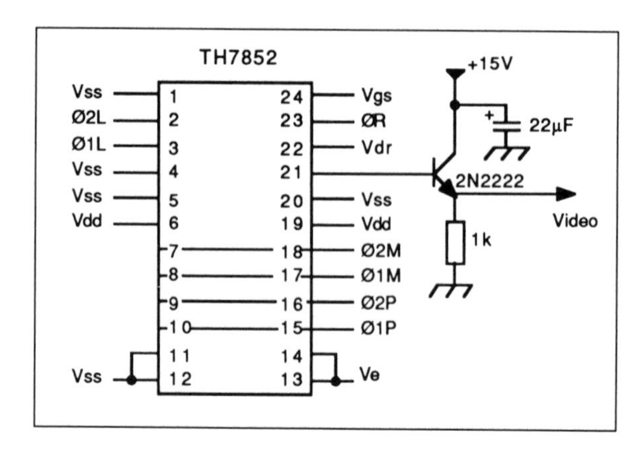

Figure 2.20 *Schematic of the CCD board. For clarity the capacitors that decouple the DC power supply leads* (V_{dd}, V_{dr}, V_{gs}, V_{ss}, V_e) *are not shown.*

2.3.7 The CCD Board

The CCD circuit board is very simple. Besides the CCD, there are several decoupling capacitors and a video signal driver transistor. This transistor is needed because the output impedance of the CCD is relatively high (1kΩ), and the video signal cannot travel, even for a short distance, without becoming distorted. The transistor is wired as an emitter-follower; the output (emitter) follows the input (base) with a shift of about −0.6V. This board is specifically designed to be small so that it can be installed inside a compact and light-weight optical head.

The pins of the CCD should not be soldered directly to the printed circuit since the MOS circuits can be destroyed by the static charges produced by some soldering irons. Further, it is almost impossible to remove a soldered CCD without damaging it. The detector is thus mounted on a traditional, integrated circuit socket.

The bias voltages V_{dd}, V_{dr}, V_{gs}, V_e and the power for the transistor should be decoupled with 22μF capacitors between each supply voltage and ground. If there is any space remaining after installing the 22μF capacitors, a 100pF capacitor should also be installed. All capacitors should be placed as close as possible to the CCD's pins.

The transistor is a 2N2222A; it can be replaced by a low noise FET transistor. Whatever type of transistor is used, its base pin should be placed very close to the CCD's video output pin. This transistor is a simple follower that does not modify the shape of the video signal.

Shorting the CCD's video output to ground will destroy the CCD. One should, therefore, conduct any measurements of the video signal at the transistor's output and not directly on the CCD's output pin. In this way, should a short circuit occur it will destroy the transistor and hopefully not the CCD.

The lines for the 7 clock signals, 6 DC voltages, and ground terminate at a connector mounted on the side of the optical head. The video signal output is via a shielded cable using BNC connectors.

The follower transistor can be replaced by an operational amplifier as shown in Figure 2.21. Both of these circuits have identical noise characteristics.

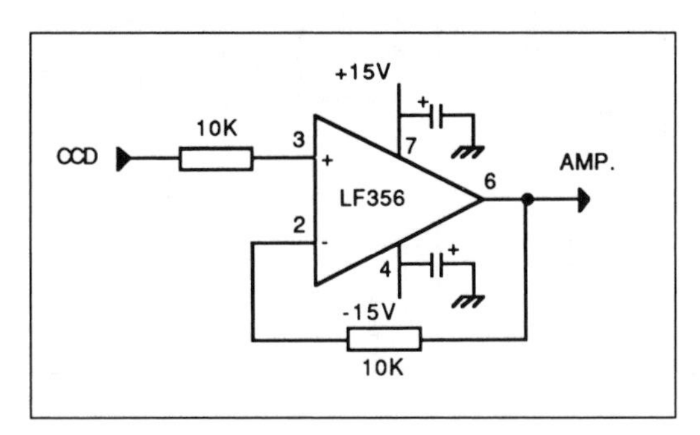

Figure 2.21 *Operational amplifier wired as a follower directly to the CCD's output.*

2.3.8 The Amplifier Board

The role of this board is to remove the ≈9V present in the video signal and then to amplify the latter.

The useful amplitude of the video signal is about 1V. To read this signal with a analog-to-digital (A/D) converter it has to be amplified in such a way that the resulting voltage amplitude ranges between 0 and 10 V.

The ≈9VDC component of the video signal is a problem. If it is not removed, it will saturate the amplifier. There are several ways to eliminate this voltage:

1. The CCD can be capacitive coupled to the amplifier, that is, a capacitor is placed in series at the output of the driver transistor (see Figure 2.22). This capacitor passes only the varying video (AC) component and blocks the DC component. Zero volts then represents the average value of the fluctuations of the video signal. This solution is not very satisfactory. The readout of the CCD must be done slowly because the capacitor has a long time constant that causes the output to vary according to whether it is at the beginning or end of a line.
2. Operational amplifiers can remove the DC component from the video signal without affecting the signal.

We have adopted this second solution. Figures 2.23 and 2.24 show two possible methods—a summing amplifier or a differential amplifier.

In the summing amplifier (Figure 2.23) the video signal plus a negative voltage approximately equal to the 9VDC video component is applied to the inverting input of the amplifier. The circuit sums these two voltages so that the real video signal is stripped of the DC component.

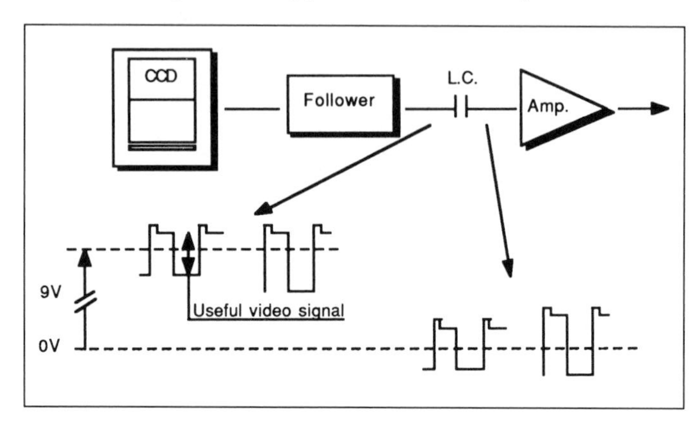

Figure 2.22 *Placing a capacitor (L.C. = link capacitor) in the video line is a simple way of removing the DC component from the signal.*

The negative voltage must be very stable over time since it will be amplified along with the video signal. The negative voltage is produced by an LM104 regulator which is designed to deliver very stable voltages. This voltage is smoothed by an RC filter (resistor of 10 ohms and capacitor of 42μF). These values have been selected to remove the maximum amount of noise. The $10\text{k}\Omega$ multiturn trimmer allows a rough regulation of the voltage (about -9V). The $1\text{k}\Omega$ multiturn trimmer can be adjusted from outside the amplifier case so that any drift of the electronics can be finely adjusted during image acquisition. We will come back to this point.

The amplifier, an LF356, is a low-noise FET op-amp. It is commonly used and inexpensive. The gain is determined by the ratio of resistances R2 and R1. Their values must be carefully chosen. There are two possible methods: either we favor dynamic range and choose a relatively weak gain that will let us make the scale of the CCD's linear response just enter into the measurement range of the analog-digital converter (0 to 10V); or we favor detectivity and select a high gain to let us measure the noise.

When using analog-to-digital converters that resolve less than 16 bits the second alternative should be chosen. We will see in Chapter 5 that by averaging several images it is possible to extract significant information which would be completely undetectable in an individual image. To be able to do this the noise has to be measurable. Typically, the gain will be adjusted so that the RMS noise at the ADC output represents 1.5 to 2 quantification units. In my circuit, I adopted a gain of −22.

Experience shows that making a variable gain amplifier is unnecessary if we digitize with an adequate number of bits (starting at 10 bits). Because the amplifier works as an inverter (the gain is negative), an input signal with a negative voltage variation will give a positive variation in the output. After amplification, the video signal will therefore present a positive variation as the CCD is illuminated.

The power supplies must be carefully isolated at a point that is as close as possible to the amplifier to eliminate any risk of interference from the 60Hz or 50Hz originating in commercial power lines. An effective solution is to use batteries to power the circuit.

The amplifier components are mounted in a small metal case which acts as a shield. The case is located close to the optical head so that the video signal travels only a short distance. The video link, which is about 20cm long, is made with shielded coaxial cable. The case and the operational amplifier should share a common ground—pin 3 of the LF356 to prevent ground loops from occurring.

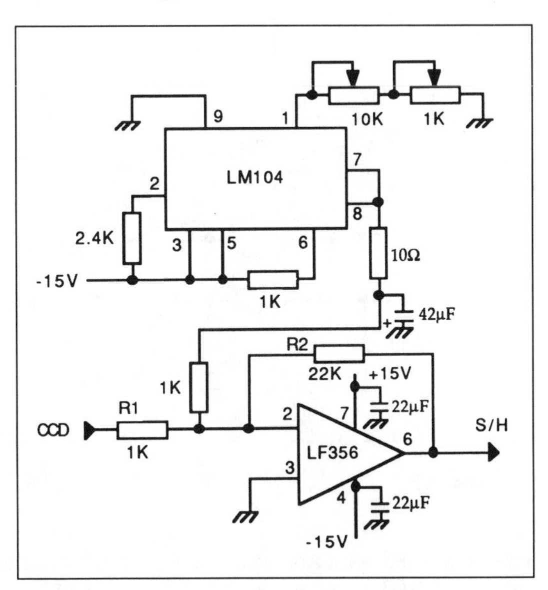

Figure 2.23 *Diagram of the amplifier wired as a summer.*

Figure 2.24 shows the circuit for a differential amplifier which has equivalent performance to our summing amplifier. Here, the amplifier creates an output that is the difference between the voltages present on the inverting and noninverting inputs. The offset voltage will therefore be positive. It is generated by an LM105 regulator, which is the positive voltage equivalent to the LM104.

The amplifier gain is

$$G = \frac{R_1}{R_2} = \frac{R_4}{R_3},$$

and the output voltage:

$$V_s = \frac{R_1}{R_2}(V_2 - V_1) = \frac{R_4}{R_3}(V_2 - V_1).$$

With R_1, R_4 each 120kΩ and R_2, R_3 each 5.1kΩ, the differential gain is about 23.

If a differential amplifier is used, follow the same shielding, grounding and decoupling techniques described above for the summing amplifier.

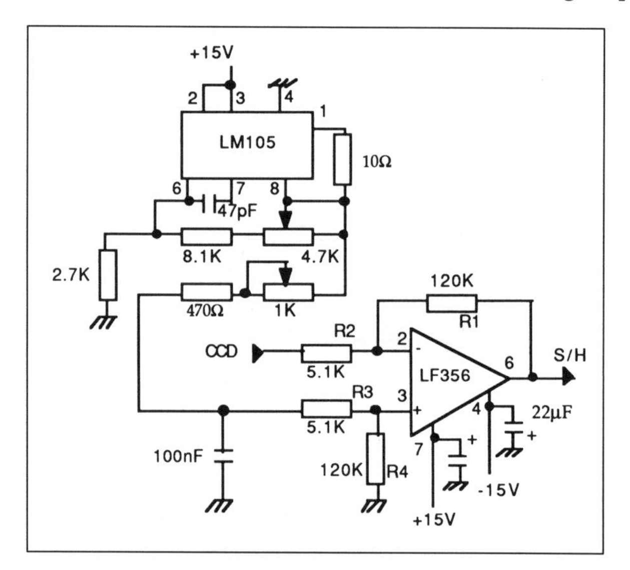

Figure 2.24 *Diagram of a differential amplifier.*

The amplifiers described here are equivalent in performance. The choice of one or the other might be determined by the ease of producing a well regulated positive or negative voltage supply.

2.3.9 The Analog-Digital Conversion Board

The function of this board is to convert the analog signal to digital levels the computer can deal with. It is essentially a sample-hold circuit and the A/D converter. Before discussing the functioning of this board we should come back to the characteristics of the video signal and understand what we are going to digitize.

We know that a point on the image will be represented in the video signal by the presence of three levels: the reset level, the reference level, and the signal level. The important information, that which translates the luminous intensity falling on a pixel, is the difference in voltage between the signal level and the reference level. This voltage difference is what has to be translated in digital form.

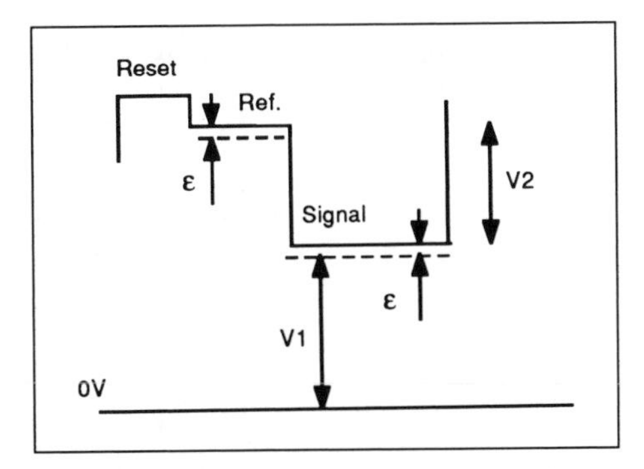

Figure 2.25 *The measurement of the amplitude (V_1) of the signal level compared to a fictional reference, here the ground, is not accurate because of the presence of reset noise (ϵ). On the other hand, the difference (V_2) between the signal level and the reference level is much more representative of the number of charges contained in the packet which is read since the uncertainty ϵ disappears in this operation.*

Measuring only the amplitude of the signal level will lead to incorrect results. Several factors cause this problem. In the first place, the zero point is not defined. It depends, for example, on the offset regulation of the amplifier. Even more serious is that the position of the zero point evolves as the circuit's characteristics change due to temperature and voltage variations. Furthermore the output diode's charge has some uncertainty, as shown by a random fluctuation of the video level (reset noise, see section 1.7.12).

These problems can be avoided by digitizing the exact difference between the signal and reference levels. Then the position of the zero point is no longer important, and at the same time the reset noise is removed since the two levels have noises that are perfectly correlated (see Figure 2.25).

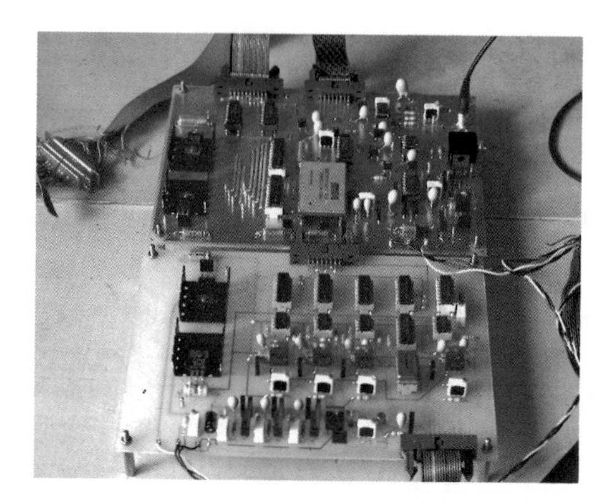

Figure 2.26 *The board at the top supports the analog-digital conversion circuits (the converter itself is located toward the center), and the board at the bottom contains the clock and bias wirings of the CCD. The latter can be set from a series of multi-turn trimmers.*

Sometimes for simplicity, the ØR clock is identical to Ø2L. Figure 2.27 shows that this solution is not valid when it is necessary to obtain a minimum readout noise.

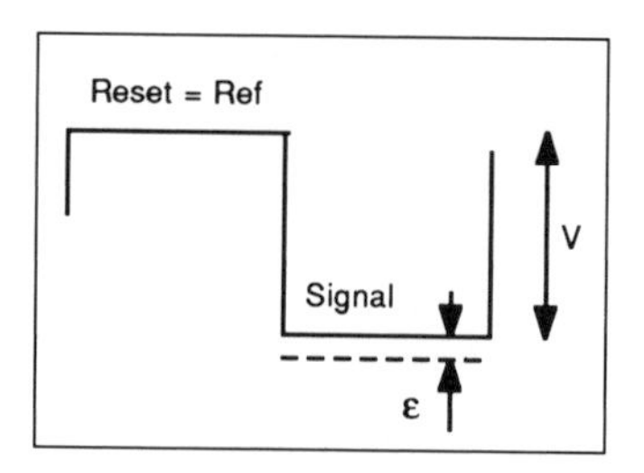

Figure 2.27 *By identifying clock ØR with clock Ø2L we remove the generation of a phase. Unfortunately, in this case the reference level no longer exists and the difference between the signal level and the reset level does not remove the noise.*

There are many techniques for isolating the signal level difference, all of which are somewhat difficult to put into practice. They are known collectively as "Correlated Double Sampling" (CDS). We will examine four: clamping, analog double sampling, double integration, and digital double sampling.

Clamping

Clamping is illustrated in Figure 2.28. As the reference level passes, switch I is closed, charging capacitor C. This switching function is usually accomplished with a transistor. At this point in the circuit, the signal is at ground potential. As the switch is opened, the output signal maintains this level. When the signal level passes, the capacitor transmits the difference in levels between the reference and signal levels (AC voltages more or less pass through capacitors depending on the amount of capacitance and the frequency of the AC). For the next pixel, the switch is again closed on the reference level and the capacitor thus recovers the slight delay that may exist with the preceding reference level.

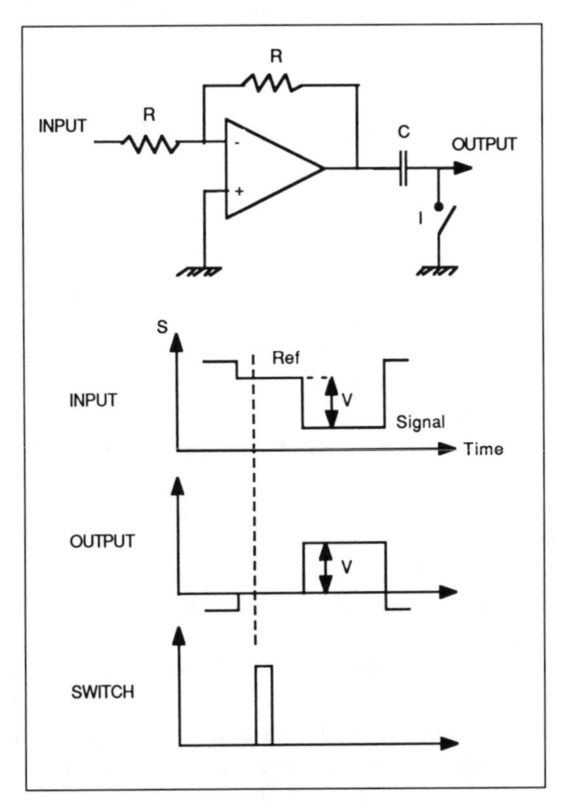

Figure 2.28 *Outline of a clamping circuit. The opening and closing of switch I is controlled by the clock signal visible at the bottom of the figure. When the clock is at high level, the switch is closed; when the clock is at low, the switch is open.*

On paper this solution is simple and efficient. In practice, things are much more complicated—the difficulty being the construction of the switch. We are faced with the contradiction: a transistor possessing very low resistance when closed *and* to avoid leakage a very high resistance when it is open. Field effect transistors (FET) are often used for this function because of their high resistance when open.

Clamping creates difficult problems. Parasitics can be created through coupling of the transistor's command with the output signal. In addition, the reference level can be brought to ground potential only with limited precision. Further, the resulting difference can vary with the amplitude of the signal—an intolerable condition for our purposes. These problems can be resolved with a highly sophisticated circuit using many discrete components. Because of this, it is not recommended to choose clamping.

Analog Double Sampling

The main feature of analog double sampling is the sample-hold circuit. This circuit detects the value of a signal at a precise moment (sampling) and retains this value in memory (hold) until the next sampling. It is hard to make this kind of circuit with discrete components for the same reasons as those mentioned in the section on clamping (switching problems). It is better to use integrated components—they are inexpensive and readily available. A sample-hold circuit has an input toward which the signal to be processed is sent, an output where we pick up the memorized voltage and a digital command input that lets us set the instant of hold. When the sample-hold circuit is commanded, its output voltage faithfully follows the variations of the signal applied at the input. Changing the command creates the hold: the output signal remains constant and equal to the value of the input signal at the instant of the change.

A sample-hold circuit is essentially made up of an analog switch controlled by the command signal, a capacitor that maintains the signal when the switch is open and a high input impedance operational amplifier follower to isolate the capacitor.

In hybrid circuits, the capacitor is usually integrated in the casing. On the other hand, with monolithic circuits it often has to be added by the user as a discrete component. In this case, it is better to use a high-quality capacitor made of polystyrene or polycarbonate which is more temperature stable.

For CCD applications the following characteristics of a sample-hold circuit are important:

1. Acquisition time. This is the period of time needed for the capacitor to be charged to a value identical to the input signal. This occurs in the sampling phase just after hold.
2. Hold time. This is the period between the issuance of a command to open a switch and the effective opening of that switch. This period

is not constant (it jitters). When the input signal is highly variable it introduces amplitude errors on the sample.

3. Charge error. During hold, stray capacitance within the switch causes a delay in signal transmission. This delay introduces noise and offsets the signal. Damped oscillations also appear on the output signal. Because of delays, offsets, and damped oscillations, the signal is only usable after it has stabilized.
4. Charge variation. The output voltage varies during the hold period due to leakage in both the switch and capacitor. This variation is linear over time. The higher the hold capacitance, the slower the variation occurs. Charge variation is also referred to as loss of memorization or droop rate.
5. Coupling. During the hold period, a fraction of the input signal appears at the output.

The sample-hold circuit's capacity should match the resolution of the A/D converter. A mediocre sample-hold circuit gives 8-bit accuracy, a good one 12-bits, and an exceptional one can give 16-bit accuracy.

Here, for example, are some of the important characteristics for Datel's SHM-20C sample-hold circuit which I used in my circuit to convert the video signal into 12 bits:

1. Acquisition time (within 0.01%) : $1\mu s$;
2. Hold time: 30ns;
3. Jitter on hold time: 1ns;
4. Charge variation: $0.08\mu V/\mu s$ with an internal capacitance of 100pF;
5. Charge error: 1mV;
6. Time needed to stabilize output signal to 0.01% after the hold order is issued: 185ns.

Double sampling requires at least two sample-hold circuits and an operational amplifier (Figure 2.29). The amplified video signal is sent to two samplers. One sampler holds a sample of the reference voltage—V_1 and is commanded by the leading edge of ØR. The other sampler blocks the video signal when Ø2L falls, giving voltage V_2. V_1 and V_2 are fed into the operational amplifier which produces the difference between these two voltages.

A third sample-hold circuit can follow the differential stage to hold the voltage at a constant level during analog/digital conversion.

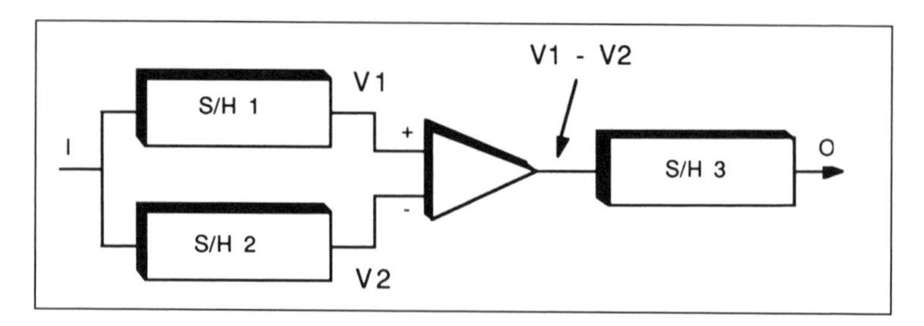

Figure 2.29 *Diagram of an analog double-sampling circuit.*

A variation of the double-sampling circuit uses a single sample-hold circuit to produce the difference as in Figure 2.30. Here the amplitude of the reference level is held as previously with a sampler whose output is connected to one of the amplifier's inputs. The video signal is also sent as it is on the other input of the amplifier. The output of the latter therefore permanently translates the difference between the signal level and the reference level. The sampler at the amplifier's output then becomes indispensable to set this difference during the analog/digital conversion.

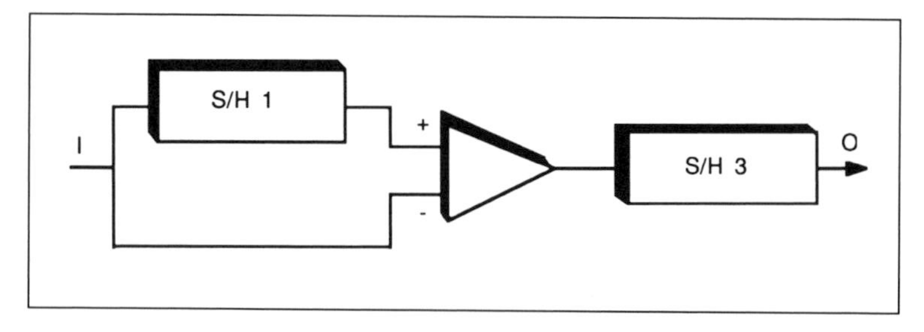

Figure 2.30 *A variation of analog double-sampling.*

The disadvantage of the second diagram compared to the first is the non-symmetry that can cause a delay on one of the tracks. This, however, is not a problem if this offset is constant.

Double integration

An alternative to analog double-sampling is double integration. After amplification, the video signal is directed to an integrating amplifier (IC3, see Figure 2.31), either through a follower amplifier (IC1) or through a unity-gain inverting amplifier (IC2). Analog switches (I1 and I2) allow the selection of one or another track. Switch I3 drives the functioning mode of the integrator in the following manner:

1. If I3 is closed, the capacitor C is at ground potential.
2. If I3 is open, the capacitor integrates the signal at the input of amplifier IC3. Then if the input voltage is V_e and the output voltage V_s, we will have the relationship

$$\frac{d_{V_s}}{d_t} = \frac{-V_e}{R \cdot C},$$

with d_t the time elapsed since the closing of switch I3.

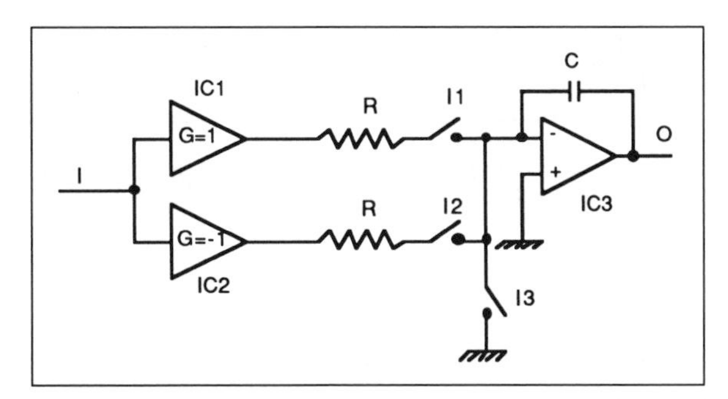

Figure 2.31 *A double integration sampling circuit.*

The processing stages are as follows:

1. Capacitor C is discharged to ground by the closing of I3.
2. When charged to the reference level, I3 is opened and I1 is closed, causing capacitor C to charge to a level proportional to the voltage applied to the integrator and to the closing time of I1.
3. The opening of I1 isolates the capacitor during the transition to the signal level.
4. When the signal level is established, switch I2 is closed for the same period as I1 was during the integration of the reference level. The capacitor is then partially discharged. At the end of the integration period, the output voltage of IC3 represents the difference between the signal and reference levels.
5. Switches I1, I2 and I3 are then kept open during the digitizing of this voltage.
6. A new cycle begins by setting the capacitor to zero.

The clock timing sequence of Figure 2.32 summarizes these operations.

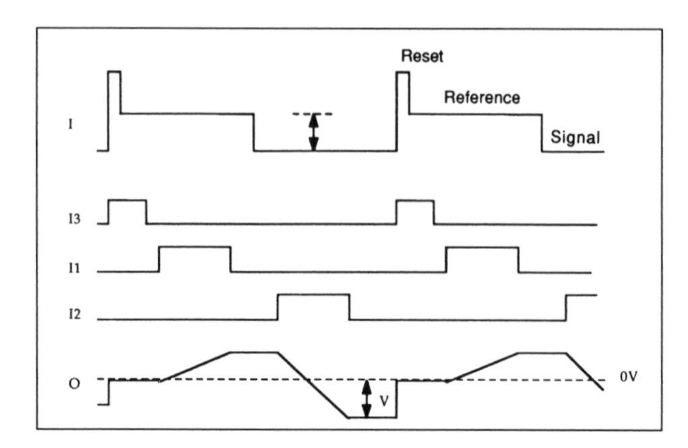

Figure 2.32 *Clock timing sequence of a double integration. The input video signal is I (we can recognize the three levels), the signal at the integrator's output is O.*

The advantage of this processing mode is the ability to integrate the signal. Its fluctuations (noise) are averaged during the integration while the signal itself is summed. The result is a notable improvement in the signal-to-noise ratio.

Double integration circuitry is difficult to construct from discrete components but high quality monolithic circuits such as Harris's HI-201HS exist. Note that the closing times of switches I1 and I2 must be perfectly constant, a situation which requires a good time base.

The double integration technique notably reduces the noise generated in the output stage of the CCD and in the preamplifier. It is often found in the cameras at professional observatories and, when used with a good CCD, reduces the global noise of the detector chain to less than 10 electrons.

Digital Double Sampling

Digital double sampling digitizes the video signal twice for each pixel—once during the reference level period and again during the signal level period. The difference is then determined digitally inside the computer. It is by far the simplest method, even though it is always necessary to use a sample-hold circuit to set the voltage to be digitized.

The time spent quantifying an image point is doubled with digital double sampling. This time increase is not really a problem, however, since today we can find fast converters at a reasonable price (12 bit conversion in one micro-second for less than $150).

The biggest problem with this method is the introduction of a quantification noise. This noise occurs because the two operands of the subtraction are quantified with a finite value and the subtraction itself produces a round-off error. To reduce this noise we either have to over-sample (digitize on 12 bits to achieve a valid 11 bit signal), or have an amplification gain such that the quantification noise is nearly negligible when compared to the analog noise. This latter approach is the one we used in our cameras.

An improvement in the signal-to-noise ratio is possible by carrying out several digitizings on the same level and by averaging the values obtained. In a first approximation, if we suppose the noise to be non-deterministic[2] and if there are N acquisitions on a level, the measurement precision is improved by a factor $\sqrt{N}$ but the readout time is multiplied by N.

Digital double sampling is used very successfully in the camera described here. It is in agreement with our objective of simplifying the electronics by assigning most of the work to the computer.

Figures 2.33 and 2.34 show the circuitry found on the analog-to-digital conversion board.

The video signal, stripped of its DC voltage component and amplified, is applied to the input of a Datel SHM20C sample-hold circuit. This device is an economical ADC sampler adapted to 12-bit conversion. Maintaining the voltage to be measured at a constant value during the conversion is essential. To understand this requirement, let us examine the functioning of an ADC.

Most converters work on the same principle of successive approximation (or successive weighing). The digital output is determined by comparing input voltages with internally generated reference voltages. Let us suppose that we apply 8V to the input of an ADC that converts a 0 to 10V signal to 8 bits resolution. The converter carries out a first comparison by internally creating a reference voltage equal to half its dynamic range, of 5V. The input signal is greater than this value, and the converter sets to "1" bit 7 of the logic output word. The second approximation is carried out by adding a voltage representing a quarter of its dynamic range to the value of the first approximation, i.e., 5 + 2.5 = 7.5V. Once again, the signal being digitized is greater than this reference voltage. Therefore, bit 6 is set at "1". Next, the device adds an eighth of 10V to the preceding comparison value: 7.5 + 1.25 = 8.75V. This time the internal voltage is greater and bit 5 is set at "0". The next step deducts a sixteenth of 10V from last comparison voltage and so on, each time dividing by 2 the value added to or subtracted from the comparison voltage. After 8 iterations the input signal is entirely converted.

[2]The digitizing is not done while the switching circuitry is rebounding

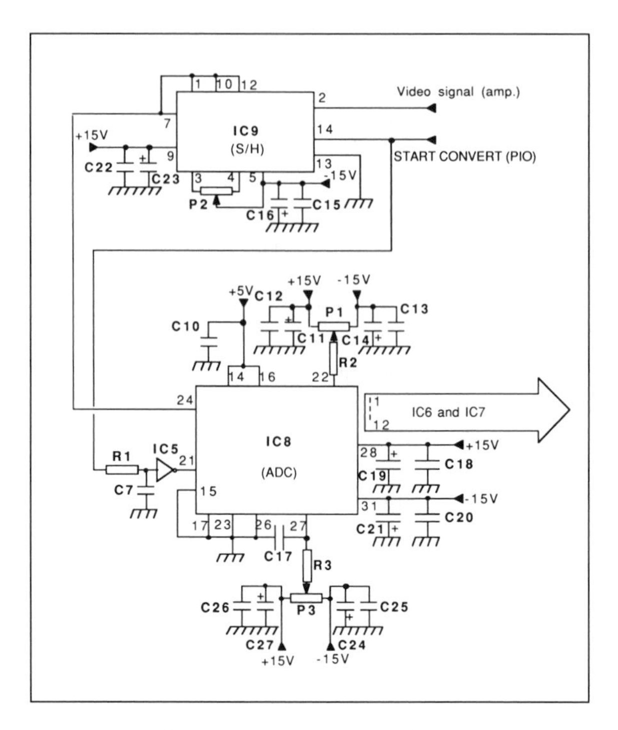

Figure 2.33 *Diagram of the sampler and the ADC portion of the analog-to-digital board.*

Components Found on the ADC Board

IC1	7915	C11,14,16,23,24,27	22μF
IC2	7812	C2,3,9,19,21	4.7μF
IC3	7805	C6	47μF
IC4	7815	C7	5.6nF
IC5,6,7	74LS04	R1	330
IC8	ADC84	R2	2.7M
IC9	SHM20C	R3	18M
C10	10μF	P1,3	22k
C1,4,8,12,13,15,22,25,26	100nF	P2	100k
C5,17,18,20	10nF		

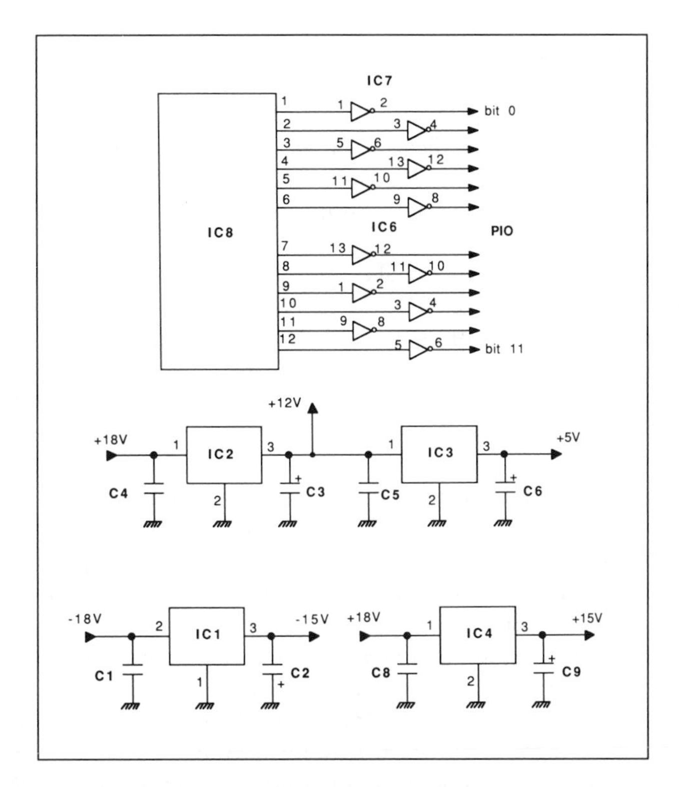

Figure 2.34 *Diagram of the ADC output and power supplies portion of the analog-to-digital board.*

A significant variation of the input signal during the conversion will cause the output to converge toward an incorrect value. This type of problem can be detected in the digital signal by the presence of abrupt jumps that have an amplitude equal to a power of 2. This phenomenon almost inevitably begins with 12-bit resolution levels if a sample-hold circuit is not used to hold the ADC's input voltage during the conversion. Since the conversion in the unit described here takes 10 microseconds, the signal is held while the conversion takes place. A hold begins when pin 14 of the SHM20C is brought high. This control signal is produced by the computer (bit 4 of B port). The trimmer, P2, is for zero adjustment.

The sampled signal is directed to pin 24 of IC8, the analog-to-digital converter. This is a 12-bit ADC84 device made by Analog Devices. The conversion takes 10μs. These devices are widely available and suitable substitutes exist (e.g. Datel ADC-HZ12B). In fact, it could easily be replaced by a more modern component such as Analog Devices' AD1678 that makes a conversion in 4μs and includes the sample-hold internally, thereby simplifying the layout even more.

The conversion starts when the logic signal on pin 21 passes from high to low. This is the START-CONVERT signal, the beginning of the conversion. A DATA READY signal indicates the end. We do not use this indicator which requires bit polling. A waitloop of sufficient duration is generated between the START-CONVERT instant and the recovery of the digitized data.

The START-CONVERT signal is produced from the signal-holding command with a small delay created in the RC circuit (R1-C7). This delay prevents the conversion from starting before the sampled signal stabilizes.

An ADC84 works with a negative logic—by applying a 0 voltage to its input, the 12 output bits become 1111 1111 1111 (i.e., 4095 in decimal). On the other hand, by applying +10 volts to the input, the 12 output bits read 0.

Trimmer P1 regulates the circuit's offset. By grounding the ADC input we should be able to adjust P1 so that the output value fluctuates between 4094 and 4095. The trimmer P2 lets us regulate the gain. By now applying +10V to the input, and adjusting P2 we should be able to get an output that fluctuates between 0 and 1.

The ADC board is mounted in a metal box, along with the clock board, and located close to the optical head. A 6-meter cable joins this box to the interface board located in the computer. The cable is multi-conductor and carries clock signals, the ADC's 12-bit output, and ground. Shielding is necessary to isolate the lines from interference. The signals (ADC) going to the computer should be isolated from the signals (clocks) coming out of the computer. If this is not the case transient signals may be produced and interpreted as additional clock signals. If interference is a problem, the clock signal lines can be bypassed to ground with 1 pF capacitors mounted on the ADC board. For severe interference a special driver IC will have to be installed at both ends of the line. These line drivers work in pairs: one is a transmitter (DS26LS31), the other a receiver (D26SLS32). If line drivers are used, twisted pair wire should be used between them, particularly if they are several meters apart.

The ADC's digital output is usually directly connected to its successive approximation register. The smallest amount of interference with this register causes the approximation to diverge. Lines connecting the ADC's parallel outputs must be short to minimize the possibility of interference.

IC6 and IC7 are inverters that buffer the ADC output. These two circuits should be mounted within several centimeters of the ADC output. In addition to buffering, IC6 and IC7 convert the ADC's negative output logic to a more traditional binary representation.

Most of the voltages are obtained with 78xx or 79xx monolithic regulators. IC2 and IC3 regulators should be mounted on small heat sinks. The amplifier is powered by ±15V produced on this board. In the example shown here, the camera's ±18V is supplied by a professional dual-voltage power supply. It is, however, fairly easy to make such a power supply since the current requirements are low: negative 0.2A and positive 0.5A.

Careful decoupling of the power supply leads is essential to minimize noise. A pair of capacitors, 4.7μF and 100nF, are mounted as close as possible to ADC input pins. The ADC board should also have a ground plane to minimize interference between the digital and analog signals. The ground potential of this plane should be established at a point located under the ADC. The ground line coming from the computer is also attached at this point. To further limit coupling between digital and analog signals the circuit board layout should physically separate these lines. The sensitive analog signal lines could be further isolated by alternating signal traces with traces connected to the ground plane. The trimmers that regulate the offset and gain levels should also be placed very close to the ADC.

Parasitic coupling is frequently a problem when analog electronics are associated with a computer. This is especially true when the computer has a switching power supply. There is no established rule for getting rid of these parasitics, but experience has led us to experiment with different ground wiring configurations. We have obtained good results by placing the CCD power supply close to the computer and directly connecting its ground to the computer's ground. At every point along its path the video signal must be transmitted via a coaxial cable until it reaches the sample-hold circuit (pin 2 of the IC9).

2.4 Setup

The camera described here is not difficult to build. Only a few circuits are required to be assembled, and it is possible to check them as they are wired. In addition to common tools such as a soldering iron and a multimeter, an oscilloscope is a necessity. This instrument is used to observe the video signal and detect noise sources. A two channel model will help to detect programming errors in the clock timing sequence. Since the camera works with relatively low frequency clocks, an inexpensive 10MHz bandwidth instrument is sufficient.

In the camera described here, the electronics (hardware) and the software are closely linked. It is at this level that the greatest difficulties arise. When a problem comes up, it is often hard to know which of the two—the electronics or the software—is the cause. The easiest thing to do in this case is to write a short program that carries out a unique function and then to observe the reaction in the electronics. We will come back to this subject with some practical examples.

2.4.1 The PIO board

The first circuit to be built should be the computer's interface. This board must work at once; otherwise troubleshooting will be very difficult. Indeed, with a simple oscilloscope, it is extremely hard to understand the signals carried on the computer's bus—a logic analyzer is required and it is a luxury few amateurs can afford.

Another difficulty in checking the operation of this board has a mechanical origin: with some computers it is very hard to measure the signals when the board is mounted on the bus—a bus extension card is needed to elevate the PIO board so that test leads can be connected to the various components.

Three simple precautions should be taken to protect the PIO and the computer from damage:

1. The power leads to the integrated circuits must be wired correctly. First, remove all the IC's, apply power to the board, and then check to see that ground and +5V levels are connected correctly. It is relatively quick and easy to trace these levels throughout the circuit board with a VOM meter.
2. A computer bus address conflict can lead to component failures—usually the 74LS244's. Only one board should be addressed at 768 to 771. This address is as a matter of practice reserved for prototype boards so it should not be a problem unless other prototype boards are simultaneously installed.
3. Because of their number, it is easy to mis-wire the address and data lines. To prevent this problem the wiring should be methodically checked with an ohmmeter before applying power.

If the board does not work, first check for the presence of the Chip-Select signal at pin 6 of the PIO by continually writing to the PIO's register (e.g., `OUT 768,0`).

To further check the PIO's operation a short program will suffice. For this purpose all the PIO ports are programmed in output and all the bits should be set successively at 1 and at 0. The program will look like the one which follows:

```
OUT 771,128  :REM all the ports in output
x=0
start:
 OUT 768,255
 x%=0
 OUT 769,255
 x%=0
```

```
 OUT 770,255
 x%=0
 OUT 768,0
 x%=0
 OUT 769,0
 x%=0
 OUT 770,0
GOTO start
```

Remember that the instruction `x%=0` is there to produce a small but indispensable delay between two successive `OUT` instructions.

Once the program is running, use the oscilloscope to confirm that the 24 bits of ports A, B and C change state cyclically. If the test is passed successfully, it is very likely that the PIO's input is also functional.

2.4.2 The Clock Board

After assembling the clock board, check the following:

1. Regulation of the +5V at the logic circuits;
2. Voltage at pin 5 of the 75361 translators. This should be adjusted to about 10.5V with voltage regulator trimmers;
3. The CCD's DC voltages (V_{dd}=14V, V_{rd} = 12V, V_{gs} = 8V, V_{ss} = −1.5V, V_e = 12V).

After these precautionary checks install the integrated circuits in their mounting sockets and connect the clock board to the PIO board.

The following program carries out a minimum clock timing sequence to make the array work. There is no attempt made here to digitize the image. The program only generates phases ØP, ØM and ØL at a high speed for easy inspection of the clocks and later of the video signal.

```
REM*****************
REM*CCD TEST PROGRAM*
REM*****************
OUT 771,144:REM initializing PIO (10010000)
start:
REM***transfer image zone---memory zone***
FOR i% = 1 TO 145
 OUT 769,3 : REM 00000011
 x%=0
 x%=0
 x%=0
 OUT 769,0 : REM 00000000
 x%=0
 x%=0
 x%=0
NEXT
REM***transfer memory zone---horizontal reg.***
FOR i% = 1 TO 145
 OUT 769,2 : REM 00000010
 x%=0
 x%=0
 x%=0
 OUT 769,0 : REM 00000000
 x%=0
```

```
  x%=0
  x%=0
  REM***readout horizontal register (signal G to 1)
  OUT 769,8 : REM 00001000
  FOR j% = 1 TO 120
  REM***hold delay of signal G at level 1***
  NEXT
  OUT 769,0 : REM 00000000
NEXT
A$ = INKEY$
IF A$="s" then END
GOTO start
```

It is important to understand how this program works. For each `OUT` order we translated the words written in the PIO into binary so that it will be easier to see the state of the bits continuously during the readout of the CCD. The purpose of the first loop is to move the charges into the memory zone. For this procedure, clocks ØP and ØM are simultaneously activated (bits 0 and 1 of port B). Index 145 of this loop corresponds to the number of lines in the image. The second loop controls the transfer of the lines from the memory zone toward the horizontal register. The internal loop (`FOR j% =1 TO 120. . .`) sets a delay during which the GATE signal is clocked high and thus during which clock ØL (at 500KHz) is applied to the horizontal register. The length of this delay is to be adjusted as a function of the computer's speed so that it lasts at least $218 \times 2 = 436$ microseconds (218 is the number of pixels contained in a line and the factor 2 comes from our reading 0.5 pixels per microsecond). In the present case, we remind the reader that the computer is an 8MHz PC AT. It is advisable to send more ØL clocks than necessary since doing so improves the emptying of the register (a good choice is a duration 20% longer than the minimum delay required). After completing the readout of the memory zone, the next image is moved into the memory zone and so on as long as we do not push the "s" key on the computer's keyboard. Here the integration time is equal to the CCD's readout time.

For the time being the CCD should not be connected. When running the program, we should observe the state of the signals on the clock board:

1. On pins 6 and 7 of IC10, clocks Ø2P and Ø1P. These clocks appear as a train of very short pulses followed by long pauses corresponding successively to the transfer of charges from the image zone to the memory zone and to the emptying time of the memory zone;

2. On pins 6 and 7 of IC11, clocks Ø2M and Ø1M, which present the same train of short, tightly spaced pulses seen at ØP (transfer from image zone "Iz" to the memory zone"Mz") along with additional clock signals that are more widely spaced and which correspond to the transfer of a line from the memory zone to the horizontal register;

3. On pins 6 and 7 of IC12, clocks Ø2L and Ø1L should be easily recognizable with their 500KHz frequency;
4. On pin 7 of IC16, we should see the ØR operating at the same frequency as ØL.

If clocks Ø1L and Ø2L are not present, check that the GATE signal produced by the computer is indeed high (pin 12 of IC18) and that the ØL signal is low (pins 9 and 10 of IC18). If in spite of everything all the Ø1L and Ø2L clocks are still not produced, check that the oscillator is running at 1MHz at pin 1 of IC 14.

Usually the ØR clock does not at first appear because the monostable vibrator, IC17, is not properly adjusted. Turning trimmer P10 up and down while watching the oscilloscope should enable one to catch sight of ØR. The trimmer is set so that the leading edge of ØR is produced at the same time as the leading edge of Ø2L. ØR's high level width is not too important; it can be set at approximately 300 ns.

2.4.3 The ADC board

This board must be tested with the CCD removed from its socket. The presence of the absent CCD is simulated by applying a DC voltage that can be varied between 0 and 10V to the input of the sample-hold circuit (pin 2 of IC9). This voltage should be stable to better than one millivolt peak to peak. The following program allows us to digitize the voltage applied at the input of the sample-hold circuit:

```
REM ********************
REM * ADC TEST PROGRAM *
REM ********************
OUT 771,153 : REM initialization of PIO
start:
 OUT 769,0
 x%=0
 x%=0
 x%=0
 x%=0
 x%=0
 x%=0
 x%=0
 OUT 769,16 : REM start convert
 x%=0
 x%=0
 x%=0
 x%=0
 x%=0
 x%=0
 x%=0
 PRINT 256 * INP (770) + INP (768), : REM ADC readout
 a$=INKEY$
 if a $="s" THEN END
 GOTO start
```

After initializing the PIO, the first `OUT` sets the sample-hold circuit command at `0`. In this case, the sampler output follows its input. After a delay to establish a stable signal, adjusted by the number `x% = 0` statements (typically 10 μs), a hold order as well as a conversion order is sent to the ADC (start-convert). A new delay is generated to leave the ADC enough time to proceed with the conversion. To provide a margin of safety this time should be slightly longer that the ADC's 10 microsecond delay (e.g., 10 + 3 μs). After the conversion the 12-bit output of the ADC is recovered and displayed on the monitor. Notice that the most significant byte is multiplied by 256 to create a 16-bit word.

Varying the input voltage from 0V to 10V, should cause the ADC output to correspondingly vary from 0 to 4095. Compare the voltage on pins 2 and 7 of IC9—they should be the same. Also, with an oscilloscope, check for the presence of the command signals on pin 14 of IC9 and pin 21 of IC8.

Normally, when the input voltage increases by about 2 millivolts, the binary word should increase by a quantification unit. It is important to check for this relationship over the entire measurement range (0 to 10V). If the readout word abnormally departs from this relationship check the following:

1. That the 12 bits of the output are correctly wired;
2. That the PIO's bits PC4 through PC7 are at ground potential (4 most significant bits of the most significant byte);
3. That the conversion time allotted to the ADC is long enough.

If the measured voltage is well regulated and filtered (better than the millivolt), the binary word readout should not fluctuate by more than a quantification unit. This underscores the importance of a clean, low noise power supply.

Trimmer P2 sets the offset of the sample-hold circuit. With the sampler's input grounded and the command line low, adjust P2 to null the voltage at the sampler's output. The ADC's offset is set by adjusting trimmer P1. Setting the ADC input at 0.000V should cause the output word to fluctuate between 0 and 1. To set the full scale of the ADC, apply 10.000V to the input and adjust P3—the output word should fluctuate between 4094 and 4095.

2.4.4 The Amplifier Board

The only adjustment on the amplifier board is the offset voltage control for the video signal's DC component. Apply about +9V to the amplifier's input. Set the fine tune trimmer mid-way and adjust the rough tune trimmer (10kΩ) for a null voltage at the amplifier output. This setting will probably have to be done again once the CCD is installed since the video DC voltage level is a function of the detector's internal characteristics and of the applied biases.

2.4.5 The CCD Board

Given its simplicity and the absence of controls, the CCD board should not be a problem if properly wired. Before installing the CCD, double

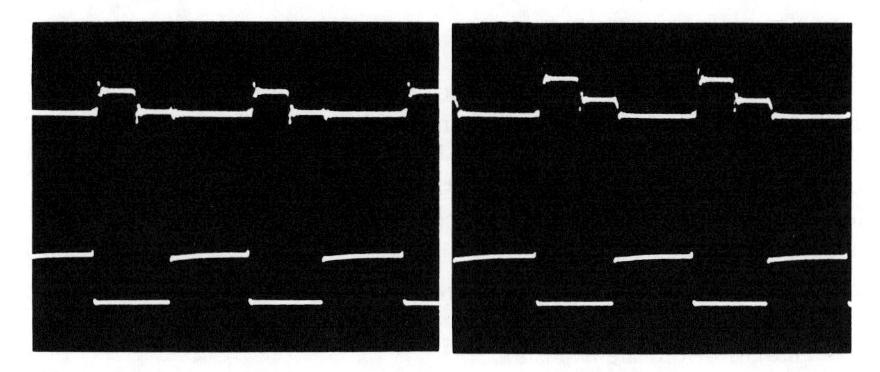

Figure 2.35 *The video signal at the output of the CCD (at the top). At the bottom we see the track of clock Ø1L. On the oscillogram on the left, the detector is placed in the dark. The reference and signal levels are then almost at the same level. On the other hand, the reset level is clearly seen. On the right the CCD is slightly illuminated, and the signal level separates from the reference level negatively. The maximum separation between these two levels depends on the array. A value of 0.7V is typical. We will remember that a positive 9V voltage is superimposed on the video signal (this offset is not visible on these oscillograms).*

check for the presence of the correct signals on each pin while running the minimum clock timing sequence program detailed above.

If everything checks out properly, place the CCD in its socket while taking precautions to prevent static charges that can destroy the detector (see section 1.7.1). Connect an oscilloscope probe to the emitter of the follower transistor, and place a black cloth over the CCD. Turn the camera on and start the minimum program—the long awaited video signal should appear. While lifting the black cloth covering the CCD observe the oscilloscope patterns. The video component should shift when compared to the reference level. This is a sign that the CCD is sensitive to light.

The saturation voltage should be around 1V. If the CCD is too brightly illuminated, this level may be exceeded, showing that the functioning is strongly disturbed.

A golden rule: never observe the video signal by connecting a probe directly to the CCD's output pin. A short circuit here will destroy the detector. The signal must always be picked up after the follower transistor where a short circuit only destroys the 2N2222, which costs about 50 cents, but protects the CCD.

The video signal is next connected to the input of the amplifier. At the amplifier's output the appearance of the levels changes noticeably because of the low bandwidth of the LF356. The offset has to be fine tuned in such a way that the reference level is about +100mV at the amplifier output.

2.4.6 Digitizing an Image

Writing a program for the acquisition and visualization of images should proceed along with the construction and testing of the electronics. Indeed, any malfunction in the hardware will be immediately visible. Information about visualization will be found in the chapter about image processing. Here we will only discuss the actual acquisition of the image. The program should include the following:

1. Several fast reading cycles of the CCD (minimum clock timing sequences already explained) to clean the detector of any residual charges before the exposure;
2. Generation of a delay corresponding to the integration time;
3. Transfer of the integrated electronic image into the memory zone;
4. Slow reading of the memory zone with digitization and storage of the image in the computer's memory. An example of an acquisition routine is shown below.

In the following routine the ability to bin is included. As a reminder, this operation consists in gathering several neighboring pixels to form superpixels, decreasing the resolution but increasing the sensitivity. The binning factor can be set in `X` (lines) or in `Y` (columns). A binning factor of `(1,1)` corresponds to a normal image, a factor of `(2,2)` corresponds to a compression rate of 2 on the two axes, and so on. The binning factor is contained in the variables `BINNX` and `BINNY`. If binning in `X` is done by summing the lines in the horizontal register, binning in `Y` is obtained by arithmetic summing within the program.

The image is stored in a table of integers `a% (i%,j%)` of size `(145,218)`. The real variable `INTEGRATION!` contains the length of the integration time and is used by the `TEMPO` routine to generate the corresponding delay.

```
REM ***********************************
REM * PART OF ACQUISITION SUBROUTINE *
REM ***********************************
FOR k% = 1 TO 7
 CALL EMPTY : REM the sequence of fast reading of the CCD is called
NEXT
IF integration! <> 0 THEN
 CALL TEMPO : REM delay corresponding to integration time
END IF
FOR i% = 1 TO 145
 OUT 769,3 : REM image zone to memory zone transfer
 x% = 0 : x% = 0 : x% = 0
```

```
 OUT 769,0
 x% = 0 : x% = 0 : x% = 0
NEXT
ii% = 0
FOR i% = 1 TO 145 STEP binnx%
 ii% = ii% + 1
 FOR k% = 1 TO binnx%
 OUT 769,2 : REM sum binnx% into horizontal register
 x% = 0 : x% = 0 : x% = 0 : x% = 0
 OUT 769,0
 x% = 0 : x% = 0 : x% = 0 : x% = 0
 NEXT
 jj% = 0
 FOR j% = 1 TO 218 STEP binny%
 s% = 0
 jj% = jj% + 1
 FOR k% = 1 TO binny%
 OUT 769,16 x% = 0 : x% = 0 : x% = 0 : x% = 0
 OUT 769,0
 x% = 0 : x% = 0 : x% = 0 : x% = 0 : x% = 0 : x% = 0 : x% = 0
 OUT 769,16
 x% = 0 : x% = 0 : x% = 0 : x% = 0 : x% = 0 : x% = 0 : x% = 0
 ref% = 256 * INP(770) + INP(768) : REM reference level
 OUT 769,20
 x% = 0 : x% = 0 : x% = 0 : x% = 0
 OUT 769,4
 x% = 0 : x% = 0 : x% = 0 : x% = 0 : x% = 0 : x% = 0 : x% = 0
 OUT 769,20
 x% = 0 : x% = 0 : x% = 0 : x% = 0 : x% = 0 : x% = 0 : x% = 0
 REM signal level -- reference level
 s% = s% + 256 * INP(770) + INP(768)
 NEXT
 a% (ii% ,jj% ) = s%
 NEXT
 OUT 769,8 : REM cleaning horizontal register
 FOR k% = 1 TO 80
 NEXT
 OUT 769,0
NEXT
```

This program fragment is difficult to follow but important to the operation of the camera so it will be discussed line by line.

The program starts with several fast readings of the array (`CALL EMPTY`); The routine `EMPTY` carries out the basic clock timing sequence of the CCD reading without digitizing (see section 2.4.2). The number of cleaning cycles is high (7) to make sure that the array has no trace of any strong illumination preceding the exposure. Having more cleaning cycles than necessary does not impair the CCD's functioning—only time is lost between each acquisition (a cycle lasts about 70 ms). The exact value of this number is not critical, but experience shows that 7 cycles is the lower limit.

A temporization routine is then called (`CALL TEMPO`). The user specified integration time is managed with this routine. With a long exposure, the seconds appear one after the other on the computer screen. After the integration period, the memory zone is read (`CALL ZAPMEMO`). This routine transfers the charges from the memory zone into the horizontal register with rapid readout of the latter (this is the second part of the basic readout program of the CCD given in section 2.4.2). This operation precedes the photocharge transfer from the image zone (Iz) to the memory zone (Mz). Its purpose is to eliminate from the memory zone any thermal charges that may have accumulated during integration. Without this precaution the first two or three lines of the digitized image could be polluted by accumulated thermal charges when they move into the memory zone during the transfer Iz-Mz.

The transfer of the 145 lines from the image zone to the memory zone then takes place. For this transfer we set clocks ØP and ØM together (`OUT 769,3`) at high level, then again at low level, still together (`OUT 769,0`). This process is repeated 145 times.

After the Iz-Mz transfer, a line of the memory zone is sent into the horizontal register during the sequence `OUT 769,2`—`OUT 769,0`. It corresponds to a time period equal to ØM. The length of this period is relatively long to ensure a good transfer in this part of the CCD (see Figure 1.25).

The 218 points of the horizontal register are then digitized. The process begins by digitizing the reference level of the first pixel. The delay between instruction `OUT 769,0` and instruction `OUT 796,16` is fundamental. It corresponds to the duration of the quiescent level at the amplifier's output. This delay is set with an oscilloscope so that the level is exactly horizontal at the instant the sample-hold circuit issues a hold command. This must occur whatever the level of the signal (within the limits of saturation, of course). Once the reference level is established, the DC component is blocked and digitizing of the video component begins (`OUT 769,16`). A delay follows to give the ADC time to convert ($10\mu s$).

Note that the digitizing sequence of the reference level starts with an `OUT 769,16`. This instruction holds the signal at the input of the sample-hold circuit to avoid transmitting a reset parasitic pulse to the amplifier, where it might cause erratic operation.

At the end of the digitizing delay, the ADC output is read. The most significant digit of the 12 bit word is located at address `770` (in decimal). We multiply it by 256, then add to it the least significant digit (address `768`). The result is stored in a temporary variable—`ref`.

The next instruction (`OUT 769,20`) sets Ø1L high. This initiates the charge packet read in the output diode and causes the appearance of a signal level at the CCD's output. The sampler is frozen during this operation.

Next the sampler is set in "follower" position (`OUT 769,4`) during a delay that is long enough for the signal level to stabilize. This delay is slightly longer than the reference level delay because the establishment time here is more dependent on the signal level. The hold order of the signal level is then sent at the same time as the `START CONVERT` of the ADC (`OUT 769,20`).

A delay is again needed for the ADC to digitize. Next the amplitude of the signal level is read and the reference level amplitude is removed (this is numerical double sampling). The result of digitizing the first point of the image is stored in the two-dimensional table `a% (..., ...)`.

The procedure continues by reading the next point in the horizontal register. After digitizing 218 points, a rapid cleaning of the horizontal register is carried out by sending about a hundred ØL clocks at 500KHz. This removes any trace of charges that may linger due to transfer inefficiencies. Next, a new line is then sent from the memory zone into the horizontal register, which is again completely read. This process continues until transfer from the memory zone is complete.

It is important not to have too much delay between each change in the state of the clock timing sequence. The delay period should be sufficient to establish stable video signal levels or to leave enough time for the converter to work. However, too long a delay can be the source of a low frequency noise in the image. With a long delay, random events (electronic instability, temperature changes of the CCD, etc.) have a greater chance of being recorded. In practice the point period (time to digitize a pixel of the image) should take about 100 microseconds.

The sequence that we have just described is repeated for each image acquisition. It is called a *single shot* sequence because only one frame is read at each acquisition.

This routine is given as an example. It will need some changes when it is moved to a machine differing from the one used in this example. Special note should be taken of the flexibility of having the computer manage the generation of the clock timing sequences. I leave it to the reader to determine the number of integrated circuits that would have to be assembled into wired electronics to produce the above functions for his computer.

At this stage of the operation, three tests have to be carried out:

1. When the CCD is placed in complete darkness and is operating at ambient temperature, the image slopes from one side to the other in the direction of the lines. This slope occurs because the charge packets of the first pixels leaving the horizontal register are affected little by the dark current, while the information contained in the most distant lines of this register goes through the detector during the whole readout period and is therefore more subject to thermal charge effects. In our camera digitizing a frame lasts about 1.5 seconds. The last digitized pixels therefore contain a surplus of thermal charges corresponding to an exposure in the dark lasting as long as the readout time, which is far from being negligible. With a TH7852 array at 20°C and a readout time of one second, the image's non-uniformity reaches 10% to 20%. The production of thermal charges is generally stronger

on the periphery of the detector, giving a characteristic appearance of a raised edge to the image. But, starting at 0°C, the contribution of thermal charges during the readout is almost non-existent, and in darkness the images become "flat".

2. With the CCD being progressively illuminated until saturation of the converter (level 4095), one should observe a steady increase of the signal in the image. An abrupt increase in the level on all or part of the image can be considered as a charge transfer problem. In this case, either the phasing of the clocks is incorrect, the high level is not high enough, or the array is insufficiently cleaned before the exposure. With the electronics described here, the full scale of the converter represents a voltage of about 450 mV at the CCD's output. The saturation voltage specified by the manufacturer and actually checked is around 1V, therefore, some leeway exists. Of course, if the detector is too strongly illuminated, the CCD's output signal will exceed the threshold value and the converter will present an incorrect reading of 4095.

3. A photographic lens is coupled to the CCD to take the first real image. This lens should have a 28 mm or 35 mm focal length to provide a reasonable field for the CCD's small sensitive surface. To prevent stray light from entering the system an adapter will have to be made to mount the lens onto the CCD's head. After taking pictures of friends and objects, the next step is to test the system using an artificial star. The star can be the filament of a small incandescent light bulb which is just barely lit. It should be positioned three or four meters from the camera. Until a signal level that saturates the ADC is reached, the star displayed on the monitor should remain a point source without excessive diffusion and with no streaks. When the luminous flux is two to three times greater than the saturation threshold of the CCD (for that, increase the brightness and/or the integration time), a characteristic streak starting from the star will start to appear. This streak is the sign that there are too many charges to move and that they have been left behind during the transfer (poor transfer efficiency because of the saturation). The voltage V_e of the anti-blooming drains can be set in such a way that the number of charges stored in a well is limited. One must be careful, however, to avoid the opposite effect, because the dynamics can be affected. This disturbance is usually shown by the appearance of "noise" that is spatially located for bright illumination, corresponding to the efficiency of the anti-bloooming device which depends on the position on the array.

Fine tuning a CCD can be tedious. Unless the CCD has very low noise and very high quality electronics (10 to 20 electrons), the settings specified by the manufacturer can be considered as correct and variations on the order of half a volt around the nominal voltages will usually have no visible consequences. It is nevertheless a good idea to carry out the supplementary tests which follow:

1. To evaluate the influence of the shape of the clocks on the transfer efficiency, the CCD is weakly illuminated with the artificial star, and we check the computer monitor to see if the base of the star is in fact round (the lens must be perfectly centered with respect to the CCD and the star aligned with the optical axis). The cosmic rays that sometimes hit the sensitive surface produce spots that rarely cover more than one pixel and can be used for this test. Professionals use X-ray sources that generate such spots on demand. An asymmetry of the image spot can be the sign of an excessive transfer inefficiency. The falling edges of the clocks should be checked to insure that they do not present two distinct negative peaks. Adjusting the clock levels and the phase crossings usually cures this problem.

   Another method to check transfer efficiency rapidly and to evaluate the influence of the settings is to send more clock signals than necessary to read the horizontal register. If the CCD is illuminated, a more or less abrupt transition in the signal between the last significant pixel and the first surplus point can be seen in the video signal. The appearance of this transition is a function of the transfer efficiency because it shows the residue of charges left behind by the last useful pixel and recovered by the points read in supplement. (This technique is called EPER—Extended Pixel Edge Response.)

   If Q_e is the sum of the intensities of the supplementary pixels (considering 2 to 4 points after the transition is usually enough), if Q_i is the intensity of the last significant point of the horizontal register, and if N is the number of columns (number of transfers in the register), then an evaluation of charge transfer efficiency (CTE) will be given by the formula:

$$\mathrm{CTE} = 1 - \frac{Q_\mathrm{e}}{N \cdot Q_\mathrm{i}}.$$

   The same procedure can be used to measure the transfer efficiency in the direction of the lines (the direction of the transfer is from the image zone toward the memory zone). To do so, one simply reads more lines than necessary and measures the streaking effect in the first supplementary lines. With Thomson CCD's, this is not straightforward because the last points read in the horizontal register act as dark references and are therefore not sensitive to light. Therefore, to create the transition, a dark current has to be generated by a short integration period and by moderately cooling the CCD. Correct measurement of the CTE requires the averaging of several CCD readings to decrease the noise.

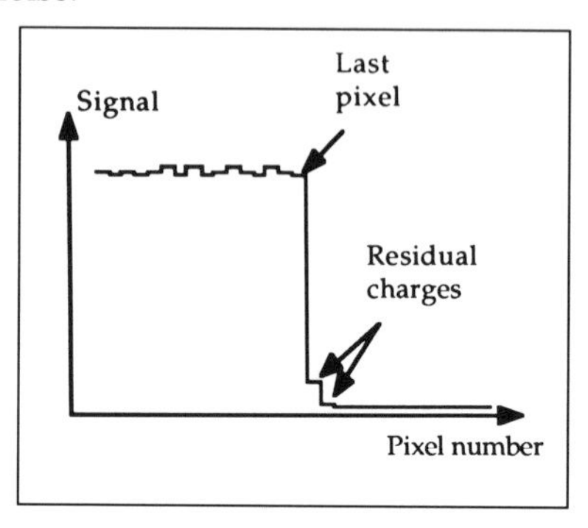

Figure 2.36 *Appearance of the video signal at the transition between the useful pixels and the supplementary pixels.*

2. The detector's linearity is affected by varying video levels during digitizing. These levels can be adjusted by varying the exposure time. Linearity is also dependent on the amplifier's operating conditions. So that the latter works correctly, it is important that $V_{dd} - V_{dr} =$ 2V to 2.5V. It is advisable to make a light transfer curve to check both the linearity and the camera's noise. Linearity can also be checked by stabilizing the temperature of the CCD around 0°C and then verifying that the dark current is indeed proportional to the integration time (which in this case will go from 0 to about 60 seconds). Several measurements are averaged for each integration time to decrease the random variation of the thermal noise.

3. The value of the voltage V_{gs} has very little influence on the CCD's operation. It always has the value of 8V.

4. A truly random noise should appear on the image when the CCD is dark and cooled below −20°C. Any trace of correlated noise must be completely removed. Correlated noise is caused when an image does not completely clear before another image is loaded. An asymmetry in the appearance of the noise from one edge of the image to the other is the sign that some of the clock frequencies are too high. If the noise has a periodic pattern, there is a grounding or a radiation problem. Problems like these may be traced to 60 Hertz coupling, or interference from signals generated in the computer. It is important to isolate the optical head and the high gain portions of the electronics (from

the amplifier to the ADC) from any source of radio-frequency radiation (television set, large transformer, or even the computer itself). If the pattern persists rf shielding of the interconnecting wires and the electronic enclosures will have to be done. At the Pic du Midi Observatory there is a very powerful television transmitter nearby which causes severe interference. With one of my first cameras I had to wait patiently until the TV went off the air—well after midnight—before my CCD could acquire good images!

Once these tests have been completed, all that is left to do is to mount the camera on a telescope and begin taking pictures.

2.5 Toward an Ever Bigger CCD

The Thomson TH7852 CCD has many positive qualities, the first of which is cost—about \$350. In spite of the fairly large size of the pixel, the resolution is excellent. It delivers excellent images of planets even in the infrared spectrum. Its anti-blooming gate allows observation of dim objects close to much brighter objects. Phobos and Deimos were recorded near the limb of Mars, which is considerably brighter, without blooming using the 1-meter (T1M) telescope at Pic du Midi during an observing run in September 1988.

The disadvantage of this array is the small sensitive surface (6.2×4.3 mm). Therefore, I developed a new camera using the Thomson-CSF TH7863 CCD. This device has 384×288 useful pixels, each 23 μm long. The sensitive surface measures 8.8×6.6 mm which is double the size of the TH7852. Because the TH7863 has no anti-blooming gate the whole pixel is photosensitive. The image and memory zone transfer is four-phase, while the horizontal register is two-phased. The output stage is very sensitive. For example to obtain a 1mV output, 1100 electrons were needed with the TH7852, but the TH7863 needs only 470.

Since the device itself is small in comparison with the sensitive surface (25mm long, 18mm wide, and 5mm thick) the optical head can be made very compact. The two rows of 10 pins are 0.6 inches apart and the pins are spaced at $^1/_{10}$ of an inch.

The nominal bias values are $V_{dd} = 15.5$V, $V_{gs} = 2.5$V, $V_{dr} = 13.3$V, $V_s = V_{ss}$ (biasing of the output stage) and $V_{ss} = -3.0$V.

Moving from a TH7852 to a TH7863 was not difficult. Much of what was learned with the first camera was directly transferrable. The clock phases are generated by the computer and the digitizing chain is identical. Only a few details were modified.

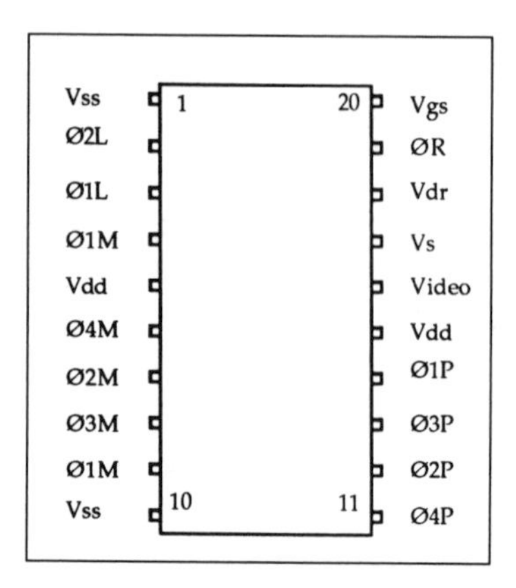

Figure 2.37 *Pinnout for the TH7863array.*

The TH7863 requires more clocks than the TH7852—11 instead of 7. However the 4 phases of a register can be easily separated into two groups of two phases. Thus for the transfer into the image zone, phases Ø1P and Ø3P are symmetrical. The same is true for phases Ø2P and Ø4P. The computer needs only make phases Ø1P and Ø2P, the other clocks can be derived from them using discrete IC's (see Figure 2.38).

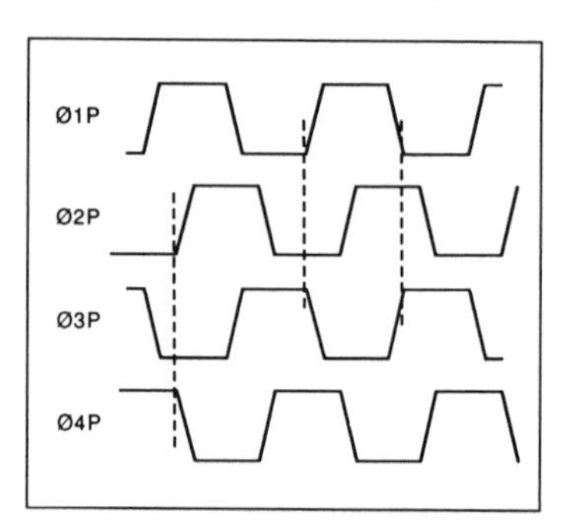

Figure 2.38 *Clock phasing for the transfer into the memory zone. The program generates phases Ø1P and Ø2P while Ø3P and Ø4P are obtained with inverters from the programmed clocks. The shift between Ø1P and Ø2P is regulated by timing software in the program.*

Generation of the clocks for transfer into the memory zone and for the image zone are identical. Figure 2.39 shows the electronic diagram to make clocks ØP (similar diagram for ØM). Note the use of a DS0026 (National) instead of the SN75361 for the conversion of the TTL to CMOS levels. The DS0026 is easier to use because it needs only one power supply. We recommend this choice.

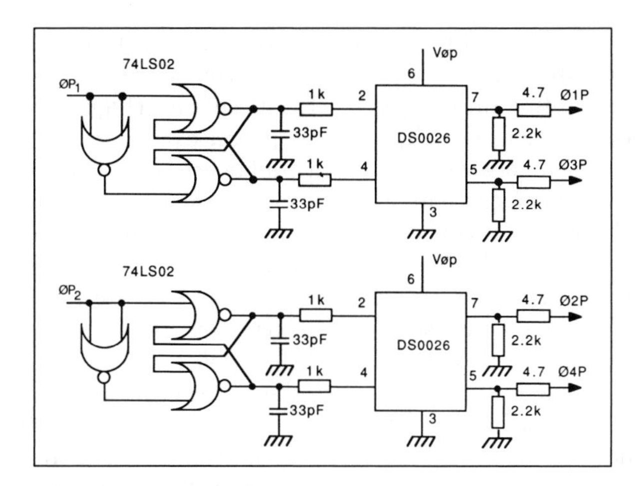

Figure 2.39 *Diagram of the electronics used to produce phases Ø1P, Ø2P, Ø3P and Ø4P.*

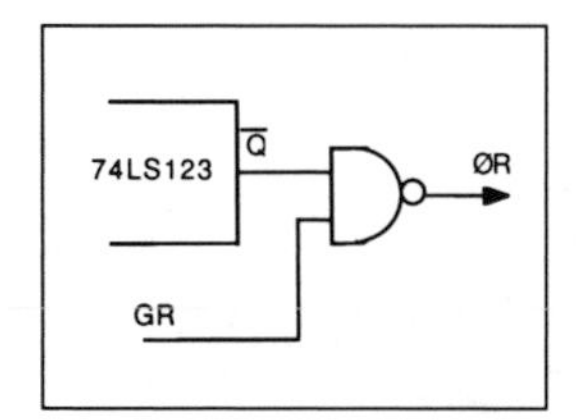

Figure 2.40 *When the signal GR, controlled by the computer, is at zero, the production of clock ØR is on hold.*

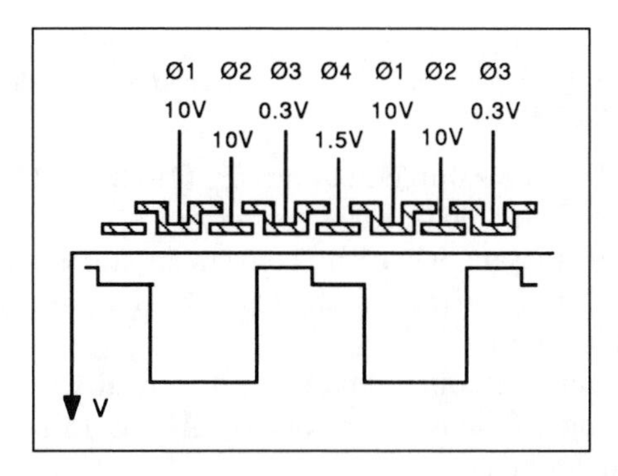

Figure 2.41 *Appearance of the diagram of the voltages during the exposure.*

The horizontal register works with two phases. Therefore, electronics similar to those described for the TH7852 array can be used for producing phases ØL and ØR. A command (GR) has been added to hold at will the generation of ØR, allowing binning in the output diode.

While integrating the charges into the image zone (Figure 2.41), two adjacent phases are clocked high, two other phases clocked low with, however, a short difference between them. This difference improves the resolution by better channeling the photocharges toward the nearest potential well.

The difference in bias between the two low level adjacent phases is obtained with the circuit in Figure 2.42. Here a shift in the negative voltage on pin 3 of the DS0026 driver controls phase Ø4P (see Figure 2.39). When the signal INTE is high, transistor T2 conducts and transistor T1 is blocked. This shifts the ground point of DS0026 by twice the diode voltage, i.e., 1.2V. This voltage in turn is added to the voltage constant of the DS0026 which is around 0.3V, so that the low level of Ø4P is 1.5V during integration. At the moment of charge transfer from the image zone to the memory zone, the INTE signal is set at 0 and transistor T1 becomes the conductor and T2 is blocked. The low level of Ø4P is then that of the offset of DS0026 (0.3V).

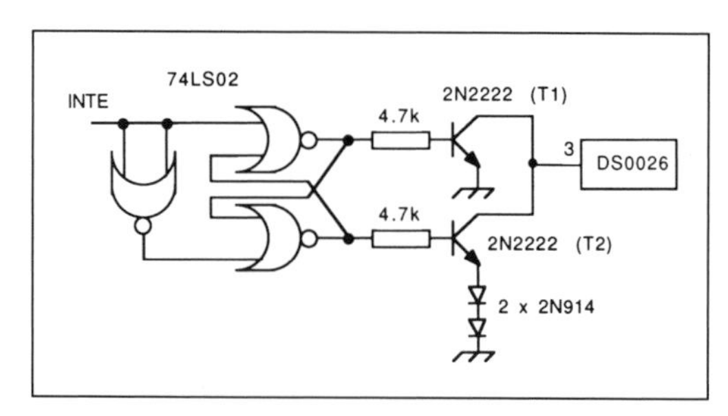

Figure 2.42 *Instead of connecting pin 3 of the DS0026 that controls phases Ø2P and Ø4P directly to ground, a circuit is introduced to shift the low level of Ø4P during the integration.*

The command of the sample-hold circuit (`TRACK/HOLD`) has been separated from that of the analog-digital converter (`START-CONVERT`) to better manage the precise instant when the conversion starts compared to when hold of the video signal starts. For the rest, the diagram is identical to the one for the TH7852 array.

The software must of course be changed to deal with more clocks and an image containing 3.6 times more pixels. About 15 seconds are needed to digitize the whole image.

Here is an example of a part of a routine for reading the CCD that transfers from the image zone to the memory zone. The allocation of the bits for port B of the PIO) is:

bit 0 = Ø1P
bit 1 = Ø2P
bit 2 = Ø1M
bit 3 = Ø2M

```
For i% = 1 TO 290
    OUT 769,5 : REM 00000101
    x% = 0
    x% = 0
    OUT 769,15 : REM 00001111
    x% = 0
    x% = 0
    OUT 769,10 : REM 00001010
    x% = 0
    x% = 0
    OUT 769,0 : REM 00000000
```

The other bits of port B have the following assignments: bit 4 = ØL (only one useful bit since the register is two-phased); bit 5 = GR (freezing of ØR for binning); bit 6 = `HOLD` (sample-hold command); bit 7 = `START` (converter command). We had to use the most significant nibble of Port C to generate additional signals. Bit 4 = `INT` (defines low level of ØP during integration); bit 5 = `GL` (rapid generation of ØL); bit 6 = reserved (MPP mode?); bit 7 = reserved Port A; and the least significant bit of port B is used for the ADC readout.

Long digitizing periods can create difficulties. Therefore, it is necessary to cool the detector, even during the setup period.

But there is another more subtle problem. An image has just been integrated, then transferred into the memory zone to be read. During the reading of the memory zone, the image zone is still "active" and registers incident photons. After a certain time T, a function of the incident flux, the photosites become saturated. If T is lower than the readout time of the memory zone, the latter's contents will be strongly affected by the surplus of charges produced in the image zone. We then see an image correctly digitized up to the line read before time T and then a complete mess (usually a strong signal that looks like a global saturation of the detector). To avoid this phenomenon, the sensitive surface has to be protected from light during the readout: the image can then be correctly digitized. The other way to get rid of this phenomenon is to illuminate the detector sufficiently so that it does not saturate during a period equal to the readout period. In fact, we are almost always doing this when we take deep sky images, and the problem described is therefore completely unnoticed with "normal" use of a CCD. All the same there are some exceptions, for example with luminous objects such as planets. But the situation is less dramatic than it seems; in practice the image zone can tolerate an oversaturation without affecting the readout of the memory zone. If the emptying of the CCD lasts 15

Figure 2.43 *A (small) portion of NGC 7000, the North America Nebula showing the area of the Gulf of Mexico. Images taken with 10 minute exposures with a 28mm telescope and a TH7863 CCD.*

seconds, we can take full dynamic range exposures lasting 0.1 second with no problem. If the integration time must be even shorter because of strong incident flux, either a filter or a mechanical shutter will have to be placed in the beam (neutral density or colored filter).

As for data processing, we have to emphasize the fact that it is impractical to define an array bigger than 64KB on a PC working under MS DOS. MS Basic 7.0 offers the possibility of working on an array occupying the memory beyond 64KB, but the execution speed is 3 times slower. To compensate for these memory limitations the image is divided into four parts which are stored in four distinct arrays, each occupying less than 64KB. The separation into four arrays is done automatically at the moment of acquisition and is thus transparent for the user. The data contained in these arrays slightly overlaps (by 10 pixels), thus allowing independent processing on the four quarters of the images and the creation of a mosaic at the time of visualization without any edge effect. The re-assembly of the image into a single piece is perfect. This four-file structure is unwieldy, but it is not bothersome if the programming is clever (using macro-commands which allow manipulation of the four files with a single command issued by the user).

A complete image from a TH7863 array coded into 16 bits represents 220K. That is a lot for a computer that has only 640K of RAM. There are solutions to this problem: use EMS or move to a different operating system—OS/2 or Unix.

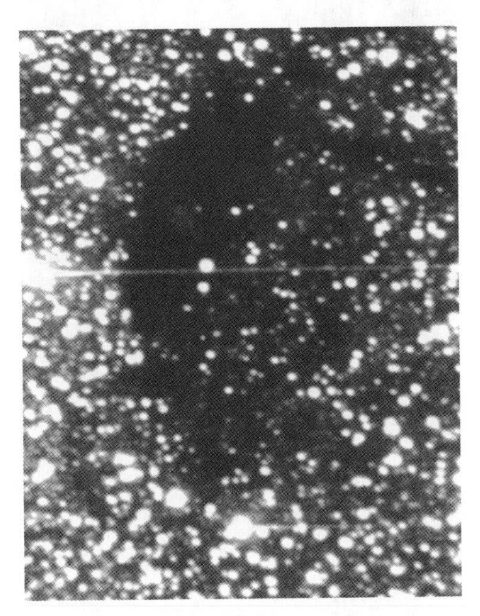

Figure 2.44 *The dark nebula LDN 323. Operating a CCD throughout its entire spectral range can result in some surprises. Here, the familiar outlines of this nebulae are missing because the CCD has detected the infrared radiation.*

Figure 2.45 *The globular cluster M10. 5 minute exposure with a 280 mm telescope.*

Chapter 3

Cooling the CCD

3.1 Introduction

Earlier we explained how thermal noise (dark signal) can saturate a CCD—even when the "exposure" is for only a few seconds and in "total darkness". This saturation is a problem since some astronomical objects require longer exposures. Obviously the CCD must be cooled to a temperature at which thermal charges do not interfere with its operation. In fact, for a CCD camera to be optimized for astronomical observation, cooling is as important as the electronics and the computer software. The overall performance of the entire system will depend upon the care taken in the design and construction of the cooling system. For example, the dark signal decreases by a factor of 2.5 each time the temperature is reduced by 10°C. A 10°C drop in temperature increases the CCD's sensitivity by one magnitude. A one magnitude gain is significant, so the quest to reduce the CCD's operating temperature will be our concern throughout this chapter.

A cooling system can be visualized as an electronic circuit in which the thermal load is a "charge". To cool (reduce the charge in) this circuit, power (watts) must be removed. To calculate the value of this thermal charge and thereby define the size of the cooling system, it is necessary to take into account not only the heat produced by the component but also that added by the environment. Thus the heat circulates through the electric wiring (conduction), is transmitted by the atmosphere (convection), and is received by the detector in the form of radiation.

3.2 Calculating the Thermal Charge

3.2.1 The CCD's Proper Heat

Proper heat is the translation, in thermal form, of the electric power used by the detector to work. The power dissipated by a component such as a TH7852 CCD is about 200 mW.

3.2.2 Conduction

Every attachment to the CCD creates a thermal "charge" pathway. While the cooling system is removing heat from the CCD, the wires which connect it to the circuit are conducting heat back into it. Ideally these electrical conductors should be thermal insulators—typically nickel, platinum and other proprietary materials. These materials are expensive and hard to find. In most cases they will have to be replaced with common copper wiring. Stranded wire should not be used. Only the finest gauge single strand wire available should be used (no larger than 0.1 mm in diameter—wire wrap is appropriate and handy). The thermal charge produced by a wire is calculated by the equation

$$Q = K\frac{A}{L}(T_{\mathrm{h}} - T_{\mathrm{c}})$$

where

K is the thermal conductivity of the material ($K = 4.1$ W/cm/°C for copper),

A is the diameter of the wire in cm^2,

L the length of the wire in cm, and

$T_{\mathrm{h}} - T_{\mathrm{c}}$ the difference in temperature between the two ends of the wire.

Figure 3.1 shows how to evaluate the thermal charge of an electric wire as a function of the material and the diameter. If the diameter D of the conductor and its length L are different from those shown in Figure 3.1, the charge is calculated by the following formula:

$$Q \text{ (new size)} = Q(\text{curve})\frac{L}{D^2}(\text{curve})\frac{D^2}{L} \text{ (newsize)}.$$

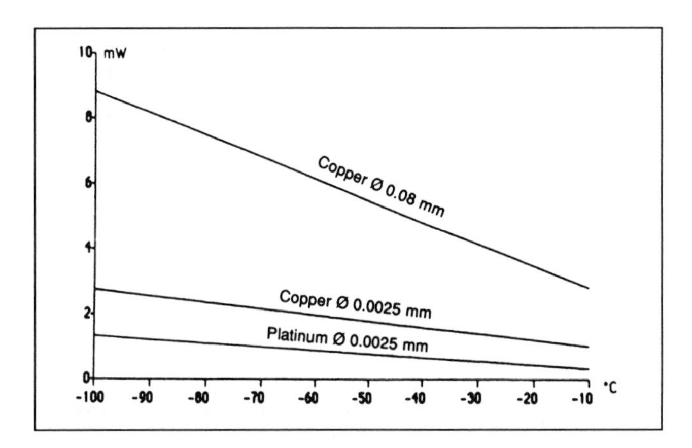

Figure 3.1 *Thermal charge produced by conduction in a power supply wire as a function of the temperature of its cold end. The curves are valid for a wire 25 mm long and for a temperature of 27° C at the hot end of the conductor.*

3.2.3 Convection

Figure 3.2 shows the value of the thermal charge per surface area caused by convection of the detector exposed in dry air and nitrogen.

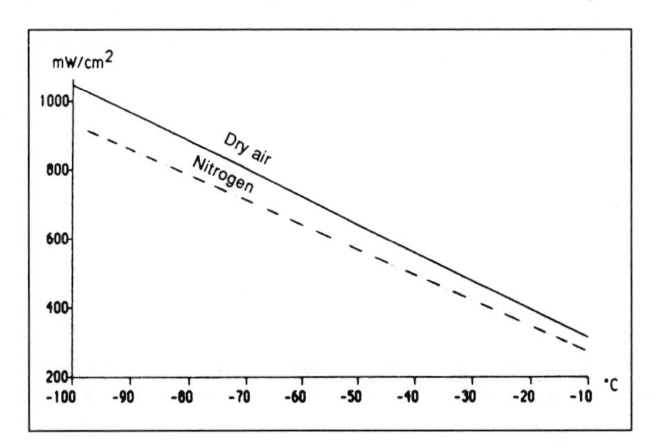

Figure 3.2 *Thermal charge per unit surface caused by convection when the detector is placed in dry air (upper curve) and in a nitrogen environment (lower line) as a function of the detector's temperature. In this case the prevailing temperature is 27° C.*

Heating caused by convection is considerably decreased in a vacuum. We will see later that this also prevents the formation of frost. The importance of convection is revealed by Figure 3.3 where we see the evolution of the thermal charge as a function of the quality of the vacuum.

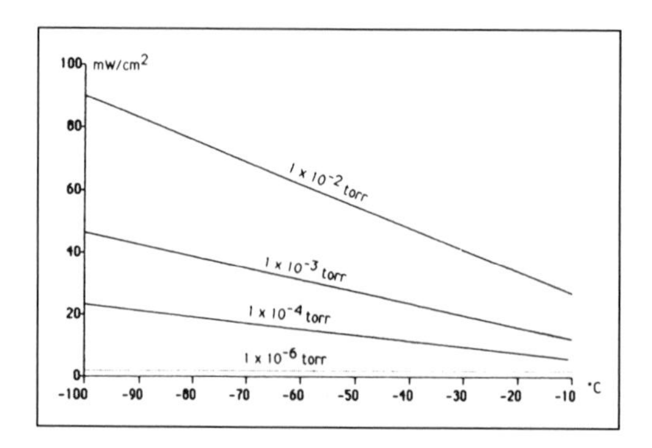

Figure 3.3 *Thermal charge per unit surface caused by convection in a more or less perfect vacuum. 1 torr = 1mm of mercury.*

3.2.4 Radiation

The detector's casing absorbs radiation which adds heat. Figure 3.4 allows us to calculate the thermal charge induced by radiation for a given area of the detector.

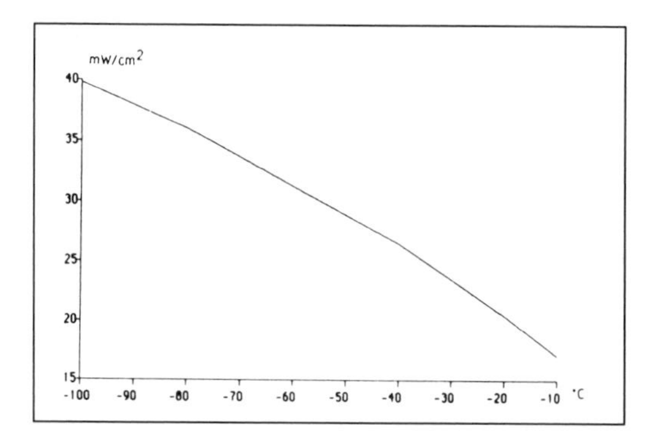

Figure 3.4 *Thermal charge caused by radiation. The emissivity of the cold surface is assumed to equal 1.0. The ambient temperature is 27° C.*

3.2.5 Calculation of the Thermal Charge

The thermal charge produced by the functioning CCD is calculated in the following manner. The detector is linked to the outside world by 24 copper single strand electric wires each with a diameter of 0.5 mm and 70 mm long. Two more identical wires connect the thermal probe, giving a total of 26 wires. The CCD is located in an enclosure where there is a vacuum of 10^{-2} Torr. The CCD's casing presents a free surface of 8 cm^2

(a part of the CCD's array is in contact with the cooling element and is therefore not exposed to air). The detector's temperature is −20°C, and the ambient temperature is 27°C. A metallic conductor called a "wick" is used to conduct the cold to the detector.

Here is the result:

1. The electric dissipation is of the order of 200 mW for the TH7852 array.
2. Copper's thermal conductivity has a value of 4.1 W/cm/°C. The section of a wire is 0.002 cm^2. Thus the thermal charge caused by conduction is calculated as follows:
$$Qc = \frac{(4.1)(0.002)}{7}(+27 - (-20)) \times 26 = 1430 \text{ mW}.$$
3. At −20°C, the power absorbed by radiation is 19 mW/cm^2 (Figure 3.4), which gives for the 8 cm^2 free surface area of the casing:
$$19 \times 8 = 152 \text{ mW}.$$
4. The convection is 35 mW/cm^2 (Figure 3.3) for a pressure of 10^{-2} Torr, giving a loss by convection from the casing of:
$$35 \times 8 = 280 \text{ mW}$$

From these calculations one can deduce the value of the thermal charge:

$$Q = 200 + 1430 + 152 + 280 = 2062 \text{ mW}.$$

To be complete, it is necessary to take into account the loss caused by the wick that brings the low temperature to the detector. Assuming that it is a piece of copper with a free surface of 15 cm^2, representing a charge of 810 mW, the total thermal charge is then seen to be

$$2062 + 810 = 2872\text{mW, or about } 2.9 \text{ W}.$$

The losses caused by the cold finger[1] are far from negligible. The other important item is the loss in the electric wires. By using single strand wire (e.g., wire wrap), we decrease by an order of magnitude the conduction losses. A CCD installed in a mounting that is even slightly optimized represents a thermal charge of the order of 1 W.

Copper is the ideal thermal conducting material (thermal conductivity of 4.1 W/cm/°C). Aluminum should be used only when absolutely necessary (conductivity = 2.0 W/cm/°C). Brass should never be used to conduct cold (conductivity = 1.1 W/cm/°C). As far as possible, the active cooling element should be placed directly in contact with the CCD, avoiding any intermediaries.

[1] A "cold finger" is that part of the wick that is in direct contact with the CCD.

3.3 Methods of Temperature Reduction

Cooling is a source of endless problems. Even staying within reasonable limits, there is much room for experimentation in relation to what is said here. We will describe some cooling methods, but this list is far from complete because we limited ourselves to techniques that we think are accessible to amateurs, except the first one.

3.3.1 Cooling with Cryogenic Liquid

Cryogenic liquid cooling is universally used in professional observatories. In this procedure the CCD is put into (indirect!) contact with a liquified gas. Astronomers' favorite cryogenic fluid is liquid nitrogen which can reach a theoretical temperature of $-196°$C. This temperature is more than low enough since the detector should not be cooled to below $-120°$C. The safety margin is used for regulation and to sponge up various thermal losses.

The nitrogen is placed in a reservoir (dewar) thermally insulated from the outside environment so that it will remain in a liquid state as long as possible. This container resembles a thermos bottle made with double walls that contain an insulating vacuum. The dewar can be made from either glass or metal. The whole assembly—composed of the dewar, electrical connectors and a window to pass light to the detector—is called a "cryostat". Once installed in a cryostat, the CCD is maintained at low temperatures by periodically adding nitrogen. A well designed cryostat is filled with 2 liters of nitrogen once every 24 hours.

The cold is conducted to the CCD by a "wick" of copper or some other braided material that soaks in the cold liquid. To avoid formation of frost on the window, the detector itself is in an enclosure in which there is a nearly perfect vacuum (10^{-6} Torr). The CCD's temperature is generally maintained to within a tenth of a degree by resistance heating elements arranged on the cold finger.

As a further precaution against frost, the outside of the window should be resistance heated. To reduce readout noise as much as possible, the amplification circuits are sometimes placed immediately beside the CCD in the vacuum enclosure. If these circuits give off too much heat, it is dissipated with a thermally conducting potting compound since there is no conducting gas in the vacuum.

Setting up a cryostat is not easy. A nearly perfect vacuum must be maintained. To accomplish this, a two stage vacuum pump is needed. These pumps are expensive and usually beyond the amateur's resources. However, it is sometimes possible to find a scientific laboratory that will agree to occasionally pump the cryostat. If the cryostat is well built and correctly pumped, it should hold its vacuum for possibly a year. To be leakproof, the assembly must be nearly perfect: thick walls, grooves precisely machined to properly hold O-rings and special connectors. It's really a professional job!

Because the telescope moves to track celestial objects, the cryostat must be able to work in any orientation, a requirement that is not easy to meet when the cooling element is a liquid. On the other hand, whatever the position, the conducting wick that brings the cold to the CCD must always be in contact with the liquid. If it is not, the CCD undergoes intolerable temperature variations.

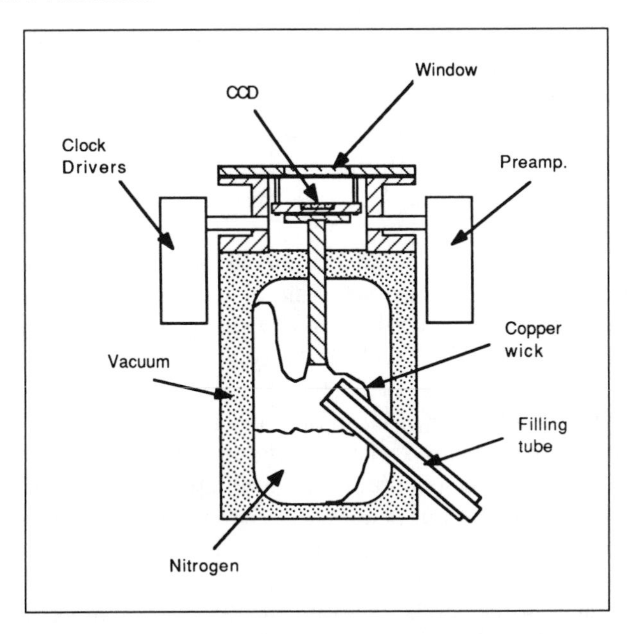

Figure 3.5 *Diagram of a cryostat. Note the orientation of the filling tube which prevents the nitrogen from being driven back out of the system. The thermal contact between the nitrogen and the CCD is carried out by an assembly of copper wires and a massive bus also made of copper. The sensitive electronics are placed very close to the CCD. Leakproof couplings allow the conductors to pass through the wall of the cryostat.*

The CCD should not undergo frequent or abrupt thermal change. Indeed at $-100°$C, the differential expansions between the silicon chip and its support are far from negligible, a situation that can lead to failure (ungluing, etc.). The temperature, therefore, must be lowered progressively (several degrees per minute), and then held at its working level as long as possible. This constraint is a problem for the amateur who will have to get liquid nitrogen regularly. This fluid is not expensive, but its storage is difficult because it must be maintained at very low temperatures. Most professional observatories have equipment to extract liquid nitrogen from the air since they are usually far from commercial sources.

Using liquid nitrogen to cool the CCD is an exotic solution, whose difficulties could discourage most amateurs. They will probably prefer to moderate their enthusiasm and choose solutions that are less effective but more realistic.

3.3.2 Cooling with Dry Ice

Dry ice is well suited for simple cooling of the CCD. The temperature of dry ice is about $-80°$C. It is easily obtained by suddenly releasing pressurized carbon dioxide through a specially shaped nozzle. The nozzle is machined in such a way that its end has a diameter of the order of a millimeter, the other end being screwed directly onto the end of a bottle of CO_2 (no relief valve). A cloth bag is attached solidly at the end of the nozzle to recover the dry ice. Although this home-made apparatus works, the output is low.

Devices to make dry ice are commercially available. For instance Bioblock in France and Fisher Scientific in the United States market an efficient device that can make bars of dry ice. Also some companies can directly supply dry ice, which will keep very well for several days in a home freezer.

Figure 3.6 shows how dry ice can be used. It is confined in a metal container whose inside is covered with thermal insulation (polystyrene foam). The tank is filled by the spoonful through an opening on the side. This is an old-fashioned method, but it works. A piston system compresses the dry ice onto a piece of copper which in turn conducts the cold to the detector. The dry ice must be kept in constant contact with the copper conductor, which is difficult to accomplish because the dry ice evaporates at this contact where it is relatively warm. This evaporation forms pockets of gas that causes the device's efficiency to drop significantly.

The device shown in Figure 3.6 can make the temperature drop without too much difficulty to about $-50°$C in a clean, economical way. There are, however, a few problems with dry ice:

1. It is difficult to regulate the CCD's temperature. The temperature is in fact a function of the way the contacts between the dry ice and the thermal bus are made. But these contacts change as the dry ice sublimates.

2. Even though making dry ice from a bottle of CO_2 is simple, doing so has to be considered a last ditch resort. It is much better to buy dry ice through commercial sources. Handling big bottles of

compressed gas is tiring in the long run. I still remember my first CCD observing run with the 60 cm telescope at the Pic du Midi Observatory. At that time I used a camera working partly with dry ice, and I carried an enormous 50 kg bottle of CO_2 to the summit with great difficulty. It was all the more tiring since the bottle was empty after two nights of observation while the run was supposed to last for about two weeks. Fortunately there was a back-up cooling method which enabled completion of the run.

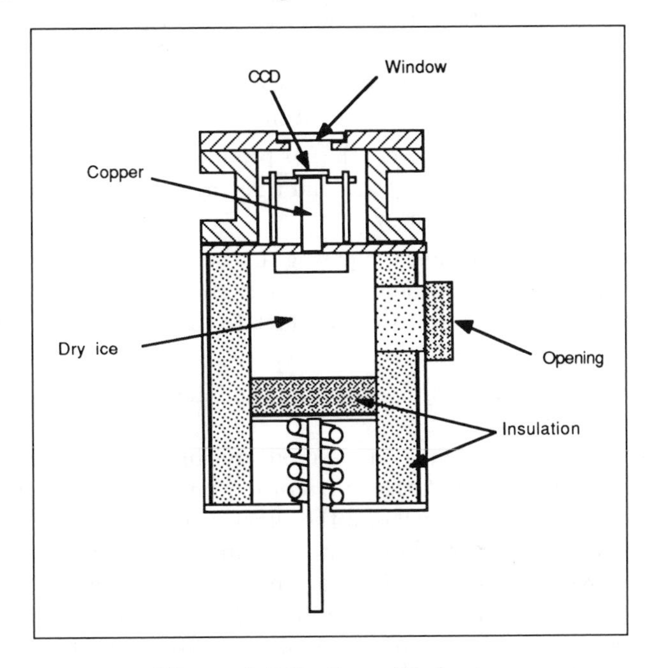

Figure 3.6 *Cooling with dry ice.*

3.3.3 Cooling by Mechanical Means

The simplest mechanical means is to use Joule-Thomson expansion: a highly compressed gas is expanded in a fine tube ending in a little opening. This expansion decreases the temperature in a way which is similar to the method of producing dry ice described in the preceding section. To increase the efficiency of the device, some of the cold gas is diverted back onto the outside of this fine tubing. To further increase efficiency the tubing is usually finned. As the pressure of the gas is reduced, its temperature decreases progressively. With compressed nitrogen, this process is capable of reducing temperatures by 77°K. Nitrogen is readily available in the air and can be collected with a compressor having a capacity of 14.7 psi/atm. × 200. The Joule-Thomson method is classified as "open cycle".

Another possibility is the "closed cycle" method which is used by home refrigerators. These machines generally use the principle of compressing vapor. Figure 3.7 shows the cycle of such a machine in a very simple way. The cooling circuit includes the following components:

1. An evaporator in which a fluid passes from a liquid to a gas under high pressure. As the liquid changes to gas it picks up heat from the environment. This absorption of heat produces the cooling. Either the evaporator is placed directly in contact with the detector, or it cools an intermediary liquid (glycol, alcohol, etc.) which is pumped into the radiator that is in contact with the detector (cryodiver principle).
2. A compressor which recovers the gas produced in the evaporator. The compressor compresses the gas and thus increases the pressure in the circuit by furnishing energy.
3. A condenser that ensures the passage of the cooling fluid from gaseous to liquid state. The condenser is usually constructed of finned tubing which enables the heat in the coolant to dissipate into the environment.

The cycle begins again by evaporating the fluid, lowering the temperature and so on.

The most common cooling fluids used are known under the DuPont trade name Freon.[2] So that the refrigeration unit will have a maximum output, the fluid should not contain the slightest trace of oil (which might come from the compressor's lubricants), of air (leaks), or of water (poor drying of the system when initially charged with the coolant). To maintain these ideal conditions, an oil separator and dehydrator are usually installed in the system. The presence of traces of water is critical because it can freeze and block the expansion unit or decompose the cooling fluid into acids that cause corrosion throughout the system.

Gas compressing machines can effectively cool a CCD. With these machines it is possible to achieve temperatures below −50°C. They can have great reserve cooling power—perhaps too much because one of the problems that has been observed is the difficulty of closely regulating the output temperature. The system is compact and, once it is perfected, can work for years without any problems.

Optimizing a cryogenic machine is a job for a professional. Suitable commercial units are available but they are expensive. If, despite everything, one wishes to build a machine from salvaged refrigerator parts, it is highly advisable to seek the assistance of a professional.

[2]Various grades of this coolant are available.

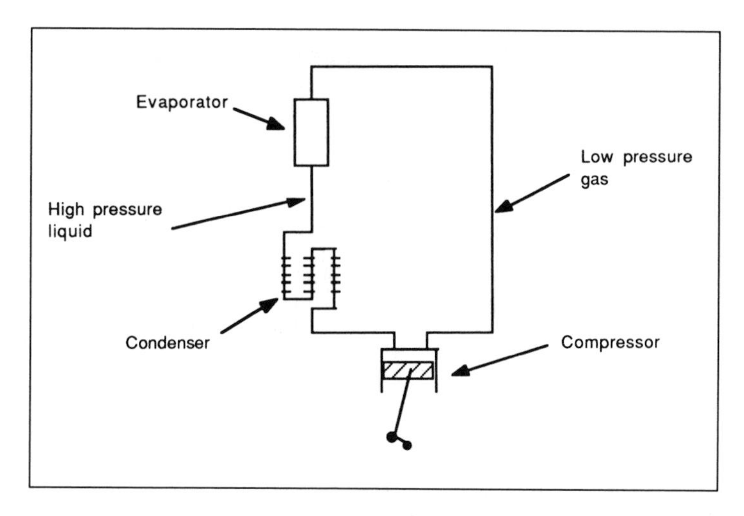

Figure 3.7 *Diagram of a gas compression circuit.*

3.3.4 Thermoelectric Cooling

Thermoelectric cooling is based upon the principle that when two different metals are put into contact and if there is a difference in temperature between these two, an *electromotive* force is created (Seebeck effect). This phenomenon is put to use in, among other things, thermocouples whose role is to measure temperature.

The Peltier effect is the opposite of the Seebeck effect; it is the absorption or the release of heat at the junction of two different metals when an electric current passes through them in a specific direction. The junctions are formed by two semiconducting materials, one N doped; the other, P doped (usually lead or bismuth telluride) and joined by copper bridges.

An external potential sets the electrons in motion. When the electrons travel from the P material to the N material, they go from a low energy state to a high energy state. This increase in energy corresponds to an absorption of thermal energy at the level of the junction and therefore to a lowering of the temperature. Then the electrons go from material P to material N, passing from a high energy state to a low energy state. They give up their energy in the form of calories on the warm side of the module.

In practice several couples are mounted between two ceramic plates. The ceramic allows the electrical isolation of the couples and gives good thermal conductivity and excellent mechanical stability. The couples are electrically linked in series, but thermally they are in parallel so that their individual actions will be added together. With such a device it is possible to build Peltier elements with a large cooling surface. The dimensions of the modules go from a few square millimeters to several tens of square centimeters. The thickness is usually 5 millimeters.

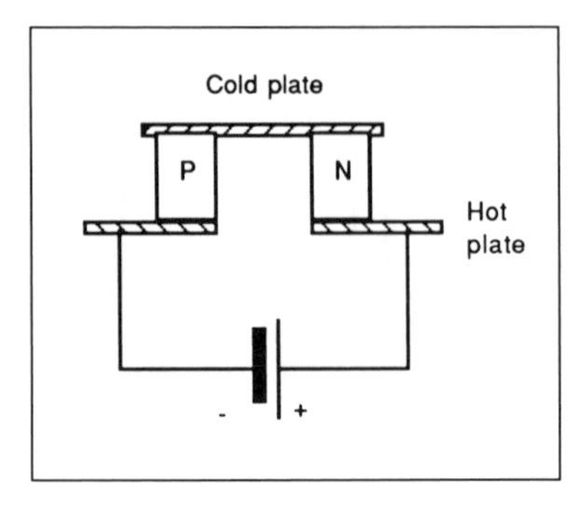

Figure 3.8 *Diagram of a thermoelectric couple. The two semiconductors are joined by pieces of copper. Note that by inverting the direction of the current, it is possible to invert the hot side and the cold side.*

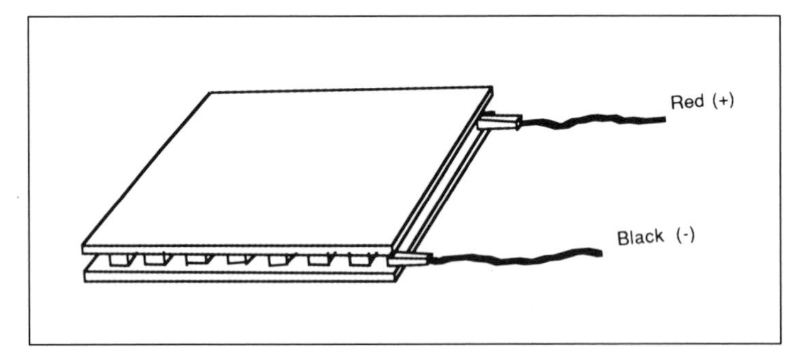

Figure 3.9 *A Peltier module. The power supply wires are always attached to the hot side.*

A thermoelectric module acts like a heat pump. It absorbs heat on the cold plate and radiates heat on the hot plate. The hot plate must have an efficient means of radiating this accumulated heat *plus* the electric power applied to the module to create the Joule effect. There are two basic characteristics of a thermoelectric module. First, heat pumping is maximum when the temperature difference between the plates is nil. The maximum temperature variation (ΔT) occurs when the thermal charge equals 0. A ΔT of 70°C can be achieved when no thermal charge is applied to the cold plate if the calories produced internally by the Joule effect are properly radiated away from the module. This last point is quite difficult to achieve. Second, pumping capacity is maximum when $\Delta T = 0$. The temperature difference between the cold plate and the hot plate varies as a function of the temperature of the hot plate as shown in the following table.

Hot Plate (°C)	Cold Plate (°C)	ΔT (°C)
50	−25	75
25	−42	67
0	−59	59
−20	−72	52

The relationship between hot and cold plate temperature directly affects performance and must be considered when the system is being designed (see the calculation examples below).

When comparing different modules the COP (Coefficient of Performance), which is the ratio between the power pumped and the electric power used, is useful. Depending on the specific type of module, the maximum efficiency is reached when the module is operating between 30% and 50% of its maximum current rating.

In practice the manufacturer's ΔT is impossible to reach. As soon as a thermal charge is applied to the cold plate, the temperature difference between the hot and cold elements decreases since the Peltiers act like heat pumps whose power is, unfortunately, not infinite.

The efficiency of a thermoelectric module is limited by two things:

1. Heating by Joule effect resulting from the electric current crossing the semiconducting elements;
2. Thermal conduction from the hot plate toward the cold plate.

To reach very low temperatures, several Peltier modules have to be cascaded, with the cold plate of the first in contact with the hot plate of the second, and so on. When cascaded, successive Peltier modules must have increased pumping power because a successive module must recover the heat given up by all the preceding stages. Cascaded units resemble a pyramid with heat absorbed at the top and radiated at the base. Ready-to-use cascades are available from manufacturers.[3]

Typically at 0 thermal charge, when the temperature of the cold face is −28°C with a single level, it is −56°C with two levels, −68°C with three levels, and −75°C with four levels. These performances are reached only when the characteristics of each level are perfectly optimized.

We have seen that the hot plate of a module absorbs the heat pumped from the cold plate plus the heat generated by Joule effect. To keep the hot

[3]See Appendix B for a list of manufacturers.

plate at a reasonable temperature (in principle, the ambient temperature) we have to be able to evacuate this calorific power. If not, the heat will rise throughout the entire module and reach the cold plate, which is the opposite from the effect sought.

To dissipate heat (calories) we use a radiator which is brought into mechanical contact with the module's hot plate. An ideal radiator would be able to absorb an infinite quantity of calories without its temperature rising. Of course, such a radiator does not exist in practice, but it is important to get as close as possible. The thermal dissipator (the radiator) should be considered as an integral part of any thermoelectric cooling system. Its ability to dissipate heat often determines the temperature reached on the cold plate.

The performance of a radiator is defined by its thermal resistance:

$$Rt = \frac{\text{Rise in temperature of radiator with respect to ambient temp. in } ^\circ\text{C}}{\text{Total power absorbed by the radiator in watts}}.$$

Therefore a radiator with a thermal resistance of 0.15°C/W will rise to a temperature 15°C above the ambient temperature when a thermal charge of 100 W is applied to its surface.

The radiator is often finned and made from aluminum. Aluminum is lightweight, a good heat conductor, and the fins increase the surface area from which heat can be radiated to the air through convection. There are two types of convection:

1. Natural convection (the air does not move), has a thermal resistance from 0.6°C/watt to 5.0°C/watt according to the shape and the size of the radiator;
2. Forced convection (the air close to the radiator is set into motion with a ventilator), has a thermal resistance of 0.07°C/watt to 0.5°C/watt.

The radiator's efficiency is greater when it is hotter than the surrounding air. Thermal resistance is therefore a parameter that depends on the particular condition in which the radiator is used. However, since the pumping capacity of a module decreases as ΔT increases, the temperature of the radiator should not be too high, or the cold plate of the thermoelectric module will heat. Generally, the difference between the surrounding air and the radiator should not be greater than 10°C. The thermal resistance indicated in the manufacturer's data sheets is a sufficiently precise basis for calculations.

More efficient evacuation of heat can be obtained by replacing the air radiator with a fluid heat exchanger. The easiest fluid to use is a glycol-water mixture which will not freeze on cold nights. The water circulates in a closed circuit passing through a small tank that is in contact with the

Peltier's hot plate. The tank has two openings, one for the entrance of the liquid, the other for its exit. The water is set into motion by an electric pump. The pump's speed should be variable to assure a sufficiently rapid flow of the liquid (be careful with some special fluids that are particularly viscous and therefore need greater pumping power to be moved). An easy way to dump the heat from this type of system is to pump the coolant through a coil of tubing submerged in a container holding several liters of liquid.

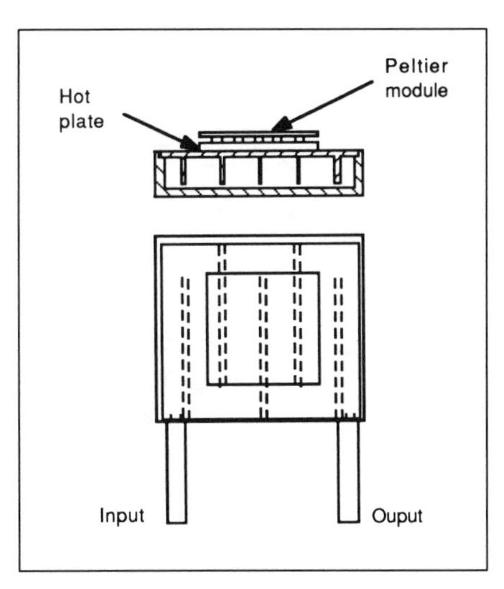

Figure 3.10 *Diagram of fluid heat dissipator. The indentations shown in the side view (top illustration) are machined into the material to increase surface area and heat transfer to the circulating fluid.*

Fluid circulation radiators with thermal resistances in the order of 0.02°C/watt are commercially available. Although copper is expensive and not easy to machine, heat exchangers made of copper are much better thermal conductors. There are also available commercially very practical little connectors for fluids with rapid linking and automatic closing of the openings that allow the decoupling of the liquid circuit from the optical without creating a mess.

Thermoelectric modules are our favorite way of cooling our cameras. They let us reach relatively low temperatures (−50°C) with a minimum of complications. They are reliable and fairly cheap (the price ranges from $18 to $35 according to the power). In addition we use fluid heat exchangers such as the one shown in Figure 3.10. The liquid is moved by a little pump whose flow is about 5 liters/min. To avoid having to prime the pump every time it is turned on, it runs in a closed circuit. The water reservoir is about

6 to 7 liters. The pump should not be a submersible type placed in the reservoir since it would heat the water and an increase of 4 or 5 degrees in the water circuit will correspond to almost a doubling of the CCD's dark signal. The water circulates from the CCD located at the telescope's focal surface through a flexible 6 mm ID tubing to a pump positioned near the base of the telescope.

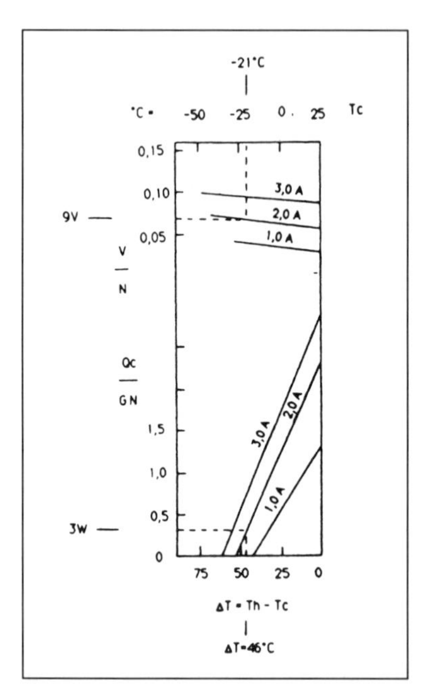

Figure 3.11 *Nomograms for calculating the performance of the module CP 1.0-127-05L for various configurations (from the Melcor catalog). The family of curves chosen corresponds to a 25° C temperature on the hot side (T_h). Each straight-line corresponds to a value of the power supply current. N and G are constants. For the model studied $N = 127$ (number of thermo-elements), $G = 0.078$, and $G{\cdot}N = 9.91$. The lower network gives the relation between the current, the charge, and the ΔT. The upper nomogram shows the relation between the current, the voltage, and the temperature of the cold side (called T_c).*

Calculating a One-level System

To define the characteristics of a thermoelectric module, almost all manufacturers offer two types of nomograms that cover a range of hot plate temperatures. One supplies, for a given thermal charge and electric current, the temperature difference that can be reached between the hot plate and the cold one. The other supplies, for given electric current and temperature of the cold plate (or the ΔT, since the temperature of the hot plate is set), the value of the voltage at the module's terminals.

As an example, we will look at the performance of a Peltier element made by Melcor.[4] The specifications for this module are: 1.) current load under 2 amperes, 2.) the thermal charge is 3 watts, and 3.) the hot plate is at 25°C. The problem is to determine the temperature reached on the cold plate and the power to be dissipated on the hot plate. Figure 3.11 gives the numbered elements for this model.

When selecting a module you should have some reserve—when $I = I_{\max}$, the module's efficiency is poor. Working at a module's extreme limits may present serious problems, should the actual system require a marginally higher capacity than originally calculated.

We begin by drawing a horizontal line at the level of $Q = 3$ W as far as the intersection with the straight-line $I = 2$ A. We then draw a vertical line and find ΔT, here 46°C. By going up in the upper nomogram, we get $T_c = -21$°C (or $T_h - \Delta T$). The straight-line $I = 2$ A shows that the model has a voltage at its terminals of 9V. The electric power used is therefore:

$$Q_e = U \cdot I = 9 \times 2 = 18\text{W}.$$

The total power dissipated on the hot plate is 18 + 3 = 21 W. This is not an insignificant amount and a good radiator will be required.

The efficiency (COP) is:

$$\text{COP} = \frac{3}{18} = 0.17.$$

Calculating a Multi-level System

The calculation for a multi-level system is slightly more complicated than for the single level system. We will describe here the device used in one of our cameras. It is composed of two levels of Melcor modules:

1. On the CCD side: model CP 1.4-31-10L, dimensions 20 × 20 × 4.7 mm, ΔT = 70°C, $Q_{\max} = 8.15$ Watts, $I_{\max} = 3.9$ A, $V_{\max} = 3.75$V. The module's current drain is under 1.7 A.

2. On the radiator side: model CP 1.0-127-05L, dimensions 30 × 30 × 3.2 mm, ΔT = 67°C, $Q_{\max} = 33.4$ Watts, $I_{\max} = 3.9$ A, $V_{\max} = 15.4$V. The module's current drain is under 2.0 A.

The specifications of this device are: ΔT = 55°C for a hot plate at 31°C (and therefore a cold plate at −24°C). The ambient air temperature is

[4]ref. no. CP 1.0-127-05L

23°C. The CCD and the modules are placed in a vacuum of 10^{-2} Torr. The thermal charge is 3 W (see the estimate at the beginning of the chapter). We will check these performances by a theoretical calculation.

Using the nomogram values of $T_h = 25°C$, $T_c = 24°C$ and $Q = 3$ W, the ΔT of the first level is of the order of 16°C, and its hot plate should be about −8°C. The calculation must therefore be refined by taking the curves plotted for $T_h = -8°C$. In fact these curves do not exist in the manufacturer's data, and they had to be calculated by extrapolation from known data. Figure 3.12 shows the result.

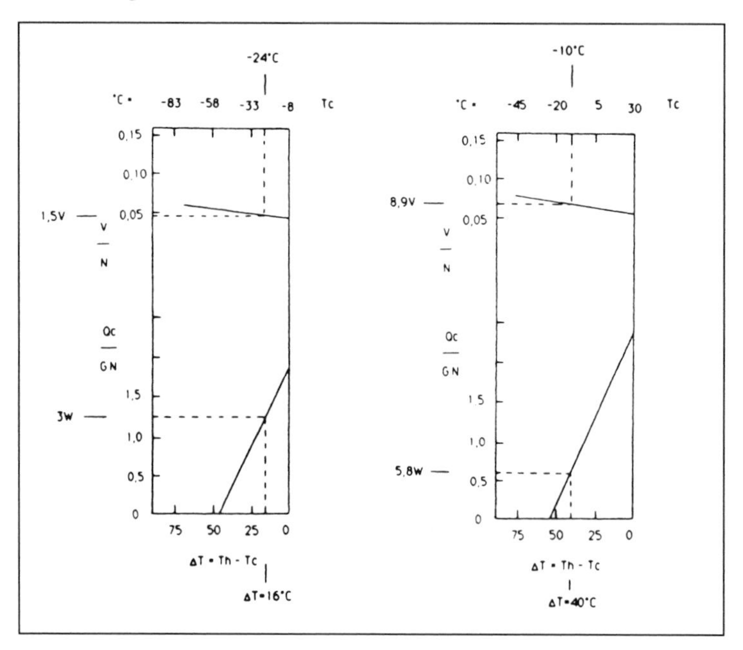

Figure 3.12 (Right) *Characteristics of module CP 1.4-31-10L for* $T_h = -8°C$, $I = 1.7$ *A.* $N = 31$, $G = 0.078$.
Figure 3.13 (Left) *Characteristics of module CP1.0-127-05L for* $T_h = 30°C$ *and* $I = 2$ *A.* $N = 127$, $G = 0.078$.

For a pumping power of 3 W, we therefore find by iteration a ΔT of 16°C. The hot plate of this first level is at −8°C. The power supply is 1.5V, from which the electric power is dissipated $1.5 \times 1.7 = 2.55$ W.

The second level "sees" a thermal charge of $2.55 + 3 = 5.55$ W to which we must add the charge induced by the free surface of the cold plate of the second level, i.e. 0.23 W (the second level is wider than the first and we get a surface exposed to the air of 5 cm^2, allowing us to deduce the thermal charge caused by convection and radiation for a surface temperature of −8°C.) The total charge applied to the second level is thus 5.8 W.

Figure 3.13 indicates that the ΔT is 40°C. If we take T_h = 32°C, the cold plate is indeed at −8°C. By calculation we predict a temperature of the radiator of 32°C while observation gives 31°C. The difference is very low and our modeling is satisfactory.

The second level dissipates an electric power of 17.8 W, and the total caloric power to be evacuated by the radiator is 17.8 + 5.8 = 23.6 W. The temperature increase of the radiator being 8°C with respect to the ambient air temperature, its thermal resistance is 8/23.6 = 0.34°C/W. This performance level is achieved by using a forced air cooled heat sink measuring 150 × 150 mm with large fins to increase the surface exposed to the air which in turn is moved by a fan.

Mounting and Optimizing a Peltier Module Cooling System

There are strict rules for correctly mounting Peltier modules. Figure 3.14 shows a basic mounting where the module is held between two metal plates, one acting as a radiator, the other conducting cold toward the element to be cooled. This is a pressure mounting.

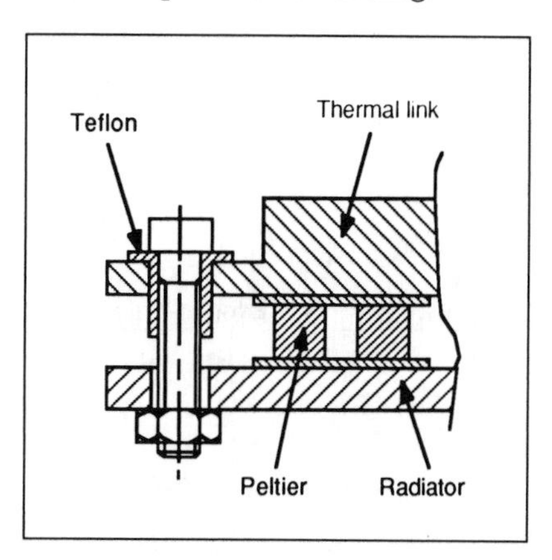

Figure 3.14 *A practical method for mounting of a Peltier module.*

For good thermal transmission, the mechanical surfaces must be smooth (roughness <0.01 mm), plane and perfectly cleaned. A silicone grease, such as that used for power transistors, is applied between the different elements to complete the thermal path. A "dry" assembling of the modules considerably lowers the system's efficiency, it should be avoided.

It is possible to mount several modules side by side. In that case, the thermal powers are set in parallel. The modules should then be placed on the same plane with a precision of 0.005 mm.

Copper should be used throughout for all thermal links. The best copper is "red copper" all surfaces of which are highly polished like glass. Smooth thermal links are more efficient in removing the heat from the unit. Rough surfaces cause internal radiation via convection and radiation.

There shouldn't be any thermal bridge between the hot and cold parts of the mounting; therefore, the holding pieces should be very poor thermal conductors. Plastic screws or metal screws with Teflon insulating shoulder washers will work, as both are poor conductors. The modules can be damaged by applying excessive torque to the screws. Screws should be tightened just enough so that the module does not slide between the two plates that are holding it—a maximum torque of 1 kg/cm^2. If thermal grease is used, it will harden over time, thus holding the mounting in place perfectly.

The CCD is usually strapped to the circuit board. The cold finger is in contact with the CCD's back (underside). Rather than threading the cold finger between the CCD's pins and the circuit board, an opening should be made in the circuit board so that the cold finger can pass directly to the back of the CCD. One must be very careful not to strain the CCD, or the ceramic chip will break. The contact pressures must be evenly distributed. If the CCD has to be strapped at its ends, the cold finger should be sized accordingly (see Figure 3.15).

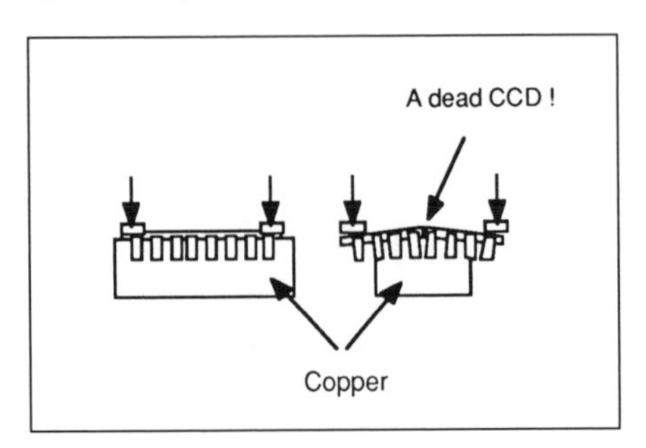

Figure 3.15 *Proper mounting of a CCD. On the left is the way it should be done; on the right, the way it should not be done. The picture on the right is not imaginary—it really happened.*

Our most recent cameras use a three-level mounting. The first level is mounted directly between the CCD's pins. That way the coldest element is in direct contact with the detector and the use of an intermediate part, which always more or less acts as thermal insulation, is avoided. Because the other elements are bulkier (30 and 40 mm long), they had to be moved away from the CCD. The copper cold finger thermal link connecting the upper module (the coldest) to the lower modules operates at a relatively high temperature (equal to 0°C). The small temperature difference between the cold finger and the surrounding environment helps diminish heat losses (see Figures 3.2, 3.3, 3.4).

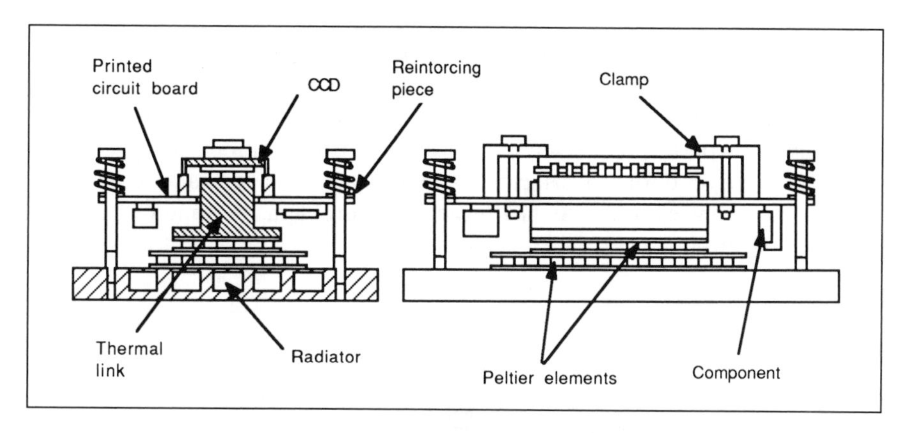

Figure 3.16 *Installation of a cooling system with three stages of Peltier modules. Special attention should be paid to the opening in the printed circuit board and to strapping the CCD with pieces made of a material that is a poor thermal conductor (Teflon). A ΔT of 50° C is feasible with this setup.*

In the arrangement shown in Figure 3.16, it is the pressure of the CCD on the first Peltier level that holds the whole assembly. The outside perimeter of the printed circuit is reinforced so that it does not bend. As always, mounting pressures must be moderate and evenly distributed. Springs can help.

Another technique of mounting the modules is to solder all elements of the assembly together—including the radiator. Low-temperature solder *must* be used or else the modules will be damaged (temperature must not exceed 130°C). The best solder for this purpose contains tin-indium. Soldering is only possible on plated modules. The last level cannot be soldered directly onto the CCD. Instead a special epoxy glue has to be used. Commercial module mounting kits are available for practically any problem that comes up. They contain solder, epoxy glue and grease. Some are even plated. Using these considerably eases the soldering work.

We strongly warn against using a module to hold a mechanical part, or even the CCD. The Peltier module is a fragile component and can only be subjected to light lateral strain, otherwise it will break.

Although it is more effective than the pressure mounting from a thermal point of view, soldering makes disassembly difficult. It also takes considerable skill to avoid damage to the module during this operation unless a temperature controlled soldering iron is used.

Use only thermal grease between the CCD and its radiator. The grease should be put only on the parts that form the thermal contact. A thin layer is enough. Silicone grease, which is very difficult to remove, should be used sparingly inside the camera (watch out for traces on the glass of the CCD!) To clean a dirty part, there is only one remedy: trichlorethylene.

We mention in passing that the pumping power required at the array (between 1 and 3 W), makes it difficult to have more than three levels of cooling. The performance of a four-level mounting prohibits pumping more than 1 W. Beyond four levels, considerable electric power has to be supplied to get just a few calories. As heat evacuation problems become very serious at the last level, it isn't worth the effort.

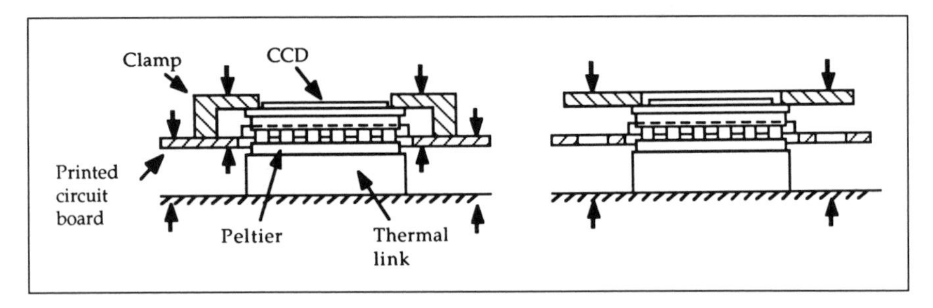

Figure 3.17 *The installation on the left is better than the one on the right. The unlabeled arrows show tightening points. The major disadvantage of the installation on the right is that the printed circuit board is held only by the pins of the CCD. The pins are forced into the support, which doesn't make a very rigid and reliable bond. A shock or rough handling during testing, such as pulling on a wire, can very well pull the CCD off the printed circuit. If this happens when the detector is on, there could be a destructive short-circuit.*

Powering Peltier Modules

The modules used in our camera are powered by laboratory grade power supplies that are stable and well regulated. It is, of course, possible to build dedicated power supplies. With today's integrated circuit regulators this is not difficult or expensive. For example, the LM138 regulator can be the basis for a 5 A supply with only a few additional parts. Building a power supply for each module enables one to "fine tune" the operation of the entire unit. Each power supply should have a small galvanometer to show the amperage passing through the module (the current and not the voltage has to be regulated).

Experience has demonstrated that a power supply ripple of more than 10% makes the output of a thermoelectric module noticeably drop. This specification can be easily maintained, even with a home-made power supply. We did, however, encounter serious problems with a power supply that had high frequency oscillations that occurred when a certain level of current was drawn from the supply. These oscillations affected the CCD, which is very close to the Peltiers, giving a considerable amount of noise in the image. We found the origin of the problem by disconnecting the power to the Peltiers and noticing the disappearance of the noise. In our case, the cause was a trimmer when set at certain positions. So, when in doubt, check the power supply (when connected to the Peltiers) with an oscilloscope to see the condition of the voltage.

The cooling of the camera should be gradual to avoid any thermal shock to the CCD and to extend the lifespan of the modules. Typically, the lowest temperature will be reached 15 minutes after the cooling begins (temperature difference of 50°C). At the end of the observing session, the unit should be gradually returned to room temperature for the same reasons.

We will see in Chapter 5 that regulating the working temperature of the CCD will improve the photometric consistency from one exposure to another. It is therefore necessary to have a temperature probe (temperature sensitive diode) mounted within the camera. This probe can be mounted on the cold finger bringing cold to the CCD if it is the coldest part (no Peltier between the cold finger and the CCD). Even better is to position the probe directly in contact with the upper part of the CCD. When the probe is placed on the cold finger space is no problem. If the probe is placed on the upper part of the CCD it will have to be small which means that it will have to be a platinum (expensive) probe or a thermocouple (not expensive but tricky to use). Temperature can be controlled by reading its value on a digital display. It is of course possible to use the output from the probe as a "feedback" to the modules' power supply and thus be able to regulate within less than a degree. To do this, the signal coming from the thermal probe is constantly compared to a reference voltage (ordered level) and any variance automatically modifies the current passing through the modules. This type of regulation is called closed loop control.

3.3.5 Protection Against Frost

Since the temperature at the CCD can become very cold, the problem of frost on the optical windows is always present.

On the first version of the camera, we solved the problem by simply placing a 40 mm thick piece of glass against the CCD's input window. Between the piece of glass and the CCD window, we inserted some microscope objective oil. This oil removed the Newton interference fringes which were formed between the two pieces of glass. The thermal conduction of this piece of glass being low, frost didn't form on the front, but this was only a stop-gap solution—the efficiency was terrible.

To avoid frost, it is necessary to draw upon time proven experience of professional astronomical systems in which the detector is placed in an enclosure filled with a dry, inert gas or even better, in which there is a vacuum. Light passes through an airtight glass window (made of silicon dioxide—optical quartz—if we are working in the UV) to reach the detector.

To let the electric signals through the walls, leakproof connectors must be used. These can be bought, but their price is very high and we preferred using standard connectors with the soldered end covered in epoxy glue (e.g., Cannon type plugs). Wiring is of course done beforehand. This method of gluing the plugs to be sure they are leakproof, although economical, does however make repair or modifying almost impossible.

It is important to weigh the pros and cons of a solution in which a vacuum is made in the enclosure compared to one where the enclosure is filled with an inert gas. Use of an inert gas does not mean that frost will not form on the detector. Figure 3.18 shows the relation between the water content of a gas and the temperature at which frost appears. It is clear that the atmosphere surrounding the CCD must be drier at low temperatures.

Several grades of bottled nitrogen gas can be purchased. Nitrogen containing about 50 ppm of water vapor is satisfactory for cooling the detector down to $-50°$C. To reach colder temperatures or to have a safety margin, it is better to have nitrogen that contains 5 ppm of water vapor.

With a CCD temperature below $-50°$C, we had difficulties with enclosures filled with inert gas. In fact, in spite of using a very dry neutral gas, traces of gasses that come from the materials used in construction of the unit always remain. One thing likely to outgas is the thermal conducting grease that is put between the cold units. To get around this problem, grease has to be replaced with soft solder. There are silicone greases that do not outgas in a vacuum, but their thermal conductivity is not very good. Other elements which can outgas are printed circuit boards, flux from soldering or the poorly cleaned walls of the enclosure.

Before the enclosure is hermetically sealed, it has to be purged with pressurized gas for several minutes to eliminate any trace of condensible materials. Packages of silica gel in the enclosure are completely useless for eliminating condensation problems. Furthermore, some packages produce dust which can settle on the CCD's window.

The gas inside the sealed enclosure eventually cools from contact with the CCD. Since this gas is also in contact with the enclosure's window, condensation will eventually form on its outside surface—this can be eliminated by placing small resistance heating elements around the window.

A vacuum eliminates any convection around the CCD, and lower temperatures can be reached than in a atmosphere of nitrogen. The difference is 5 to 6°C for a working temperature around −40°C. Leakage of the enclosure and outgassing from components inside it are always a threat to stable operation. If the enclosure will not hold a suitable vacuum it should be pumped continually during use. In any case, for safety reasons, we always do so.

The main problem to be solved is the tightness of the enclosure. This tightness can be rough if we use nitrogen because the inside and outside pressures are about the same, limiting the importance of the leaks. The casing can be made of plastic, giving a particularly light camera. Two openings are made to purge the enclosure with a moderate flow of nitrogen while pumping.

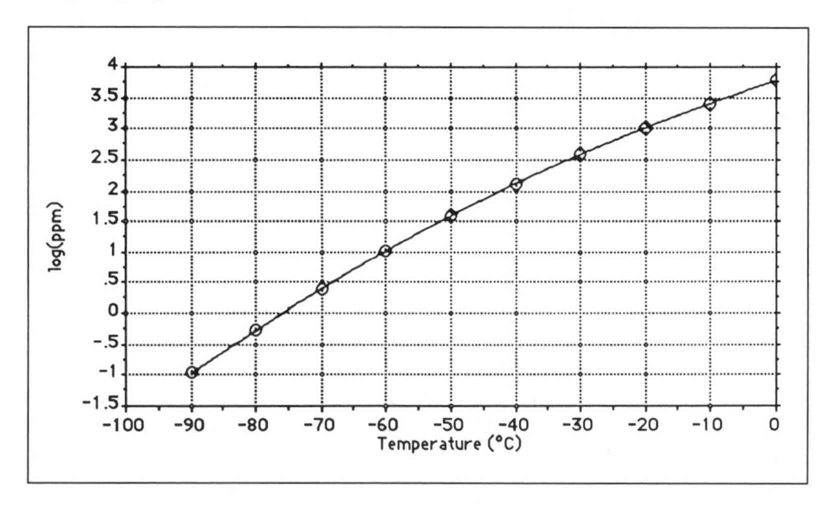

Figure 3.18 *Relation between the humidity content in log parts per million (ppm) of a gas and the frosting point.*

Seal tightness becomes much more of a problem when the enclosure must hold a vacuum. The enclosure must be made out of metal (aluminum) to limit outgassing. This increases the weight of the camera. The tank must be tightly closed and the vacuum seal between different parts foolproof. Only O-rings installed in precisely machined grooves (see the technical data of O-ring washer manufacturers) provide a practical method of achieving this objective. We have experimented with flat washers squeezed between machined flat surfaces of mating parts but this is not usually a satisfactory method. Nothing is as good as a properly mounted O-ring. The vacuum plug must be specially designed so that the valve can be opened and closed

Figure 3.19 *A CCD camera using the TH7852 array mounted on a 6-inch Newtonian telescope. The cooling system, based on Peltier modules and a liquid exchanger, enables the camera to be very compact. The dark box, in the middle of the tube, contains the main electronics. (Photograph by Olivier Gadal.)*

while connected to the vacuum pump to maintain the vacuum. A primary vacuum of 10^{-3} torr[5] is enough for temperatures from −50°C to −60°C. This level can be achieved with a small pump costing from $800 to $1000.

The CCD window must be perfectly clean because dust and traces of grease are "seeds" upon which frost develops.

It is advisable to anti-reflection coat the window. This coating limits reflections which can produce "ghost" images. These coatings are always optimized for a specific wavelength where the transmission is maximum and the reflection minimized. The window of a CCD camera requires a "wide band" treatment, which, though less efficient overall, will meet our needs since we will be observing a wide variety of objects.

In conclusion, when the CCD is moderately cooled (below −50°C) an enclosure filled with nitrogen seems the most appropriate. When the temperature is lower, caution dictates the use of a tight enclosure in which there is a vacuum.

Sometimes pure luck is on the amateur astronomer's side. With one of our cameras we noticed that the CCD, placed in an enclosure that was not very tight, never had any frost—even at a working temperature of −40°C when the enclosure was filled with air! The explanation was that the CCD was not the coldest part of the camera. That element was the cold finger between the Peltier stages and the detector. The water vapor present in the enclosure condensed on the cold finger and left the CCD window clear. Since the unit was tight enough to prevent the entry of new

[5] 1 torr = 1mm of mercury; 1 atm. = 29.92 *inches* of mercury

air (and moisture) the system maintained equilibrium once the initial air was "dehydrated". This experience shows that cooling CCD's is not a cut and dried affair.

Figure 3.20 *NGC 1360 is a large, faint planetary nebulae in Fornax. This image required two 3 minute exposures using a TH7863 CCD with the 24-inch T60 telescope at Pic du Midi Observatory.*

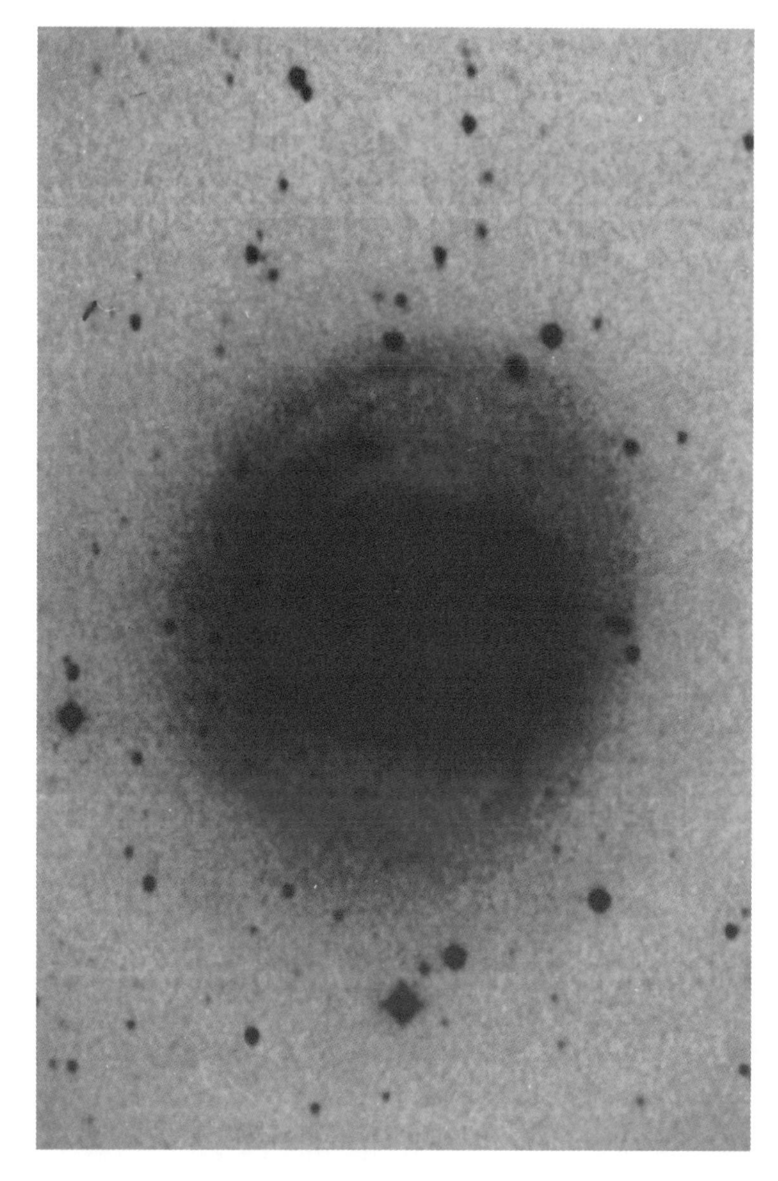

Figure 3.21 *This image of M77 shows the faint external structure of this well-known galaxy. To avoid saturation in the central regions of this very bright object five one-minute exposures were composited.*

Chapter 4

Image Processing

The potential of the astronomical CCD camera can only be exploited through effective use of computers. Most professional astronomers interact with the CCD solely through a computer keyboard, and beyond the software there is only a black box. To these users of the CCD camera the software represents the visible part of the iceberg. After laboring to create a CCD camera it is quite understandable to consider the job as nearly finished. This is far from the truth—the hours spent developing software should not be counted—for better or worse the quality of the software will, for all practical purposes, define the quality of the entire system. For the computer enthusiast, writing programs for the CCD offers the opportunity to learn new aspects of programming. For others it may be a necessary evil to be surmounted in the quest for new insights to our universe. In either case, software should be viewed as a critical necessity.

Most CCD programming efforts will be dedicated to developing routines that are related to image processing. We will use this term in a general way to include image visualizing, transformation and reduction. Image processing is a broad subject and we will therefore limit ourselves to describing what is essential for processing astronomical images. Moving software written for one type of computer to another is seldom straightforward. Another problem is that the new computer's capabilities (graphic, mouse, coprocessors, languages, etc.) will often not be fully utilized if software is just adapted to a new environment. A much sounder procedure is to assimilate the different algorithms used in image processing and to translate these into code that fully exploits your particular machine's capabilities. Therefore this chapter contains few program listings. Rather, emphasis is placed on providing algorithms because they offer the best way to be portable on past, present, and future machines. The program fragments that have been included are here only to illustrate, in a practical way, the implementation of an algorithm. To demonstrate what CCD acquisition and image processing software is, we will conceptually describe two programs (`MAT` and `IP`) that we developed. They are the product of several years' experience in the field.

4.1 Organization of the Programs

The `MAT` and `IP` programs are written for an AT compatible computer, i.e., a machine equipped with at least an Intel 80286 microprocessor. The language used is MS Basic 7.0. This is a second-generation language that compiles very fast, is extended by hundreds of instructions, error correction is easier, and it interfaces with other languages—Pascal, C, Assembler, etc. Even though MS Basic 7.0 is very fast, some routines are written in Assembler. Specific instances are image display and elementary manipulating operations (image transfer between several tables, arithmetic operations, etc.).

Scrolling menus are used as helpers (`HELP` instruction). Mouse controlled pull down menus have not been used because the MS-DOS environment makes this difficult. (Undoubtedly the new Microsoft Windows environments will make this less of a problem.) Therefore, typed commands are displayed in a reserved area located at the bottom of the screen. The rest of the screen operates in graphic mode. It is thus possible to type the commands while keeping an eye on the displayed images and the processing that is taking place. Most commands have arguments. For example, to position an image in a specific place on the screen, we type `POS 150 100`. The numbers (`150,100`) designate the screen coordinates of the lower left hand corner of the image. Sometimes there is a large number of arguments. For example, the instruction that starts the plotting of a three-dimensional view[1] of an image could be written `TRID E 120 2 .1 1`, which means that the representation must be drawn from a point of view located on the left of the image (`E` = East), on a length and width of 120 pixels centered around the last position of the graphic cursor, with a resolution of 2 lines on axis Y, a scale factor of 0.1 in intensity and a scenographic perspective. If the operator types only `TRID` then `Return` (or `Enter`), the program prompts the user for the value of the different arguments. This interface is obviously not very user friendly according to current norms. The image processing program has more than 150 commands. However, experience shows that only about a dozen commands are intensively used. We think that calling

[1] The image is placed in a three-dimensional space in which the plane (X, Y) represents the surface of the image and the axis Z, the intensity. This space is then displayed in perspective.

functions by scrolling menus reduces flexibility and speed but we recognize that this a matter that will be disputed by others.

There are two main programs: `MAT` for acquisition, and `IP` for image processing. Specialized procedures (for example, Fourier transforms) are written as sub-routines. For coherence, the programs use the same user interface and an identical internal structure (some of the routines are common to both). We will summarize the list of the principal commands of the acquisition and processing programs to give an overview of the software functions required to run a CCD camera.

4.2 The Acquisition Program

The acquisition program is called `MAT`. To explore these functions we use as an example an observer who wishes to take the image of a deep sky object with a 200 mm telescope equipped, of course, with a CCD camera.

Using a finder scope, the telescope is pointed toward the object to be studied. A short exposure is started to adjust the pointing. The object is a 14th magnitude galaxy, and a 20 second exposure should be enough to record it. Let us execute the command `LE 20` (`LE` = Long Exposure). The screen fades and the seconds counting the integration time pass one by one on the screen. After 20 seconds, plus a couple of seconds to read the CCD, the digital image appears on the monitor.

Other than a few scattered stars, apparently there is no galaxy in the field. The galaxy we are looking for, however, is not very bright and before making any quick judgements it would be a good idea to process the image so that the low luminosity levels are emphasized. To do so we are going to adjust the visualization thresholds.

There are two of these thresholds, called *low threshold* and *high threshold*. They define a section of the dynamic range of the image over which the program distributes values of gray for display on the monitor. With a standard VGA graphics board (the minimum needed on a PC), there are 16 gray levels for a screen resolution of 640 x 480 points. The values of the thresholds must fall between 4095 and 0 because the dynamic range of the analog-digital converter is 12 bits. An image pixel whose intensity is equal to or greater than the high threshold will be displayed as the lightest grey. On the other hand, a pixel whose intensity is equal to or less than the low threshold will be displayed in black. A pixel whose intensity falls between the high and low thresholds will be assigned one of the 14 levels of gray according to the position of this intensity between the two thresholds. The idea of thresholds is essential for understanding the techniques of image visualization.

When we want to visualize the entire dynamic range of the image, we write VISU 4095 0. These are the default thresholds assumed by LE. To observe faint objects, we have to reduce the threshold range. Let's try VISU 500 0. A faint light appears in a corner of the image, which could be the galaxy. To make sure, let's make the dynamic range even narrower. Running the BG (background) command informs us that the sky background is located around the quantification level 40. BG determines this value by calculating the median intensity of a sample of pixels located in a corner of the image. Let us execute VISU 100 30. This choice still allows us to distinguish the sky background and therefore the faint objects which can be confused with it. The galaxy is now visible.

The choice of thresholds can be imposed implicitly without passing through VISU with the instruction THRESHOLD (THRESHOLD 100 30, for example).

Let us re-center the galaxy by adjusting the position of the telescope. With experience it will be easy to know in which direction to go and by how much. Let's start a new exposure: LE 20. The last visualization thresholds are retained by the program, and the galaxy appears immediately after the exposure in the center of the image.

We next need to find a guide star in our guide scope, again check the centering with a 20 second integration, and then to take a long exposure: LE 300 (five minutes). We now proceed as in photography, i.e., keep the guide star in the center of the cross-hairs. At the end of the integration, the screen displays the image. Next we must adjust the visualization thresholds because the long exposure recorded more sky background and thermal charges had more time to accumulate. Let's try, for example, VISU 500 100. The stars are truly round, the tracking is fine. The galaxy looks good and we want to store the image. At this point the image is in the computer's RAM memory and it must be transferred to the hard disk with a file name: SAVE N2704 (the galaxy is NGC 2704). The red light of the hard disk lights up for two seconds. It's over—the image is saved.

We next decide to take a second exposure of NGC 2704. Later, by summing the two images, we will obtain a final image equal to a single 10 minute exposure. The second image must be saved with a different file name than the first one so that it does not overwrite the first one. If we can't recall the first name, the instruction STATUS will remind us, along with other information (exposure time, date, hour, visualization thresholds, etc.).

The AUTO command saves two files, first an image file, and second a file with key information: date, hour, exposure time, binning factor, etc. This second file may also contain information on the observing site or the characteristics of the filters used for the image acquisition—in other words data that will be of value later. This file is called a *header*. If the AUTO command is executed, the image file will be saved with a name followed by the extension `.PIC` accompanied by a header file with the same name and the extension `.HDR`. To save an image without creating a header file, we merely type `MANUAL`. Using image files and header files can present problems. For example, they may become separated when moving from one magnetic medium to another. The way to avoid this kind of inconvenience is to write the header's data to the image file. This data can be displayed along with the image by reserving a space outside the image area. Displaying this data alongside the image ensures that it will be readily available when the image is studied.

Focusing should be checked regularly. To do this we select a star of magnitude 6 or 7 and make a 1 second exposure: `ACQ 1`. Like `LE`, the command `ACQ` cleans the array, manages the integration delay, reads the array and displays the image. However, unlike `LE`, this process is repetitive. The command `ACQ` therefore is essentially a "moving picture" with the "motion" of the image display determined by the exposure time. To stop the acquisition cycle, we simply press on a key and the last image is held in memory and displayed. Continual acquisition allows interactive control, for example, to center a star.

Simple inspection will tell us about the state of the focus. For example, if the star appears in the form of a ring, we are in fact observing the image of the mirror with its central obstruction. In this case the best focus is very far from the sensitive surface. The focusing mechanism has to be moved while we observe the screen to make the star as sharp as possible. A bit of experience will ensure that the proper visualization thresholds are selected to carry out this operation successfully.

To reach a more precise setting of the focus, a new tool is needed. Let's do `TRACK 12 1`. This instruction does a normal acquisition but the image is not visualized as such. Rather, the program determines the point of the image with the strongest amplitude; then a photometric cross-section of the line passing through this point is visualized. Since the star is supposed to be the brightest object on the image, we obtain its profile. The acquisition is continuous as with `ACQ`. The cross-section always passes through the star even if it moves between two exposures. The point here is to adjust the focus to produce the finest possible profile. Argument `12` of the instruction sets the dynamic range of the visualization of the cross-section (in number of bits). The second argument is the integration time. The intensity of the star's peak is also displayed, giving an additional criterion for focusing since it is at a maximum in the best sharpness plane.

Getting a good focus is not always easy, especially when there is turbulence. Atmospheric turbulence erratically moves the star and spreads the image. The amplitude of the spot changes strongly with the distribution of energy in its disk. Under such conditions, patience is required for successful focusing.

The program has a command (MAP) that works like TRACK, but here the line or the column along which the cross-section is taken is set by the user before acquisition. This option is useful for focusing an extended object (e.g., a planetary limb).

Sometimes we want to save a series of images very quickly. This is a technique used in planetary imaging. Here many images are quickly taken with the hope that a few will have been exposed during a period when the atmosphere was calm enough to allow a superior image. The command AUTOSAVE begins the process. Each image is saved by simply pressing on a key. The file name is supplied at the beginning of the session (for example MARS1, MARS2, MARS3, etc.).

The CCD is read by numerical double sampling which requires the reference level to always have a positive amplitude. If the reference level has a negative voltage at the ADC's input, the ADC translates it digitally into a zero (our ADC is *unipolar*, i.e., it cannot convert negative voltages). In this case, even if the signal level has a positive value, the measurement of the double sampling is incorrect (see Figure 4.1).

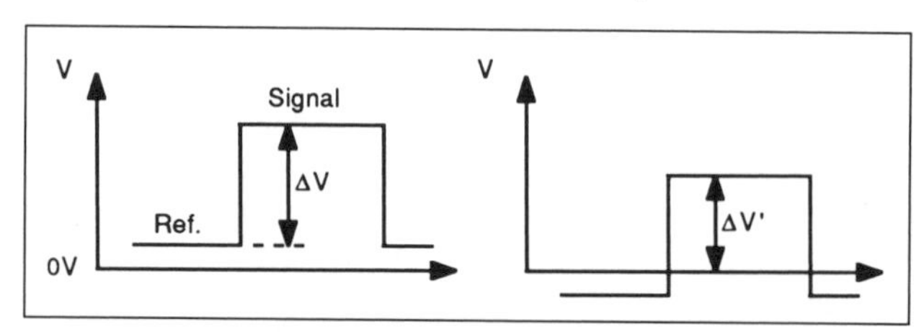

Figure 4.1 *On the left, the reference level and the signal level have a positive value at the ADC input and double sampling supplies a correct value (ΔV). On the other hand, when the reference level has a negative voltage, the signal level is measured with respect to a fictitious reference and then we find reset noise in the image.*

To insure a positive voltage reference level offset, the command TEST makes an "image" in which only the reference level is digitized. By turning the multi-turn pot that controls the amplifier offset, we adjust for a non-zero image (to check our progress, a repetitive cross-section of an arbitrary line is displayed). The reference level is positioned between 100 and 200 quantification units to provide some room for drift of the levels, which is especially helpful when the components have not yet reached their working temperature. This adjustment is necessary to obtain images without reset noise and should be checked regularly during the night.

If the reference level is set at 200, the maximum difference between the signal level and the reference level will be $4095 - 200 = 3895$. The dynamic range is therefore slightly reduced, but this is of minor consequence.

The NOISE command provides a means to check the camera's noise. We start by acquiring at least twenty images in the dark from which we calculate the average value, the variance and the standard deviation of several pixels. This test allows us to detect any possible abnormalities. With the electronics described in this text, the rms noise, given by the standard deviation, is approximately a quantification unit. Summing the lines and columns is possible with the BINNING command. The binning factor can be adjusted in rows as well as in columns.

We have just reviewed the main acquisition commands for the MAT program. Most of the commands in the IP image processing program which will be described below are shared by the MAT. In this way, the user has numerous possibilities of visualizing images and processing them on the spot. The quality of an image can therefore be judged precisely and rapidly, and another image can be taken if there is a problem.

4.3 The Image Processing Program

The IP image processing program anticipates almost every possible need (the compiled code is about 250Kb). The commands for IP can be grouped into four categories:

1. Service commands. Among other things, these commands manage the image files in the mass storage device.
2. Visualization commands. These include visualization in levels of gray and in false colors, changes in palette, three-dimensional display, isophotes, cross-sections, etc.
3. Image processing commands. These include filtering (high-pass, low-pass, median), histogram modification operators, cosmetic correction options, etc.
4. Processing commands. These are limited because this type of instruction is usually specific to a given research program, whereas the IP program is intended for general use. We have included what is strictly necessary, in particular, measurement of the position of objects, and aperture photometry.

Before going any further, let us examine the internal structure of the program. To carry out an elementary operation, such as efficiently summing two images, the two operands must be in the memory at the same time. The program will therefore have to manage at least two zones of image storage. One of these will be called the principal zone, the other the *buffer*. In IP there are in fact two buffers called BUFFER1 and BUFFER2. Commands have been created to transfer the contents of an image in the principal zone to a buffer (TBUF) and to transfer the contents of a buffer into the principal zone (VBUF). The argument of these commands designates the number of the buffer on which we are working. Figure 4.2 shows the layout of the memory organization with the different instructions allowing image transfers. There is no provision for transferring from buffer to buffer because in practice such transfer is useless.

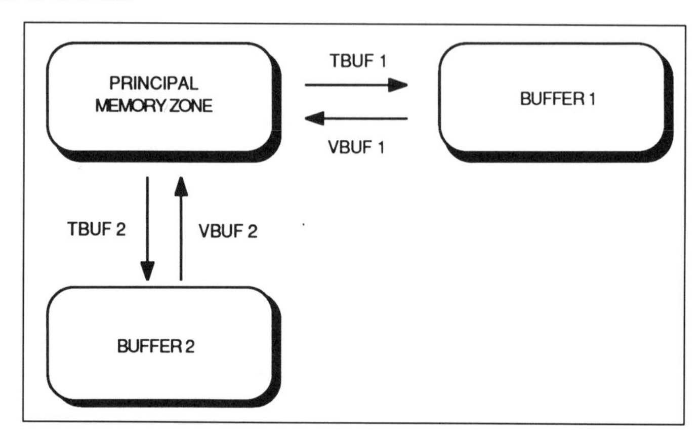

Figure 4.2 *The three memory zones reserved by the* IP *program to store the images.*

With the SWAP command the contents of the principal zone and the contents of BUFFER1 can be permuted without having to pass through BUFFER2.

The images are organized into lines and columns. We will use this systematic structure for all handling operations in the program. Thus the transfer from the principal zone to BUFFER1 will be treated in the following way:

```
FOR i% = 1 TO imax%
    FOR j% = 1 to jmax%
      buf1% (i% ,j% ) = a%(i% ,j% )
    NEXT
NEXT
```

The variables imax% and jmax% represent respectively the number of columns and lines to be treated in the image. The integer arrays a% () and buf1% () designate respectively the principal memory zone and BUFFER1. We defined them in the dynamic memory at the beginning of the program by the following instruction sequence:

```
xmax% = 145
ymax% = 218
DIM a% (xmax% ,ymax% ), buf1% (xmax% ,ymax% )
```

The variables xmax% and ymax% define the size of the image. It is possible to treat only a part of the image by modifying the value of the variables imax% and jmax%. Figure 4.3 shows how this is done. This is the windowing technique that is sometimes used when only part of the image contains useful information—the gain in calculation time can be significant when this technique is used.

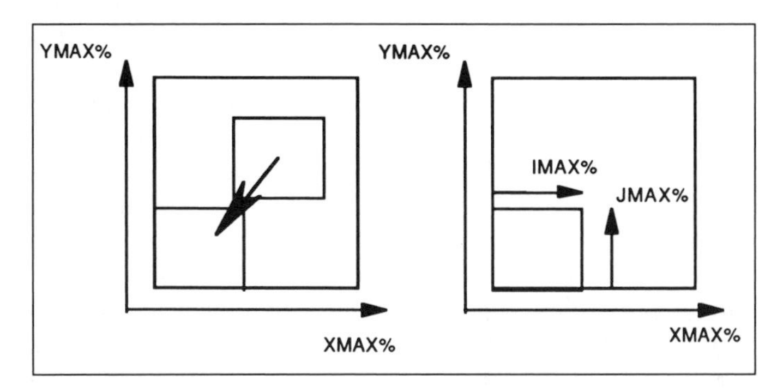

Figure 4.3 *A part of the image is defined and then translated in the lower left corner of the principal memory zone. Processing is then done on an image of size* imax%,jmax%.

Saving and loading an image to, or from a disk are operations that are easy in MS Basic 7.0. The instruction BSAVE makes a direct copy to disk from a memory segment as defined by the user. The instruction BLOAD does the same for loading. The outline of a save routine is given below:

```
length! = 2! * (xmax% + 1) * (ymax% + 1)
DEF SEG = VARSEG (a% (1,1))
BSAVE name$ + ".PIC",VARPTR (a% (1,1)), length!
```

First, the size of the image in bytes is determined and stored in the variable length!. This variable is real (as denoted by the exclamation mark) since the result of the calculation can be greater than 32768, the maximum value allowed by simple length integers in MS Basic 7.0. The intensity of an image point is coded on two bytes, which explains the multiplication by two in the evaluation.

The instruction DEF SEG sets the segment number of the first element of the image array. The image is then saved with the name stored in the variable NAME$ (defined previously) with the extension .PIC. The instruction VARPTR returns the value of the offset of the first element of the array with respect to the beginning of the segment. Loading the image is done by an equivalent but simpler procedure because the size of the image does not have to be supplied:

```
DEF SEG = VARSEG (a% (1,1))
BLOAD name$ + ".PIC", VARPTR (a% (1,1))
```

Saving a 64K image on a standard PC AT hard disk takes about one second. Loading is even faster. These instructions are thus especially fast. However, it is impossible (under MS-DOS) to save at one time a memory zone larger than a 64K segment. If the image extends beyond 64K, it must be processed in several separate pieces. This is what we do with images produced by the TH7863 CCD.

The command `EXPORT` writes out an image file from memory while `IMPORT` reads an image file into memory. These commands are simpler in structure than `BSAVE` and `BLOAD`. This simplification facilitates data exchange between different types of computers. Further, it also makes it easier to save or load a part of an image.

`IP` was designed to be flexible. For example it accepts macro-commands and it is an integrated language. The macro-commands are series of orders that are edited inside `IP` (`EDIPMACRO`). These orders are identical in syntax to those sent from the keyboard. Each time that the `GO` command is started, the series of instructions contained in the macro is executed. This vastly simplifies repetitive procedures. The macro-command can be saved (`SAVEMAC`), reloaded (`LOADMAC`), and visualized (`PRINTMAC`).

`IP`'s integrated language makes it possible to write series of instructions that are much more elaborate than those in a simple macro-command. In fact the language has loop control structures, tests, variables, and input/output instructions. It also facilitates sub-programs, mathematical calculation instructions, and more. All the `IP` commands can be executed from its program mode, and several programs can follow each other. An editor has been included to write programs. However, programs may be imported into `IP` from practically any text editor. The program is started by the command `RUN`. An extensive set of tools is available to help the user create new commands designed for a particular application not initially foreseen in `IP`. Other possibilities for simplifying usage include calling back the last n commands typed appended with their edition or a generalization of assumed parameters.

4.4 Visualizing Images

Restoring an image on a computer screen is not an easy task. The overall visualization strategy will depend on the capabilities of the graphics board. Two cases can be distinguished:

1. The screen is binary, i.e., the graphics point can be only be turned on or off;

2. The screen is analog, meaning that a graphic point's intensity can vary over a range of brightnesses or colors.

4.4.1 Visualizing With Binary Graphics Boards

This category includes the PC compatible Hercules type boards or the Macintosh screen (Mac +, Mac SE). Most printers (impact, thermal, laser) output binary type images. The ability to reproduce an image on paper justifies studying the binary representation of images.

Visualizing is problematical. On the one hand the camera records a magnitude varying continuously within a "range of grays"; on the other hand, the screen (or the printer) is only capable of binary representation. In what follows we will discuss several methods which create the illusion that an image contains levels of gray.

Reproduction With Predefined Patterns

An image's range of intensities can be divided into a finite number of intervals. Each of these intervals is assigned a pattern composed of an array of $M \times N$ screen points. A variety of patterns is obtained by turning on a greater or lesser number of points in this array. Depending on where an image element belongs on one of these intervals, the corresponding pattern will be drawn on the screen. If the pattern has been well chosen it is possible to move back from the image to a point where the artificial structure fades away and is replaced by a grey shaded image. Figure 4.4 shows a range of ten grays created from a 3×3 matrix.

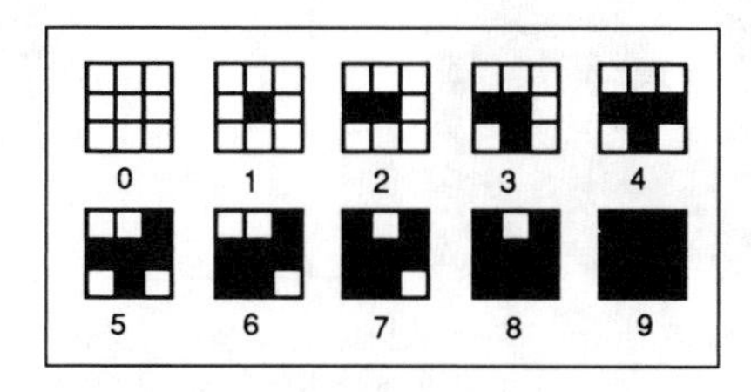

Figure 4.4 *Examples of patterns obtained from a 3×3 matrix. If the size of the array is $n \times n$, the number of gray levels simulated is $n^2 + 1$.*

The matrices are not created at random. Thus, in the range shown in Figure 4.4, if the third pattern is replaced as indicated below, horizontal lines that look artificial will appear in the otherwise uniform segments corresponding to this level of gray.

At a transition between two levels of gray states, an obvious contour may appear. To minimize these contours, a screen point lit for level j should also be lit for level $j+1$. This technique requires screens that have numerous points. In fact, in the example given, each pixel of the digitized image needs 9 screen points. In general, if the screen point is not very small, the structure of the array can be seen in the final image, giving a rather artificial texture. Laser printers with a 300 ppi (points per inch) resolution work well with this method. The algorithm is easy to program and is usually sufficient for rapid image reproduction.

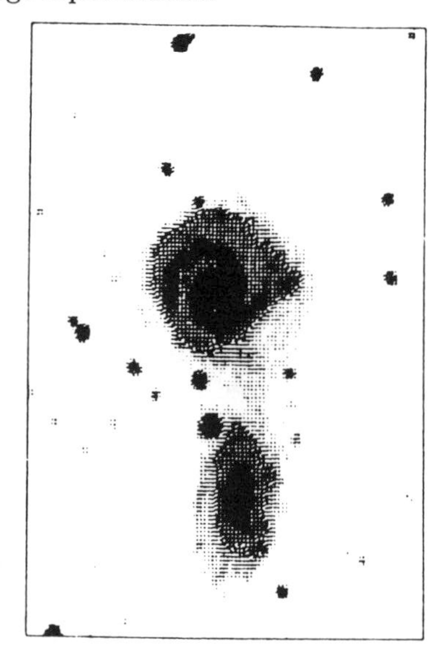

Figure 4.5 *Example of images printed with a 300 point per inch laser printer; the gray levels were simulated by a 4×4 matrix. The image is of the galaxy pair ARP271 taken with a TH7852 array. The exposure was 20 minutes with the T60 telescope at Pic du Midi Observatory diaphragmed at 40 cm. The negative representation on the right is often used in astronomy because it is easier to analyze details that may blend into the sky background if printed as a positive.*

Reproduction by the Adaptative Average Method

A local average is calculated from the image points surrounding the pixel $P_{x,y}$. Let this average be M. A quantity $P'_{x,y}$ is determined by applying the following equation

$$P'_{x,y} = M + k(P_{x,y} - M).$$

In this relation k is a parameter that acts upon the rate of contrast of the reproduction.

A threshold T is chosen based upon the following:

Screen point lit if $P'_{x,y} > T$;
Screen point turned off if $P'_{x,y} \leq T$
The reproduced image is binary if $k = 1$.

The parameter k and the threshold T are chosen by trial and error to obtain realistic visualization of the image. This procedure is known as the Jarvis-Roberts algorithm. It is particularly good at reproducing images with a large dynamic range. Another advantage is that a single screen point can represent a CCD pixel.

Reproduction by Dithering Threshold Matrices

An $n \times n$ matrix is compared with the image by a step value n. This matrix is known as a halftone screen. For a given position of the screen, the coefficients of the matrix are compared to the amplitude of the underlying pixels. According to the result of the comparison, the corresponding points on the computer screen are turned on or off. The halftone method is almost universally used to reproduce photographs in books, magazines and newspapers.

Depending on the value of the matrix's coefficients, we can define screens of special shapes. Thus the following matrix gives a screen that looks like a double spiral:

13	7	9	15	18	24	22	16
11	1	3	5	20	30	28	26
4	2	0	10	27	29	31	21
14	8	6	12	17	23	25	19
18	24	22	16	13	7	9	15
20	30	28	26	11	1	3	5
27	29	31	21	4	2	0	10
17	23	25	19	14	8	6	12

However the matrix that gives the best result with a screen of moderate resolution is a Bayer screen. Here are the coefficients:

0	32	8	40	2	34	10	42
48	16	56	24	50	18	58	26
12	44	4	36	14	46	6	38
60	28	52	20	62	30	54	22
3	35	11	43	1	33	9	41
51	19	59	27	49	17	57	25
15	47	7	39	13	45	5	37
63	31	55	23	61	29	23	21

The coefficients of the matrices can be multiplied by a coefficient whose value is dictated by the dynamic range of the image. Also, a constant may be added to adjust the position of the gray scale within the image's dynamics.

The screening technique with a matrix of so-called "ordered" coefficients produces visualizations that show details especially well. Unfortunately, the artificial texture of the screen is a bit too obvious. This defect can be remedied by randomly permuting the coefficients of the matrix at the same time as it passes in the image. However the reproduction then appears noisy and imprecise.

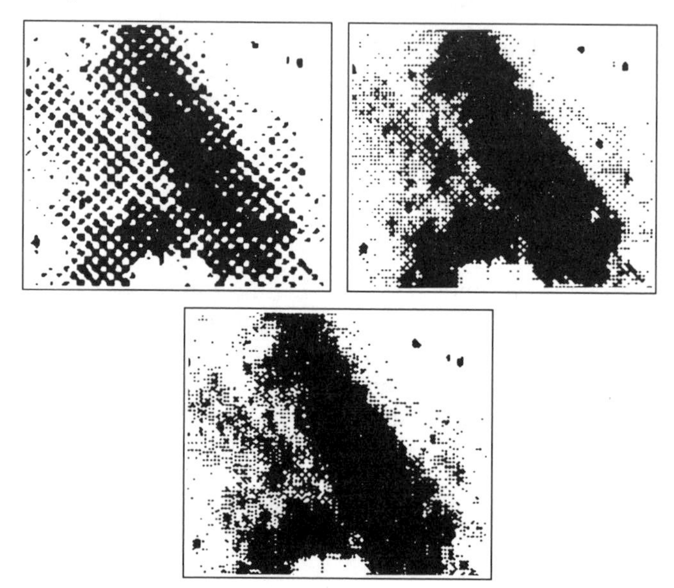

Figure 4.6 *Restoration of nebula M17. Double spiral matrix at the top left. Bayer matrix at the top right, and at the bottom a matrix whose coefficients are randomly permuted.*

Reproduction by Error Diffusion

Thresholding an image is a simple operation. After setting a threshold, the following decisions are made for each pixel:

1. If the pixel's intensity is higher than the threshold, the corresponding screen point is turned on in a light color;
2. In the opposite case, the screen point is turned off.

As the intensity of the pixels moves away from the threshold the processed image looks less and less like the original—therefore an error has been made. The purpose of error diffusion is to distribute the error introduced during the thresholding of the first pixels onto the points that remain to be plotted. For each pixel a binary decision is made that depends on the intensity of the pixel and the error introduced during the thresholding of the preceding pixels. This procedure generally uses about a dozen pixels preceding the plotted point.

If $N_{x,y}$ is the intensity of the pixel and E the error, a quantity $N'_{x,y}$ is determined from this relation:

$$N'_{x,y} = N_{x,y} + \sum_{i=1}^{N} \mu_i E_{x+a,\ y+b}.$$

In this equation the sum, called an error diffusion matrix, represents the total of the errors weighted by the coefficients μ_i distributed in the matrix.

0,1	0,1	0,6
0	0,2	$N(x,y)$

0,03	0,06	0,10	0,03
0,06	0,10	0,15	0,06
0,10	0,15	$N_{x,y}$	

Figure 4.7 *Two examples of error diffusion matrices. The coefficients are addressed by incrementing two variables (a, b) as a function of i. The value of the coefficients is determined empirically to obtain the best result.*

The special shape of the matrices in Figure 4.7 makes it possible to take into account past processing of the image provided that it is processed from left to right and from top to bottom. Starting from the quantity $N'_{x,y}$, a binary decision is made as a function of the threshold. The result is stored in a table $B_{x,y}$. An image coded in 8 bits has the following values:

$$B_{x,y} = 255 \text{ if } N'_{x,y} > \text{ threshold},$$

$$B_{x,y} = 0 \text{ if } N'_{x,y} \leq \text{ threshold}.$$

The error produced on the studied pixel is calculated by:

$$E_{x,y} = N'_{x,y} - B_{x,y}$$

The calculation is initialized by giving a zero value to the error (upper left corner of the image).

The appearance of the processed image depends on the shape of the matrix, on the value of the coefficients, and on the value of the threshold. If the coefficients have low values, the image will have high contrast and will be very sensitive to noise. On the other hand when the coefficients have high values, a low contrast image is produced with a loss of fine details.

Error diffusion produces a characteristic stringy texture (avalanche effect). The continuity of grays is excellent and details are shown without aliasing. However the relative position of objects in the reproduction is not guaranteed because position depends on the local context. The error diffusion algorithm burdens a PC and the plotting is slow.

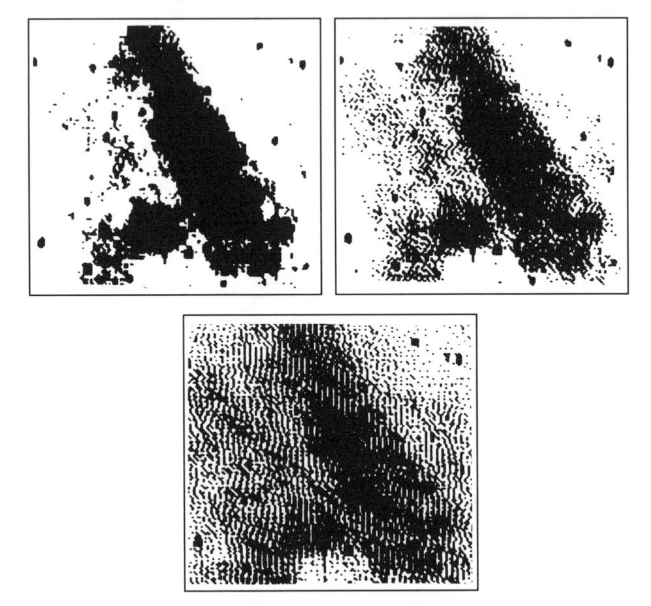

Figure 4.8 *The nebula M17 visualized with the error diffusion method. From clockwise from top left to bottom the value of the error coefficients increases.*

4.4.2 Visualizing With Graphics Boards And Analog Monitors

This visualizing mode comes closest to the appearance of the real image. The accuracy of the reproduction is limited, however, by the number of levels of gray that can be simultaneously displayed. In practice, there is a point beyond which additional levels of gray contribute nothing to the image.

By making a scale of grays (uniform segments placed side by side and of increasing "grayness"), it is possible to see the number of simultaneously displayed grays at which the eye can no longer distinguish the difference between two neighboring shades. Good graphics boards simultaneously displaying 256 gray levels are sufficient. The dynamic range of many monitors does not allow us to benefit from all these shades at one time but sometimes by adjusting the monitor's brightness and contrast controls, a more realistic image is possible. To about 64 levels the eye can clearly distinguish the gray segments. An image visualized under these conditions will show iso-contours. The effect is of course enhanced when the number of grays decreases. With fewer than 16 gray levels, the contours become a hindrance when interpreting the image. The eye easily distinguishes the iso-contours in image zones that are practically uniform such as sky background or a faint galaxy. They are less obvious when the image has a wide contrast range (for example, a moonscape).

An EGA board on a PC can simultaneously display 16 colors. They are not analog, but the 16 shades may be chosen from a range of 256 colors to create a range of 16 consecutive grays.

VGA boards have analog outputs. Sixteen true grays can be displayed at the resolution 640×480, and 256 colors at the resolution 320×200. In the latter mode, only 64 true grays can be simultaneously displayed. Additional graphics modes are regularly added to this type of board, and some display 256 colors on 640×480 points, or even 1024×768 points, producing images that approach photographic quality.

As an example we will describe a visualization procedure in a 256 color mode and 320×200 points on a PC (mode 13). First the colors that will be used to display the image are selected from among the range of colors available on the board. This range is called the palette. Graphic mode 13 offers a palette of 262,144 colors, from which we can create a sequence of 64 grays. The levels of gray are created by evenly mixing the basic colors of red, green and blue. The palette used for the display (256 colors) is validated by the instruction **PALETTE USING**, which points toward the first element of a table containing the color numbers. For example

```
FOR i% = 0 TO 63
  pal & (i% ) = i% + i% * 256 + i% * 65536
NEXT
FOR i% = 64 TO 255
  pal & (i% ) = 0
NEXT
PALETTE USING pal & (0)
```

The color table is of the long integer type (coding on 4 bytes). The first 64 elements of this table are filled with the numbers corresponding to a range of grays (the three additions correspond to the mixture of the three basic components, progressing from the dimmest to the brightest luminous intensities). In the second loop the table is completed up to a limit of 255 by zeros, i.e., by the color black.

When the image is plotted on the screen, the intensity of the image points must be linked to the gray levels of the palette. This linking is carried out by a transcoding table called a LUT (Look-Up Table). This table contains as many elements as there are possible quantification units in the image (in our case, 4096). In its simplest form, the LUT is uniformly divided into a number of zones equal to the number of colors that can be displayed (64 here), and these zones are filled by numbers representing the number of the slice. Thus from row 0 to row 63, the table will be filled with zero; from row 64 to row 127, it will be filled with one; and so on until the last slice that goes from 4032 to 4095 is filled by the number 63. When plotting an image, the LUT is read for the level of gray corresponding to a row equal to the intensity at that point. For example, if the intensity of the pixel is 72, element 72 of the LUT contains the number 1, and the screen will display a point with the intensity of the first level of gray above the black. We will see later that using LUT's allows easy modification of the screen's appearance, including adding false colors.

The fragment of the following program shows how the LUT is filled.

```
xdyn! = 63 /(ish% - isb% )
FOR i% = 0 TO maxdyn%
  color% = (i% - isb% ) * xdyn!
  IF color% <0 THEN
     lut% (i% ) = 0
  ELSE IF color% >63 THEN
     lut% (i% ) = 63
  ELSE
     lut% (i% ) = color%
END IF
NEXT
```

The variables ISH% and ISB% contain respectively the high and low thresholds. These values are entered by the user. The variable maxdyn% contains the value of the maximum dynamic range of an image (here 4095).

Plotting the image is now straightforward.

```
m% = maxdyn% + 1
j1% = screeny% - ypos% - 1
FOR i% = 1 TO imax%
  x1% = xpos% + i%
  FOR j% = 1 TO jmax%
    if a% (i% ,j% ) <m% THEN k% = a% (i% ,j% )ELSE k% = maxdyn%
    PSET (x1% ,j1% - j% ), lut% (k% )
  NEXT
NEXT
```

The array containing the image (principal memory zone) is swept by a double loop—first rows and then columns. The instruction PSET (X,Y),I writes to a point located at the screen coordinates (X,Y) with gray level I. The variables xpos% and ypos% contain the screen coordinates of the lower left corner of the image. These variables are modified by the POS command,

and it is therefore possible to have several images simultaneously on the same screen. The variable `screeny%` contains the value of the vertical resolution of the screen (200 in mode 13). In our example, the image pixels would have an intensity higher than 4095.

The speed of the above routines can be significantly increased by writing them in assembler. This is done by writing a byte representing the value of the gray level into the computer's image memory which starts at `& HA000` in mode 13.

4.4.3 VGA Boards for PC's

Serious problems appear when we wish to work in a graphic mode that is not documented in the VGA standard (beyond mode 13). This is an unfortunate situation since the extended modes offered by current VGA boards are very attractive: 640×480 in 256 colors, 800×600 in 256 colors, etc. Generally, the only information available to the buyer of such a board is the value of the interrupt parameter `10h` which allows one to change to the extended mode. Writing to the video memory is another matter. A solution is to work continually with interrupts. In fact, the BIOS compatibility of all the VGA boards is excellent. The disadvantage of this otherwise reliable method is the slowness of the display (one minute to display an image in the format 380×290 with a PC AT 8MHz, compared to 2 seconds when writing directly to memory).

Standard PC BIOS contains some graphics routines. These are accessible through the software interrupt `10h` by loading into the microprocessor's `AH` register the parameter corresponding to the function desired. Some functions require other parameters that will also be passed by specific registers.

As an example, let us study the function that enables activation of a new video mode. This is the function `0h` of the interrupt `10h`. The additional parameter, which we write to the `AL` register, defines the graphic mode. Therefore, to change the graphic mode to 640×480 with 16 colors, `AL` has to contain the value `12h` when the interrupt is called. In assembler, the initialization will be written as follows:

```
MOV AH,0
MOV AL,12h
INT 10h
```

Changing the VGA display to a 320×200 mode, with 256 colors requires that the `AL` register contain `13h`, etc. To determine the parameter for the higher modes, it is necessary to consult the board's data sheet. For example, on a Paradise board, the 640×480 mode with 256 colors is obtained by loading the value `5Fh` into register `AL` during initialization.

In any case, whatever the graphic resolution, all the functions and sub-functions of the `10h` interrupt work because a program written in ROM on each graphics board ensures the compatibility with the BIOS. This compatibility is fundamental because, among the functions of the `10h` interrupt, we find some that allow the display of a point on the screen and modify the color palette; in other words, there is a straightforward way to visualize our CCD images.

We need only 3 basic functions, besides the initialization function (`0h`) already studied, to cover all of our requirements.

1. Writing a point on the screen—function `0Ch`. The registers will take the following values:
   (a) `AH` = `0Ch`
   (b) `AL` = color of the point (0...256)
   (c) `CX` = screen X-coordinate
   (d) `DX` = screen Y-coordinate
2. Reading the color of a point on the screen—function `0Dh`:
   (a) `AH` = `0Dh`
   (b) `CX` = screen X-coordinate
   (c) `DX` = screen Y-coordinate
   (d) `AL` = return value of the color
3. Definition of the color palette—function `10h`, sub-function `10h`:
   (a) `AH` = `10h`
   (b) `AL` = `10h`
   (c) `BX` = number of the color
   (d) `CH` = percentage of green (`0...63`)
   (e) `CL` = percentage of blue (`0...63`)
   (f) `DH` = percentage of red (`0...63`)

Of course this sub-function only works in the graphic modes that allow the display of 256 simultaneous colors.

Other Visualizing Modes

To facilitate the study of an image, it is helpful to have a wide range of visualizing tools. The `IP` program makes it possible to enlarge certain details of the image during visualization. The commands `ZOOM1` and `ZOOM2` will enlarge a part of the image by a factor of 2 or 4. The enlarged area can be specified implicitly (`CENTER` command) or defined from a point designated by the `CURSOR` command. The zoom effect is obtained by interpolating additional points between those that already exist.

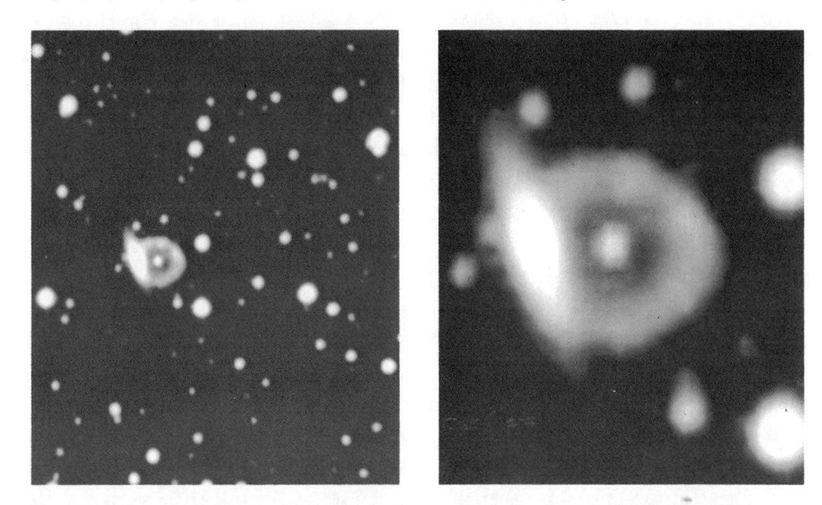

Figure 4.9 *On the left, a full-sized image that shows the field of the small planetary ring-like nebula PK 38-25.1. On the right, an enlargement of this image permits a better view of the ring. Note that it is superimposed on a galaxy in the background. 10 minute exposure with the T60 telescope at the Pic du Midi Observatory.*

The commands `EXPAND` and `SCALE` scale image files loaded into RAM memory. `ZOOM1` and `ZOOM2` only enlarge the actual image displayed on the screen. This situation will be discussed again in the section about geometric transformations.

To enhance the colors of the images, the active color palette can be modified at any moment with the `PAL` command. Traditionally, the low levels of the image are arbitrarily represented by cold colors (blues) and the high levels by warm colors (reds), but variations are possible.

A spectacular option of the `IP` program allows us to visualize an image in true colors starting from three images taken through different filters. If the filters isolate the red, green, and blue radiation, we will have an image whose appearance is close to the visible one. We can also use an infrared filter and have an image whose chromatic balance is shifted with respect to what the eye sees, but whose content is just as rich.

However, direct visualization in true colors is impossible with a computer with limited graphic capabilities. We have to be able to mix three-color components on the screen while keeping a reasonable range of intensities. To get a reproduction worthy of photography, we have to have a considerable number of simultaneous colors. The ideal is to use a graphics board with 3 color planes of 8 bits each, i.e., 256 shades of blue, 256 shades of green, and 256 shades of red. Such a board has a palette of 16,777,216 colors (24 bit graphics board). A standard VGA board offers 6 shades in each of the basic colors, representing $6 \times 6 \times 6 = 216$ colors for the three-color process, a figure which is just below the 256 simultaneous colors available with the 24 bit graphics board. This figure is sufficient to provide a good idea of the colored appearance of the image. The three monochromatic components should be present in the dynamic memory at the same time; they are stored in the two buffers and in the principal memory.

Things do improve if, instead of representing an image point by point on the screen, we use four screen points forming a little square. With a 256 color VGA graphics board, we can assign 64 shades of red to one of these four points, which will act as the representative of the "red" three-color component, 64 shades of green to the second screen point to represent the "green" component, and 64 shades of blue for the "blue" component. The easiest thing to do with the remaining screen point is to define a range of 64 grays and then to assign to the extra screen point a level of gray that corresponds to the average intensity of the 3 three-color components. Colors are rendered spectacularly with this method, especially if we use a VGA graphics board that can display 256 colors simultaneously at a resolution of 640×480 or 800×600. Obviously the image is enlarged by a factor of two along the two axes with respect to a traditional visualization (one image point = one screen point).

For an even more accurate representation of true colors, relatively sophisticated techniques have to be used. The idea is to display three trichromatic components in their basic color one after the other, while photographing the screen with color film. The camera's shutter is set on B during this exposure. Remember that a VGA board allows restoration of the blue component with 64 shades of blue, the green component with 64 shades of green, and the red component with 64 shades of red. The final photographic film image will therefore contain $64 \times 64 \times 64$ shades of color, or a palette of 262,144 colors. It is not technically difficult to obtain this kind of effect. The three components are carefully superimposed so that they appear in the same place on the screen (we will see how in section 4.5.9); then they are displayed successively for a precise length of time (to be adjusted for a correctly exposed final image). To check chromatic rendering on the film, the image should also have a scale of colors that cover the spectrum from blue to red. This scale can be placed along one of the image's borders. The images should also appear and disappear very quickly, which means writing directly to the video memory (use assembler or some evolved language that allows such manipulations). With this procedure, an image can be displayed about ten times a second. Moreover, if this frequency or one that is even a bit higher can be obtained, seeing the image in true colors can be achieved by sequentially displaying the three components—the persistence of the eye will blend the three colors into a chromatic image.

However, if for some reason the display cannot be cycled at these rates we can, for example, display the image slowly from left to right and then erase it at the same speed and in the same direction. The screen points will then be lit during the same period, no matter what their position is on the image. Another technique is to display the image in a palette where all the colors are black. Once the image is displayed (but invisible), the palette adapted to the processed trichromatic component is changed and the image suddenly appears.

Here is some advice on taking photographs of images on computer screens:

1. During the shot, the room containing the screen must be completely dark to eliminate parasitic reflections and to get good contrast.
2. Use a lens with a medium focal length to limit distortion caused by the curvature of the screen. A focal length from 80 mm to 135 mm is a good choice with 35mm film. A macro lens is ideal since it allows you to exactly frame the image whatever its size (extension rings on a normal lens will also work).
3. Don't expose at less than one-fifteenth of a second to avoid the appearance of the vertical blanking bar in the middle of the image.
4. Do not push the contrasts on the screen to extremes, or the final result on the film will tend to be too "harsh".
5. With a 100 ISO film, the usual exposure time is 1/2 second at f/4.

Three-dimensional Visualizing

Images can be processed to give them a three-dimensional appearance. If we pass a plane (X, Y) through the image to represent the geometry and an axis Z that defines the intensity, three different and useful images can be created:

1. false relief obtained by a perspective view of the image;
2. isophotal curves;
3. cross-sections.

The `TRID` command of the `IP` program makes it possible to show the image as though it has relief. This representation is very useful for judging the relative brightness of the various parts of an image.

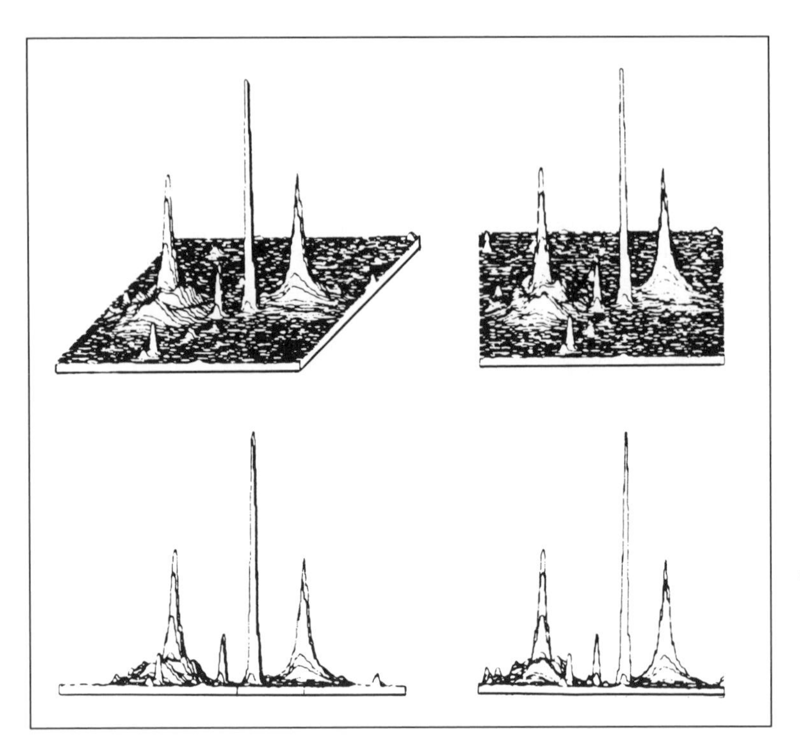

Figure 4.10 *Three-dimensional representations of the system Arp271. The appearance of energy distribution as a function of distance from the core is obvious. Different observational viewpoints are possible.*

By making a slight horizontal and vertical shift between the different plotted profiles, we create an image that gives a respectable impression of depth. This "three-dimensional" representation of the image is obtained by the following process:

1. First plot the intensity profile of the first line (or column) of the image along the x-axis of the screen (horizontal axis). This curve will be "closest" to the observer. At the same time initialize an array, called a *yt*-line or Z-buffer, with the curve's intensity values (screen y–coordinate) as a function of the screen x-coordinate (which is the index of the array).

2. Then draw the second profile which is located in the second plane. During this plotting compare the screen coordinate y_1 of the plotted profile with the coordinate y_2 contained in the peak line for position x. If $y_1 > y_2$, the screen coordinate point (x, y_1) of the new profile is displayed. In addition, position x of the crest line is given the new value y_1. In the case where $y_1 \leq y_2$, do not plot anything on the

screen since the point is hidden by a line located further forward with respect to the observer, the crest line is not changed and we go on to the next point.

3. Proceed in the same way for the following profiles.

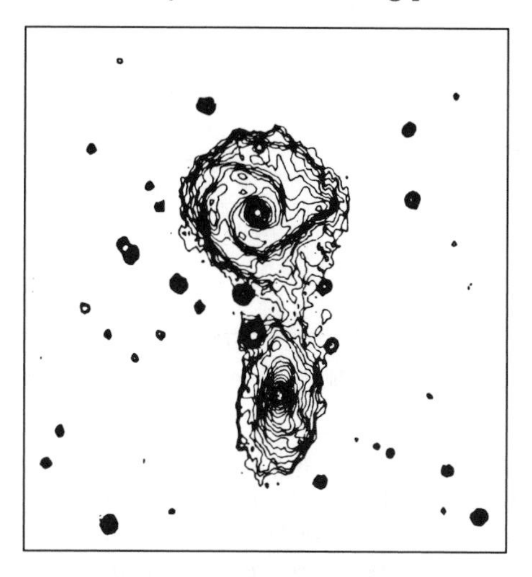

Figure 4.11 *The* `ISO` *command of the* `IP` *program lets us represent the image of the galaxy pair Arp 271 in the form of isophotal curves.*

Isophotal Curves

Figure 4.11 shows an example of plotting isophotal curves. The "level" curves link points of equal luminous intensity. The interval between curves can of course be adjusted as can the maximum and minimum levels between which the isophotes will be plotted.

Following is a method for obtaining isophotal curves. The image is explored in groups of four contiguous points. These four points make up the four corners of a square of screen coordinates (x_i, y_i), (x_{i+1}, y_i), (x_i, y_{i+1}), and (x_{i+1}, y_{i+1}). We plot the two diagonals of the square, which allows us to define four triangular zones inside it (see Figure 4.12A). A fictitious additional point is defined in the center of the square and therefore is located at the intersection of the diagonals. Its intensity is equal to the average intensity of the four points marking the square.

If we refer to a three-dimensional space (coordinate z represents the intensities), we can join our five points with triangles. Four triangles will thus be plotted. One of them is shown in Figure 4.12B.

Furthermore, we have defined intensities through which the isophotal curves will pass. These curves represent the plotting of the intersection of the image surface and of the planes located at a height z equal to the intensity of the isophotes.

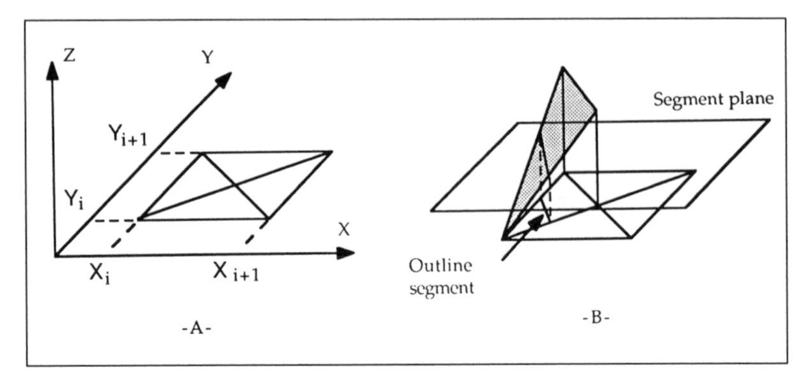

Figure 4.12 *Method of tracing isophotal curves.*

Let us study one of these planes in Figure 4.12B. It might cut one or several triangles drawn in the three-dimensional space. Several cases may occur: the intersection is on a line, on a point or following a plane. The first configuration is the one of interest to us and the one the program should detect. The small segment that results from the intersection is nothing but a part of the isophotal curve defined by our cross-section plane. The coordinates of the ends of the segment, projected in plane (X, Y), are deduced from elementary geometric relationships.

The search for the intersection is done for the four triangles as well as for all the user defined planes. Once a search is finished, the next group of four image points is processed and so on. The isophotal curves are thus plotted gradually as the routine passes over the image. The size of an elementary square counted in screen points can be varied to create a zoom effect. To improve the legibility of isophotal images, the level curves can be colored as a function of their intensity.

Cross-sections

The command `CUT` creates photometric cross-sections along the lines or the columns in the image. The cross-section plane is positioned by the keyboard's cursor keys or mouse. The photometric profile is updated in real time. At the same time, an image is displayed on the screen on which a highlighted line or column shows the position of the cross-section. The scale of the photometric profile can also be adjusted to detect faint intensity variations inside the image. Programming this kind of command is straightforward. A collection of visualizing routines should be rounded out with help screens that provide layout instructions: scales of gray, texts, image mosaics, plotting of indicators in the image itself to emphasize an object, etc.

4.5 Image Processing

This section will examine several procedures to assist in the interpretation of CCD images. These procedures are necessary because the combination of atmosphere, telescope, the CCD itself, and its associated electronics tend to degrade the image. Strong atmospheric turbulence, poor focusing or the presence of charge diffusion in the detector can cause image spread. If the object is a point source, such as a star, the point spread function will be a good criterion for appreciating the quality of the spatial resolution. This point spread function is also called instrumental function or impulsive response of the acquisition system. Knowing the point spread function allows, under certain conditions, restoration of the original appearance of the object. This is an important part of image processing.

The optical system can also deform the image (geometric distortion). Image processing helps to compensate for these deformations and to restore the original geometry. Moreover, the CCD is affected by different noises which are superimposed on the image. Extracting useful information from this kind of environment is one of the main purposes of image processing.

An image may contain too much information, which can confuse the analysis. Image processing can minimize confusion and isolate the useful information. There is almost an infinite number of applications of image processing and once the techniques are mastered, it allows us to get the maximum from an observation. This is a decisive advantage with respect to other detectors such as photography. This power can be a double-edged weapon if it is not well understood. Indeed, you can so manipulate the image that it has nothing to do with reality. Common sense and moderation are qualities that are appreciated in anyone who wishes to do image processing.

4.5.1 Low-pass Filtering

To introduce the concept of filtering, we will deal with noise in the image. An especially efficient noise reduction tool is *convolution.* This operation calculates a new intensity for an image point, taking into account the weighted intensity of neighboring points. To reduce the noise we will use a spatial averager which resembles a matrix. Called a convolution matrix, it moves by one pixel unit across the image. The elements of the matrix are the weighted averages. The intensity of the image point located at the center of the matrix is replaced by the weighted average intensity of all the points located inside the matrix. Let us take an example by numbering the image points in a 3×3 matrix as follows:

$$\begin{array}{ccc} I_1 & I_2 & I_3 \\ I_4 & I_5 & I_6 \\ I_7 & I_8 & I_9 \end{array}$$

If the coefficients have the values

$$\begin{array}{ccc} 1 & 2 & 1 \\ 2 & 4 & 2 \\ 1 & 2 & 1 \end{array},$$

the new intensity of the point located in the center of the matrix will be:

$$I_5 = \frac{I_1 + 2 \times I_2 + I_3 + 2 \times I_4 + 4 \times I_5 + 2 \times I_6 + I_7 + 2 \times I_8 + I_9}{16}.$$

Convolution matrices sized 3×3 or 5×5 are often used. There are 4 low-pass filters pre-defined in the `IP` program. They are given below in the order of increasing power:

$$\begin{array}{ccc} 0 & 1 & 0 \\ 1 & 16 & 1 \\ 0 & 1 & 0 \end{array} \qquad \begin{array}{ccc} 1 & 1 & 1 \\ 1 & 12 & 1 \\ 1 & 1 & 1 \end{array}$$

$$\begin{array}{ccc} 1 & 1 & 1 \\ 1 & 4 & 1 \\ 1 & 1 & 1 \end{array} \qquad \begin{array}{ccc} 1 & 1 & 1 \\ 1 & 1 & 1 \\ 1 & 1 & 1 \end{array}$$

Sweeping the image with the convolution matrix is programmed with the following code:

```
FOR i% = 2 TO imax% - 1
  FOR j% = 2 TO jmax% - 1
    im% = i% - 1
    ip% = i% + 1
    jm% = j% - 1
    jp% = j% + 1
    s& = a% (im% ,jm% )*m1& + a% (i% ,jm% )*m2&
         +a% (ip% ,jm% )*m3& + a% (im% ,j% )*m4&
         +a% (i% ,j% )*m5& + a% (ip% ,j% ))*m6&
         +a% (im% ,jp% )*m7& + a% (i% ,jp% )*m8&
         +a% (ip% ,jp% )*m9&
    buf1% (i% ,j% ) = CINT(s& /sum% )
  NEXT
NEXT
CALL vbuf(1)
```

The coefficients of the convolution matrix are contained in the double integer variables `m1&, m2&, m3&,` etc. The calculations are double precision since the convolution product can in certain cases be higher than 32768 (maximum value of a number coded in signed single precision on two bytes). The result of the convolution is written to one of the buffers. At this stage the main image cannot be modified since the amplitude of a point in it is used several times as the convolution matrix moves. The variable `sum%` contains the sum of the matrix's coefficients. When programming in assembler, this sum should be equal to a power of two so that the division is reduced to simple integer binary right shifts. The instruction `CINT` returns the rounded-off integer part of the expression located between the parentheses. After complete convolution of the image, the result is transferred into the principal memory by calling the `VBUF` routine. The points located on the edge of the image cannot be evaluated with this algorithm since it requires knowing the amplitude of pixels that do not belong to the image (negative coordinates or spillover). A row of points is thus lost around the outside edge of the image, but this loss is not really a problem with small-sized convolution matrices.

Convolving an image of 144×218 pixels by a 3×3 matrix takes about 6 seconds on a 8 MHz PC AT programmed in MS Basic 7.0.

In mathematics, convolution is expressed in the one-dimensional case in the form

$$G(x) = \int_{-\infty}^{+\infty} A(x-e) \cdot H(e) de$$

with $G(x)$ the convolved function, A the original function, and H a convolution vector.

Let us consider the function $A(x)$ at a point x. At a "distance" (number of pixels) e from this point counted back toward the beginning, we multiply this function by the weighting coefficient $H(e)$ which only depends on this distance. We integrate the result over all e. The processing is started again for the point $x+1$ and so on. The convolution product can be represented in the symbolic form

$$G = A * H.$$

The sign $*$ represents the convolution and not a simple multiplication.

In two-dimensional image processing, convolution is a function of two variables of space, x and y. We then write

$$G(x,y) = \int\int_{-\infty}^{+\infty} A(x-e,\ y-f) \cdot H(e,f)\ de\ df.$$

To increase the action of a convolution matrix, we can pass it over the image several times.

The effect of low-pass filtering can be expressed in terms of spatial frequencies. The fine details, i.e., the high frequencies, are removed in this operation. This is why it is called low-pass filtering—only the frequencies below a certain value are retained. The amount of filtering is a function of the convolution matrix's coefficients. The difficulty with this technique is controlling the amount of filtering to remove the noise while keeping the significant details of the image. In the case of convolution in the spatial domain, it is hard to find the right balance because this kind of filtering is not very selective.

Figure 4.13 *No, it's not a distant galaxy. On the left, the original picture of the author! In the center, processing with a low-pass filter allows us to remove the fine details (command* `FB4` *in the* `IP` *program). On the right, passage of a high-pass filter (*`FH4`*) greatly emphasizes the details.*

4.5.2 High-pass Filtering

High-pass filtering emphasizes fine details with respect to slow variations in the image's intensity. For high-pass filtering we use convolution matrices with negative coefficients. The following are pre-defined in the `IP` program:

$$\begin{matrix} 0 & -1 & 0 \\ -1 & 20 & -1 \\ 0 & -1 & 0 \end{matrix} \qquad \begin{matrix} 0 & -1 & 0 \\ -1 & 10 & -1 \\ 0 & -1 & 0 \end{matrix}$$

$$\begin{matrix} -1 & -1 & -1 \\ -1 & 16 & -1 \\ -1 & -1 & -1 \end{matrix} \qquad \begin{matrix} -1 & -1 & -1 \\ -1 & 10 & -1 \\ -1 & -1 & -1 \end{matrix}$$

These matrices modify the average level of the image. To find a level equivalent to the original, the convoluted image is divided by the sum of the weighting coefficients: 16, 6, 8 and 2 respectively.

While high-pass filtering accentuates the fine details, it also emphasizes noise. In spite of the noise, however, high-pass filters are the best method available to sharpen a slightly fuzzy image. They are often used to improve the visual aspect of planetary or lunar images.

4.5.3 Cross Convolution

Up to this point we have utilized a technique that convolves from rectangular matrices in a single operation. In some cases it may be better to convolve the image first by following the lines with a one-dimensional filter and *then* following the columns (or vice versa). This procedure is called cross convolution. Following is an example with a convolution by a Gaussian surface with the meridian profile:

$$z = \exp\left(\frac{-x^2}{2\sigma^2}\right).$$

The Gaussian surface has a bell shape:

$$z = \exp\left(\frac{-r^2}{2\sigma^2}\right)$$

with $r = \sqrt{x^2 + y^2}$.

This surface of revolution is not easy to model using a traditional matrix. In addition, to get a good representation, we need a large-sized matrix. Since the calculation time is approximately proportional to the square of the size of the matrix, it is better to use cross convolution. The equation of the exponential surface is separable and allows cross convolution by first convolving a plane function along one axis. Then the result of this first operation is convolved along the other axis. The calculation time is linear as a function of the extent of the function.

Gaussian convolution is extremely useful in image processing because, among other things, the Gaussian curve is a shape often found in nature—the profiles of stars spread by atmospheric turbulence are very similar to the Gaussian function. By adopting a large σ we can very effectively smooth an image. We will see an application of Gaussian convolution when we explore unsharp masking.

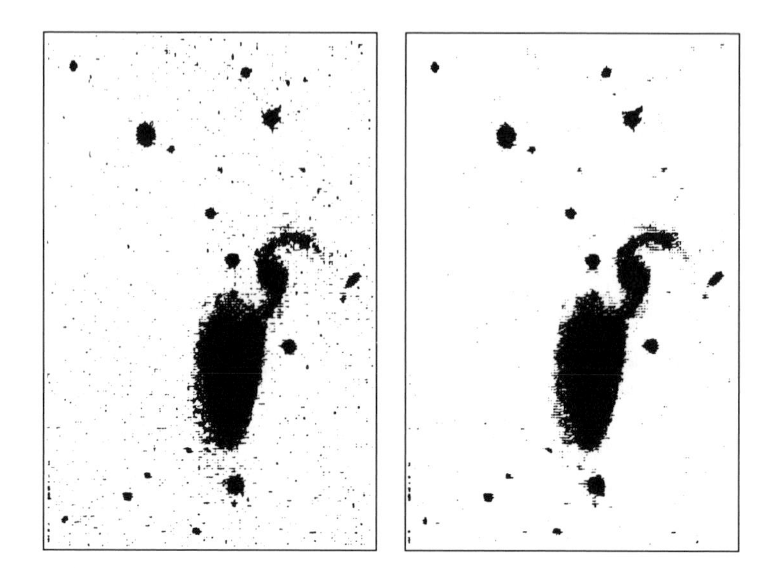

Figure 4.14 *On the left, an image of the galaxy pair Arp84 (T60 telescope stopped down to 40 cm) strongly disturbed by a noise of electronic origin. The image on the right shows the effect of a median filter. Most of the noise has been removed without significantly degrading the resolution.*

4.5.4 Statistical Filters

The disadvantage of the filters studied up to now is their linearity. They process images as an invariant domain, i.e., the action of the low-pass filter equally affects the noise and the real information, which results in a loss of resolution.

A method of selective filtering that can separate the noise from the useful information is therefore necessary. Non-linear filtering techniques must be used. A good example of a non-linear filter is the median filter.

A matrix, or convolution mask, is moved in the image. The coefficients of this matrix are taken to be equal to one. The amplitude of the point situated in the center of the mask is replaced by the median value of the sampling points located under the mask. The median value is determined by sorting the sampling points according to increasing intensity. The median value is then equal to the intensity of the point at the center of this sorted sample.

As an example we will consider a sample of 5 elements called a, b, c, d and e. Sorting by increasing intensity gives us the classification

$$b < a < d < c < e.$$

In this case, the median value is given by the intensity of point d. The median filter is classified as a statistical filter because of this sorting operation. Note that the sample must contain an odd number of elements (3×3 point matrix in the IP program).

The remarkable property of the median filter is that it evens out any pixel with an abnormal intensity with respect to its neighborhood, which is often the sign of noise, more particularly electronic, impulsive noise (parasitics), or of a cosmic ray. In fact, the deviant point will always be classified at one of the ends of the processed sample and is therefore erased. The details of the image are well preserved, however, except in the cases of very small stars that only occupy one or two pixels (take care!)

The algorithm of this filter can be refined by adding a criterion of sensitivity. Let P be a parameter chosen by the user that can take a value between 0 and 1. Whether or not to modify the pixel located in the center of the array is then determined in the following way:

```
IF ABS(a% (i% ,j% ) - sam% (5)) > p*(sam% (8) - sam% (2)) THEN
   buf1% (i% ,j% ) = sam% (5)
ELSE
   buf1% (i% ,j% ) = a% (i% ,j% )
END IF
```

The array **sam% ()** contains the amplitudes of the sorted points; **sam% (1)** contains the lowest amplitude; and **sam% (9)**, the highest. The median value is **sam% (5)** since we are working with a 3×3 matrix. If the parameter P has a value close to 0, the filter's action is maximum. On the other hand, it is almost zero if $P = 1$. Experience with many images will eventually enable the operator to easily select appropriate values for P.

The sorting can be done in a traditional way:

```
DO
  change% = 0
  FOR k% = 1 TO 8
    IF sam% (k% +1)<sam% (k% ) THEN
       SWAP sam% (k% +1), sam% (k% )
       change% = 1
    END IF
  NEXT
LOOP WHILE change% = 1
```

Two other statistical filters are maximal and minimal filters. In a maximal filter the value of the mask's central point is determined by the point with the highest amplitude in the sample while the minimal filter takes the point with the lowest amplitude. These operations correspond to an expansion or a compression of the image. To illustrate these mechanisms we will treat an example in the one-dimensional domain. Assume the following signal:

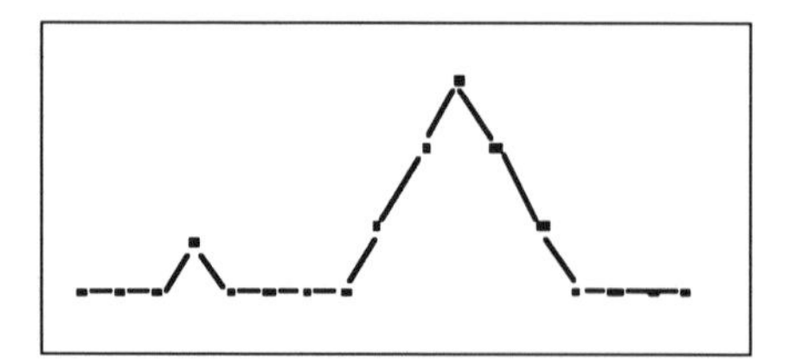

Note that there are two peaks in the above drawing. The one on the left, of low amplitude, is attributed to noise, and we will attempt to remove it by superimposing on this line a vector of 3 points representing a mask which in this case is called a structuring element. For each position of the mask in the line, the point located in the center of the mask takes the minimal value found in the sample of 3 points. We thus erode the line, and here is the result:

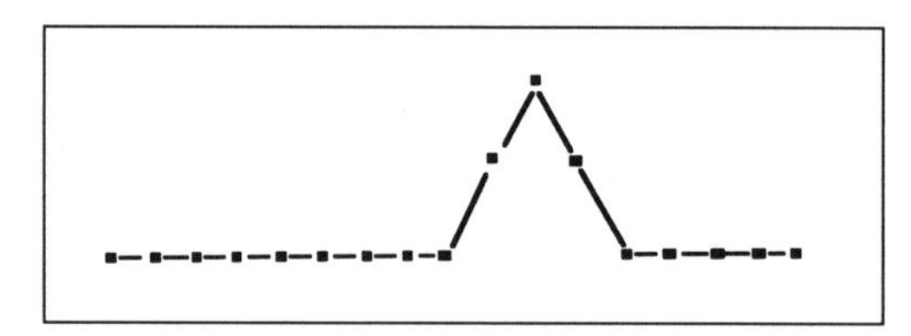

The noise has disappeared, but the object on the right is thinned (we say eroded). We will expand this eroded image to restore it to its original size. The structuring element is again moved in the line, but this time the central point takes the maximum amplitude found in the sample. The result is an image identical in size to the original model with, however, a modification in the amplitude. The noise has completely disappeared:

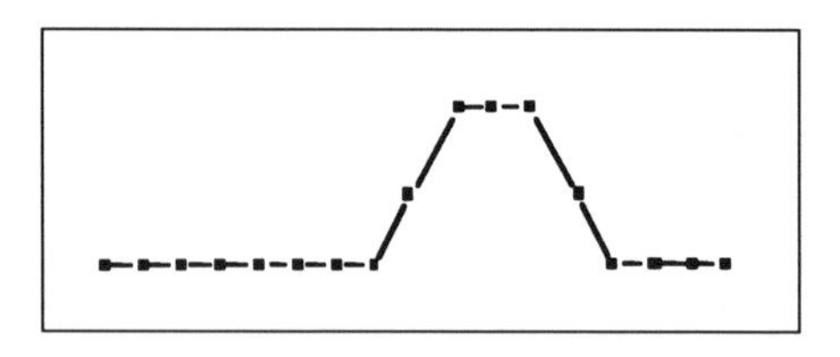

To strongly erode or expand an image, the structuring element can be applied in succession. However, the structuring element's shape has an effect on the final result—after several consecutive passes, the image tends to acquire curves that are close to the shape of the mask. Because of this, it is better to use a mask with non-angular geometric forms, or to use masks of different shapes for successive passes, such as those that follow:

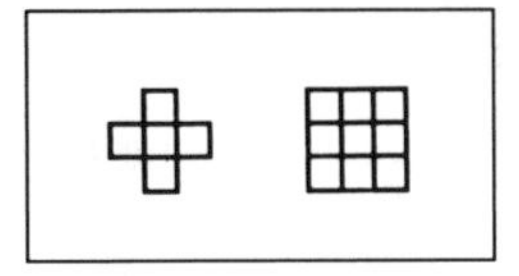

Technically an erosion followed by an expansion is called closing, while an expansion followed by an erosion is called an opening.

Figure 4.15 illustrates the technique of erosion-expansion applied to an image of Halley's comet obtained on April 2, 1986. The rough image is presented at top left. The presence of many stars may hinder image processing. Using an erosion operation we removed the small-sized objects (top right). Pushing the processing quite far leaves only the central part of the comet's tail visible (bottom left). The loss of information is too great. By expanding the eroded image to an intermediate stage, we get a comet without stars and noise but nonetheless strongly smoothed (bottom right).

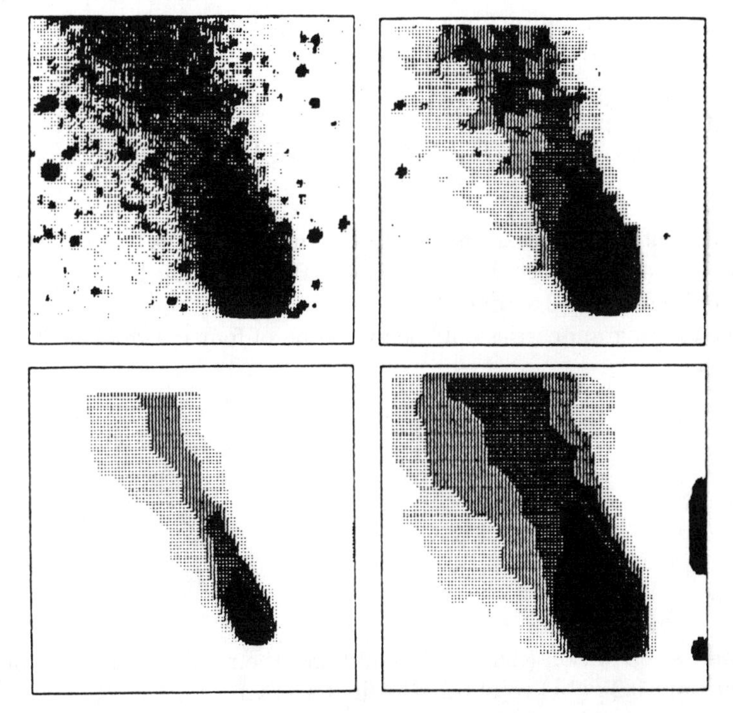

Figure 4.15 *Processing an image of Halley's comet.*

4.5.5 Arithmetic Operations on Images

The IP program allows us to apply the four elementary operations to images.

The first IP program application is addition. Addition creates a composite image made up of a number of different images. The composite is therefore an average image with less noise than any individual image. The images are first stacked and centered with respect to each other so that they are exactly superimposed. This is a registration operation. The resulting composite image shows a decrease in noise because the noise is uncorrelated from one image to another. Since noise is a random fluctuation of the signal around an average value, its effect on the image is reduced as the number of images in the composite increases. On the other hand, the useful (repetitive) signals add up. When N images are composited, the signal-to-noise ratio increases by a factor $\sqrt{N}$ with respect to a unique image.

The two arguments of the addition are placed in the principal zone and in BUFFER1. The result is found in the principal memory zone allowing the image to either be displayed or immediately stored.

```
FOR i% = 1 TO imax%
  FOR j% = 1 TO jmax%
    a% (i% ,j% ) = a% (i% ,j% ) + buf1% (i% ,j% )
  NEXT
NEXT
```

This double loop is executed in less than two seconds on an 8 MHz PC AT processing an image 145x218 pixels. Programming in assembler, the time is reduced to 0.3 seconds.

The IP program carries out the averaging of four images A1, A2, A3, A4 as follows:

```
LOAD A1
TBUF1
LOAD A2
SUM
TBUF1
LOAD A3
SUM
TBUF1
LOAD A4
SUM
MULT .25
```

The last operation (MULT .25) multiplies the result of the additions by the constant 0.25 ($1/4$) to obtain the final average image.

Subtracting two images (SUBT) is similar to that of addition but in subtraction the order of the operands is important. The subtraction is carried out in the direction of the "principal memory"—BUFFER1. Subtraction is useful for removing the dark signal from an image.

Division of two images (DIV) is often used for radiometric correction. A normalization coefficient is introduced so that the dynamic range of the dividend is not too severely modified. In practice, if an image is divided by another image that is practically the same, the result will be close to 1, and all the information is lost since we store the data in integer arrays. Usually the normalization coefficient is the average value of the divider which is obtained by the command STAT.

```
FOR i%  = 1 TO imax%
  FOR j%  = 1 TO jmax%
    IF buf1% (i% ,j% ) = 0 THEN
       buf1% (i% ,j% ) = 1
    ELSE
      x! = norm!*a% (i% ,j% )/buf1% (i% ,j% )
      IF x!> maxdyn%  THEN
        a% (i% ,j% ) = maxdyn%
      ELSE
        a% (i% ,j% ) = CINT(x)
      END IF
    END IF
  NEXT
NEXT
```

This routine provides for both division by zero and a spillover of the tolerated dynamic range (variable maxdyn%). The division is computed in real numbers (note the exclamation mark), then the resulting rounded off integer is stored in a% () (instruction CINT).

Suppose that we want to divide image A1 by image A2. The STAT instruction determines that the average value of the intensities of A2 is 1550. We then write

```
LOAD A1
LOAD A2 1
DIV 1550
```

The particular syntax of the second command allows direct loading of the image into BUFFER1 without going through the principal memory.

Other arithmetic operations are possible:

1. The OFFSET command adds a constant to (or subtracts it from) a whole image;
2. The command ZAP removes the images placed in the principal memory;
3. The command CLIP sets at zero any point in the image that is below a certain threshold. This command is useful for removing negative values from an image after it has been processed.

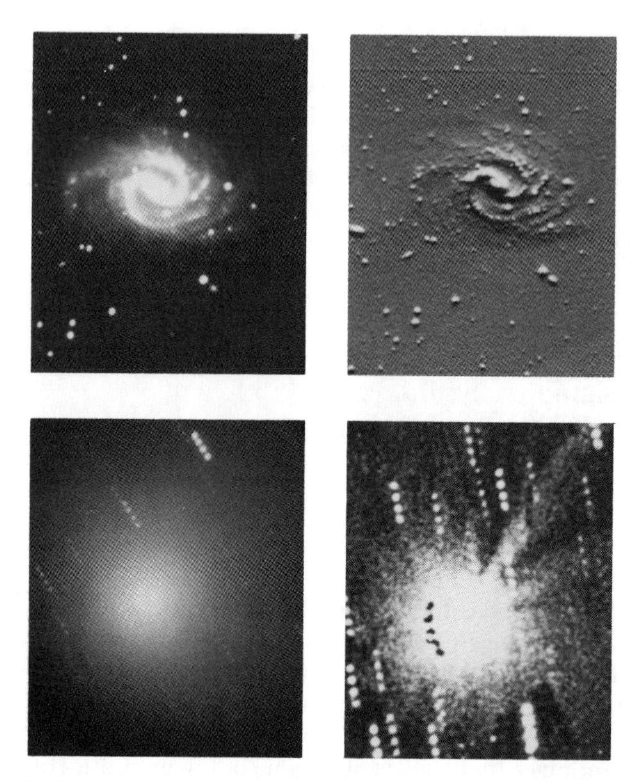

Figure 4.16 *At top left, we have a typical application of the gradient technique. We can see a standard presentation of the galaxy NGC 4535 (60 cm telescope with a TH7863 CCD) and another view after a directional gradient filter treatment, which enhances the spiral structure of the galaxy. In the lower left hand picture a radian and rotational shift algorithm was applied to the central region of comet Levy (280 mm f/6.1 telescope with a TH7863 CCD). In the initial image (lower left hand picture) the coma is so dense that it is virtually impossible to detect the presence of a jet structure. The final image was created with an algorithm introduced by Larson and Sekanina (*Astron. J., **89**, *571, 1984). The principle is simple: two images which have identical radial and rotational shifts but in the opposite direction are added. Using polar coordinates (α, R), which respect to the comet nucleus, the relationship between the initial picture B and the final B' is*

$$B'(\alpha, R, \Delta\alpha, \Delta R) = 2 \cdot B(\alpha, R) - B(\alpha - \Delta\alpha, R - \Delta R) - B(\alpha + \Delta\alpha, R - \Delta R)$$

The processed picture reveals many jets. In this example, $\Delta\alpha = 10°$ and $\Delta R = 1$. This technique is very powerful for studying weak structure within the coma vicinity.

4.5.6 Contour Filters

The purpose of a contour filter is to extract the contour of objects present on the image. They fall within the domain of shape recognition and are also known as morphological operators.

The application of contour filters to astronomy is limited. However, one type, the one-directional gradient filter, can help to better visualize certain details in an image. It is obtained by carrying out the difference two-by-two between consecutive points on the lines or columns. The result is the one-directional derivative of the image. The contours are emphasized because these are the areas of the image where we find abrupt intensity variations (the derivative is then maximum).

In practice the image is convolved by a mask with the coefficients $(-1, 1)$ or, to obtain a more symmetric shape, $(-1, 0, 1)$. The difference between two images, where one is slightly shifted with respect to the other, also gets the gradient filter. The derivative produces negative values and a constant must be added (`OFFSET` command) to obtain a completely positive image.

The derived image calculated in this way appears as a bas-relief with side illumination. The segments of the image with the highest contrast appear in white or black according to the sign of the derivative (direction of the slope). On the other hand, zones with no variations in intensity appear gray. With the `IP` program we can modify the position of the "illumination" (direction of the derivative) to optimize the visualization.

An isotropic derivative, i.e., one that is identical in all directions, is obtained from a *Laplacian* filter.

$$\frac{\partial^2}{\partial x^2} i + \frac{\partial^2}{\partial y^2} j = \text{Laplacian}$$

The Laplacian filter is made from sweeping the image with a convolution mask whose coefficients can be determined by the matrices below.

$$\begin{matrix} 0 & -1 & 0 \\ -1 & 4 & -1 \\ 0 & -1 & 0 \end{matrix} \qquad \begin{matrix} -1 & -1 & -1 \\ -1 & 8 & -1 \\ -1 & -1 & -1 \end{matrix} \qquad \begin{matrix} 1 & -2 & 1 \\ -2 & 4 & -2 \\ 1 & -2 & 1 \end{matrix}$$

Note that the sum of the weighting coefficients is equal to zero. The disadvantage of Laplacian filtering is the strong increase in image noise. Anisotropic filters such as *Sobel's* are often preferable. Image contours are detected for a horizontal direction then for a vertical direction with the matrices:

$$H_1 = \begin{bmatrix} 1 & 0 & -1 \\ 2 & 0 & -2 \\ 1 & 0 & -1 \end{bmatrix} \qquad H_2 = \begin{bmatrix} -1 & -2 & -1 \\ 0 & 0 & 0 \\ 1 & 2 & 1 \end{bmatrix}$$

Assume $K_1 = H_1 * \text{Image}$ and $K_2 = H_2 * \text{Image}$.

The point situated in the center of the matrices is given the intensity

$$I = \sqrt{K_1 \times K_1 + K_2 \times K_2}.$$

The Sobel filter is a powerful tool for extracting contours but it favors the horizontal and vertical axes.

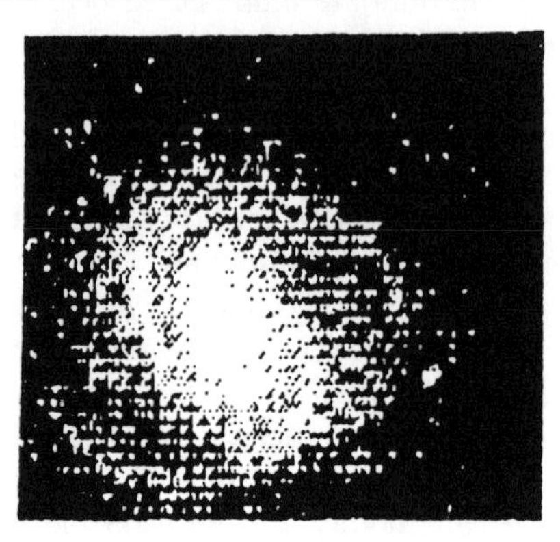
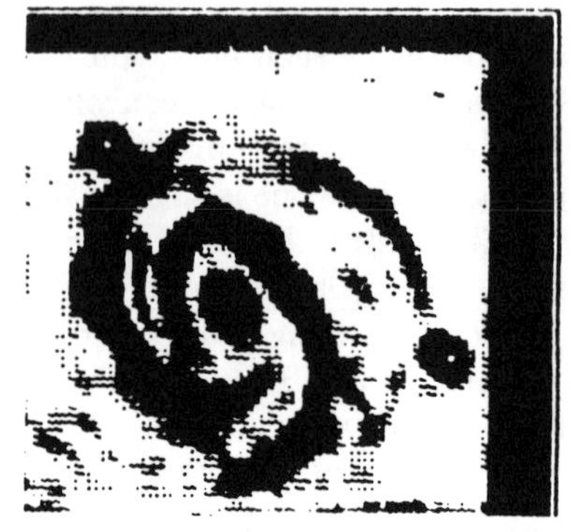

Figure 4.17 *A very noisy image of the galaxy M 100 obtained with one of our first cameras. Applying the Sobel filter (on the right) strongly emphasizes the spiral arms.*

Directional filtering can be obtained with a Sobel operator. For example, the following matrix gives a shadow effect with low-angled illumination coming from the upper left corner of the image.

$$\begin{matrix} -2 & -1 & 0 \\ -1 & 0 & 1 \\ 0 & 1 & 2 \end{matrix}$$

The IP uses matrices of this type to produce bas relief effects for different illumination directions.

Another filter that is similar to the Sobel, is the *Prewitt* filter:

$$H_1 = \begin{bmatrix} 1 & 0 & -1 \\ 1 & 0 & -1 \\ 1 & 0 & -1 \end{bmatrix} \qquad H_2 = \begin{bmatrix} -1 & -1 & -1 \\ 0 & 0 & 0 \\ 1 & 1 & 1 \end{bmatrix}$$

4.5.7 Unsharp Masking

Unsharp masking is a powerful technique allowing spectacular enhancement of the finest details of an image. The principle of unsharp masking is well-known to photographers who use it to emphasize details that are drowned in the strong density gradients of the film. The method consists in making a contact negative copy of the original with an interposed transparent spacer—a glass plate or the emulsion of the film itself. The resulting negative is an unsharp shadow of the original. This unsharp mask is then stacked and registered with the original and a print is made. The details are nonexistent in the mask, but on the other hand the mask retains the wide light gradients that cancel those of the original negative. Consequently the final result shows considerably improved details.

An identical method is used for digital processing. The image is first smoothed by a low-pass filter to create a mask. Then the resulting mask (an unsharp mask) is subtracted from the original image. To illustrate, we will examine the processing of a photometric equatorial cross-section of the planet Saturn taken with a linear array CCD in June, 1984, using the 1-meter telescope at the Pic du Midi. Our purpose will be to extract as clearly as possible the Cassini division.

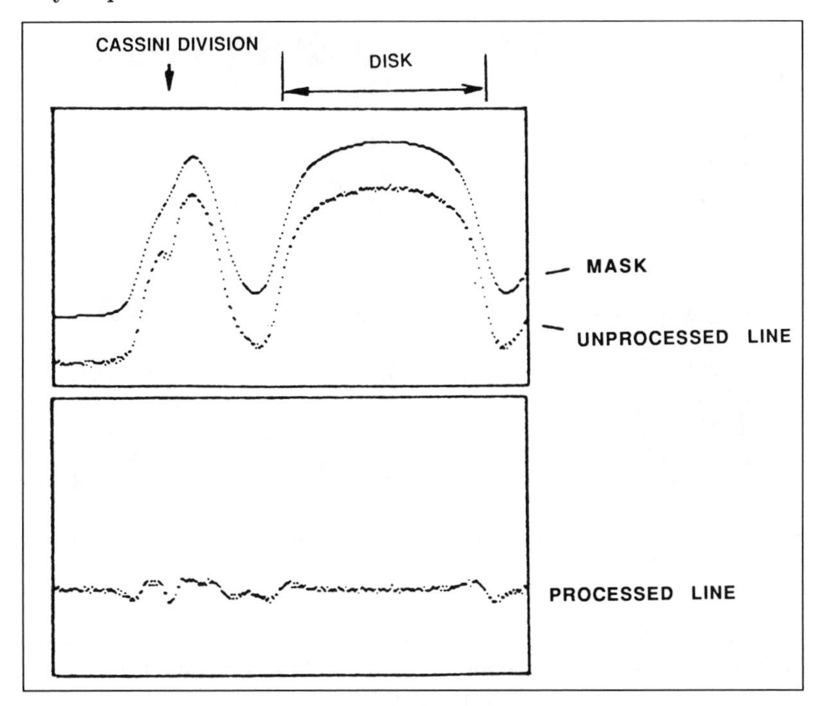

Figure 4.18 *Example of unsharp masking. Equatorial cut of the planet Saturn.*

Because the sun's rays strike the "spherical" planet at varying angles there is an obvious darkening along the limb. This effect is clearly visible on high-contrast planetary photographs and it makes it difficult to correctly visualize the edge of the planet. In CCD imaging the center-edge phenomenon is just as visible and annoying. In our example of the planet Saturn, the center-edge effect is enhanced by turbulence and a focusing error that contributed to a spreading of the image.

The cross-section is first smoothed by a low-pass filter (local average). The high frequencies are thus removed, creating an unsharp mask. We then determine the difference between the original image and the mask, i.e., the difference between an image containing both high and low spatial frequencies and an image containing only low frequencies. The result is an image in which the high frequencies dominate. The Cassini division stands out particularly well as an indentation in the ring. Note however that there are several artifacts in the high contrast zones. These are rebound phenomena and are also called Gibbs effect. These artifacts are inherent in this technique and they should not lead to false conclusions when interpreting results at the level of zones with very strong brightness gradients.

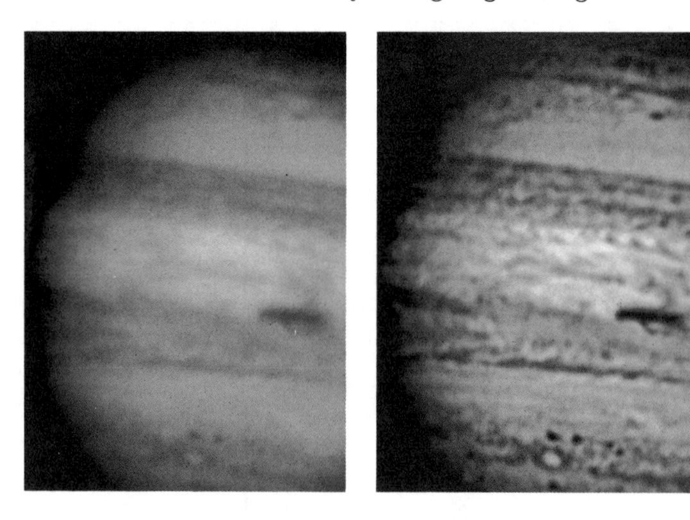

Figure 4.19 *On the left, the untreated image of the planet Jupiter obtained on December 26, 1988, with the 1-meter telescope at Pic du Midi Observatory. On the right, the image processed with unsharp masking.*

An unsharp mask is a very powerful high-pass filter. It is so powerful that in practice we have to moderate it by adding some low frequencies to make the final image look more realistic.

The unsharp mask's ideal application is in planetary imaging. As an example, here is how we would process an image of the planet Jupiter with the IP program.

`LOAD jupiter`	loading the image
`TBUF 2`	setting aside the image
`GAUSS 7`	making an unsharp mask with a Gaussian filter that is 7 pixels wide for 4 sigma
`TBUF 1`	storing the unsharp mask in `BUFFER1`
`VBUF 2`	recovering the original image
`SUBT 200`	subtracting the original image and the mask but adding a shift constant of 200 to avoid having negative intensities
`MULT 8`	multiplying the high frequencies by a factor of 8
`TBUF 1`	storing the result in `BUFFER 1`
`VBUF 2`	recovering the original image
`SUM`	adding the original image with high frequencies to retain a slight edge to the final image
`SAVE result`	saving the result

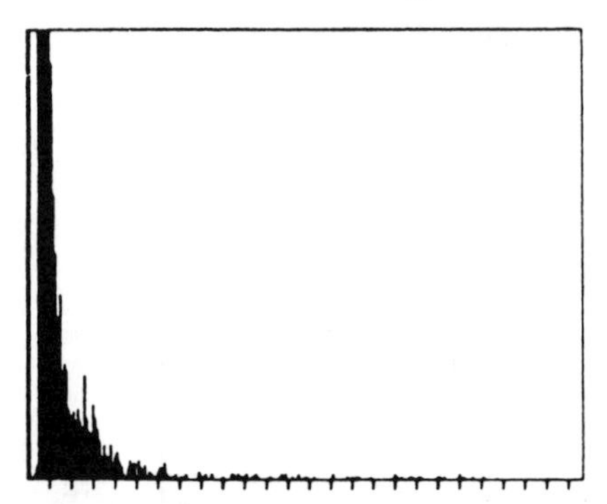
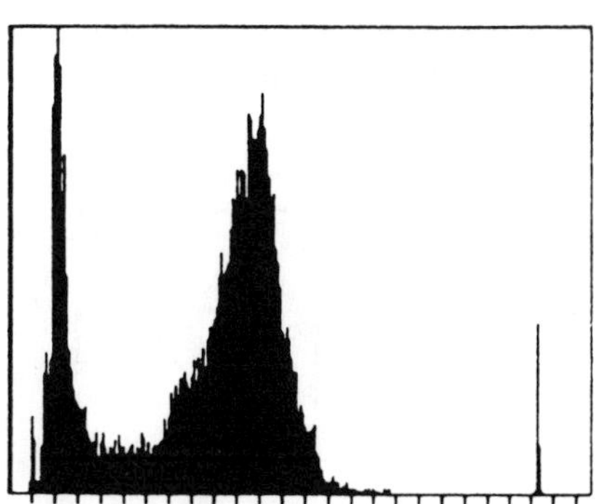

Figure 4.20 *On the left, typical histogram of a deep sky image. The sky background fills most of the image, which gives a heavy population around the quantification steps corresponding to this background. However, the histogram has an extension toward the right that represents the few objects on the image (stars, nebulae, etc.). On the right, a histogram of a lunar landscape. We can distinguish two peaks (modes) that correspond to two brightness populations in the image: the illuminated areas and the shadowed areas. The peak located on the right represents a population of saturated pixels.*

4.5.8 Changing the Histogram

In astronomy the useful information is confined to an amplitude of several dozen quantification units. Let's take the example of a faint galaxy located in the middle of a field of bright stars. Correct visualization of the galaxy requires very tight visualization thresholds bracketing the level of the sky background. The stars will then appear as large spots that may even touch. The ideal would be to increase the galaxy's brightness without significantly increasing the size of the stars.

This ideal is possible with techniques that modify the distribution of the intensity of the image points, i.e., a histogram. A histogram is a function that conveys the distribution of intensity levels in an image. In a graphic representation of the histogram, the axis of the abscissa is usually marked in quantification units; on the ordinate we put the number of pixels belonging to each of the quantification units.

Examining the histogram can facilitate the choice of visualization thresholds since with one glance we can judge the image's dynamic range. The IP program has a HISTO command with an abscissa and ordinate to zoom to allow visualization of the histogram of a memory resident image.

The most efficient way to modify a histogram is to use a transcoding table or LUT (Look Up Table). Remember from our discussion of display techniques that a LUT establishes a relation between the initial state and the final state of an image. This time the LUT is used to modify the digital image itself, not merely its representation.

The LUT of Figure 4.21 transforms the image in such a way that the intensity of the points situated below a certain threshold is set to zero.

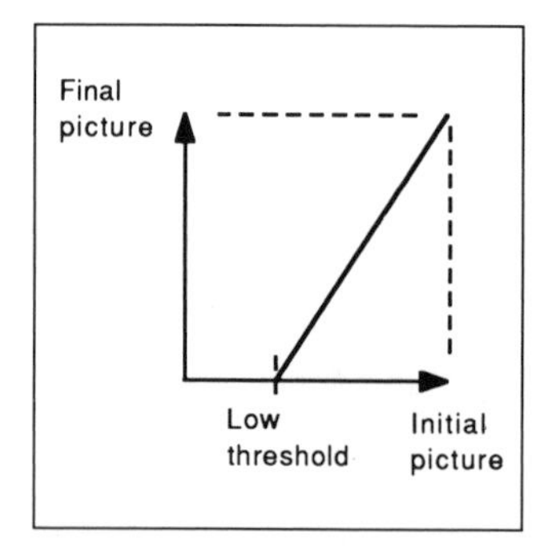

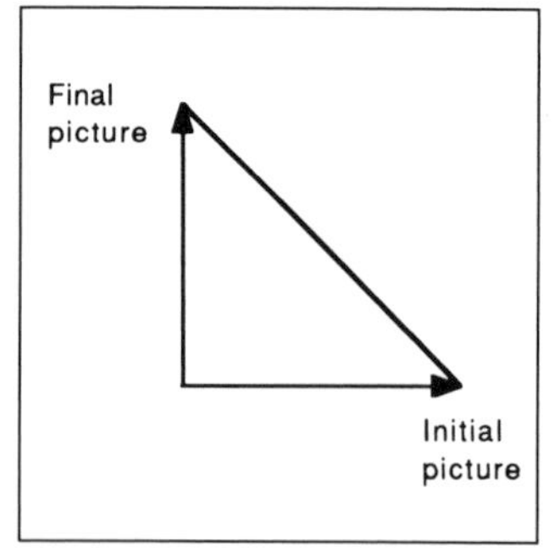

Figure 4.21 (left) *Thresholding by the bottom of an image.*
Figure 4.22 (right) *Transcoding table to obtain the negative of an image.*

The LUT of Figure 4.22 produces a negative of the image. The element of zero row of the LUT has a value of 4095, and any image point with an intensity of 0 will be replaced by a point of intensity 4095. A point of intensity 1 will be given the intensity 4094 (value contained in element 1 of the LUT) and so on.

To enhance the appearance of a faint nebula drowned in the middle of bright objects, we will use a logarithmic LUT. This is a very common technique for adapting the long dynamic range of the CCD image to the much more limited range of reproduction methods. The shape of the logarithmic LUT (Figure 4.23) shows clearly how low illumination levels are amplified to the detriment of strong illumination. The image processed in this way appears more natural and less digital. The technique also strongly enhances background noise in the sky, giving it a distinctly "pixelated" look.

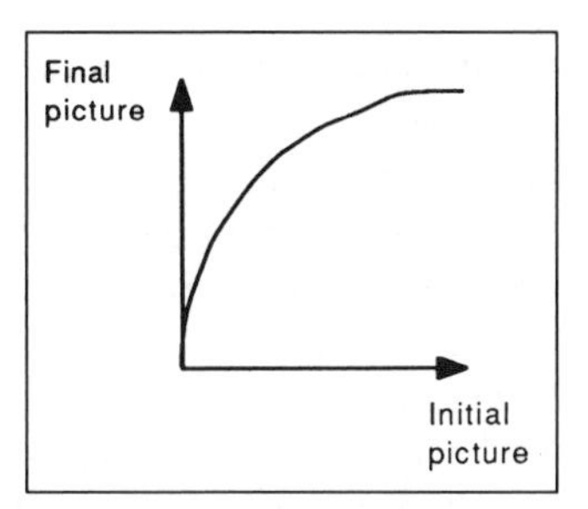

Figure 4.23 *A logarithmic LUT.*

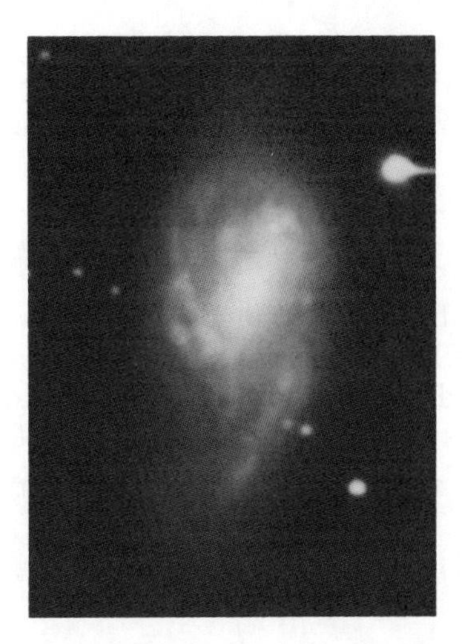

Figure 4.24 *Galaxy M66 visualized after a logarithmic transformation of the intensity scale. The details of the region close to the core are then visible at the same time as the arm extensions. Image taken with a 20 minute exposure with a 280 mm telescope and a TH7863 CCD.*

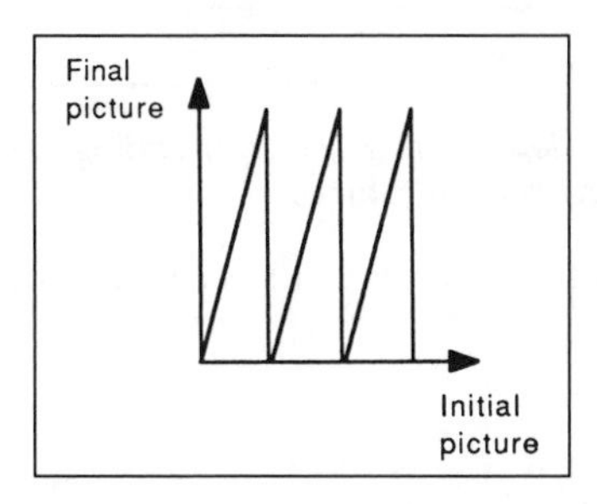

Figure 4.25 *A saw-toothed LUT.*

A wide range of possibilities exists in defining the shape of the LUT (exponential functions, power function, etc.). The LUT of Figure 4.25 produces a beautiful mixture of levels of gray and iso-contours in the image.

A transformation that is often used to enhance the contrasts in an image is called histogram leveling. The purpose of the method is to find a transformation function that, once it is applied to the image, allows the distribution to become uniform. In other words, after processing, the different levels of gray in the image have the same probability of existing.

The formulation of the histogram can be written

$$P(x_i) = \frac{n_i}{N},$$

with $P(x_i)$ the discrete probability of intensity x_i in the image; n_i the number of pixels with amplitude x_i; and N the total number of pixels in the image. The histogram is leveled by creating a transcoding table y_i such as

$$y_i = \sum_{j=0}^{i} P(x_j).$$

This LUT is built from the accumulated histogram of the image. Its effect is to spread the levels of gray represented by a strong density in the original histogram and to compensate for this spreading by a compression in the less dense zones of the histogram.

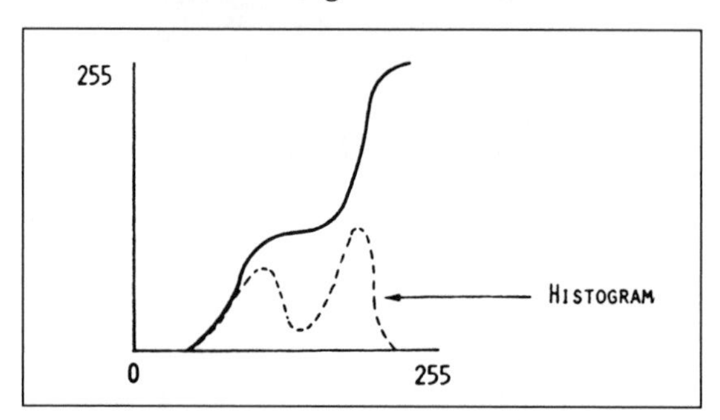

Figure 4.26 *Method of determining the transcoding table (solid line) for a histogram equalization (e.g., here in 8 bits).*

The first operation to be carried out is calculating the histogram (array HISTO% ()):

```
FOR i% = 1 TO imax%
  FOR j% = 1 TO jmax%
    intensity% = a% (i% ,j% )
    histo% (intensity% ) = histo% (intensity% ) + 1
  NEXT
NEXT
```

The accumulated histogram is then determined and stored in the array LUT%():

```
sum% = 0
FOR k% = 1 TO maxdyn%
    sum% = sum% + histo%(k%)
    lut%(k%) = sum%
NEXT
```

The accumulated histogram is normalized so that the element of row 4095 represents an intensity of 4095 (the variable sum% contains the maximum value of the LUT, calculated during the last passage in the preceding loop):

```
FOR k% = 1 TO maxdyn%
  lut% (k% ) = CINT(maxdyn% *lut% (k% )/sum% )
NEXT
```

Finally the image is transformed by the new LUT:

```
FOR i% = 1 TO imax%
  FOR j% = 1 TO jmax%
    a% (i% ,j% ) = lut% (a% (i% ,j% ))
  NEXT
NEXT
```

On a typical astronomical image, histogram leveling expands the scale of grays in the zones that are strongly represented in the original image, i.e., the zones corresponding to the sky background. Objects that are usually lost in the sky background (faint nebulae) have their amplitudes disproportionately amplified when compared to pixels belonging to other intensity populations. This amplification is the reason why this technique is so powerful.

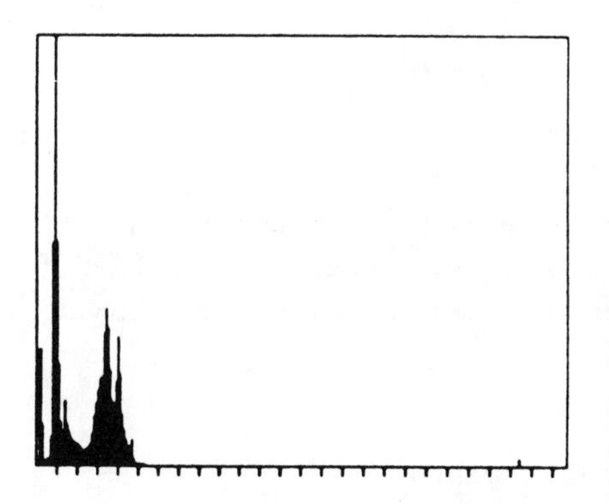

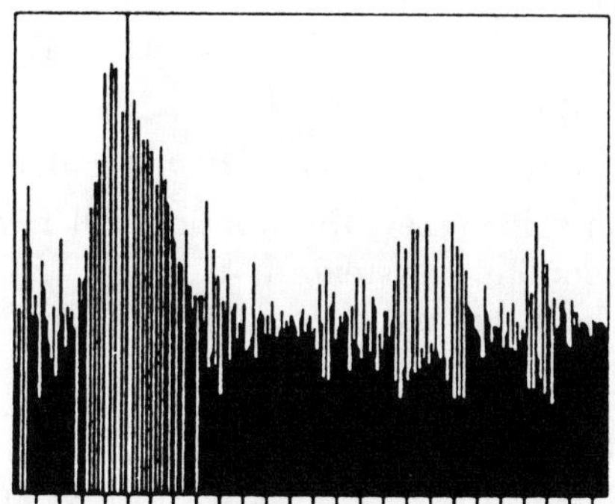

Figure 4.27 *Appearance of a histogram before and after equalization.*

4.5.9 Geometric Transformations

Transformation Equations

To be able to compare images taken under different conditions, we often have to subject them to geometric transformations. We saw an example of this previously in the discussion on compositing. Images have to be perfectly recentered with respect to each other (registered) before addition can be carried out. The most common geometric transformations are shift, scale changing, and rotation.

Shift along the axes (x, y) by a distance (a, b) is defined by the equations:

$$x' = x + a$$

and

$$y' = y + b.$$

Scale changing by a factor (r, s) is written

$$x' = rx$$

and

$$y' = sy.$$

Finally rotation by an angle α in the trigonometric (i.e., counter clockwise) sense is written

$$x' = x \cos\alpha - y \sin\alpha$$

and

$$y' = x \sin\alpha + y \cos\alpha.$$

If the rotation, instead of taking place around the origin of the coordinate system, is carried out around the coordinate point (A, B) we will get

$$x' = A + (x - A)\cos\alpha - (y - B)\sin\alpha$$

and

$$y' = B + (x - A)\sin\alpha + (y - B)\cos\alpha.$$

In some cases, the rotation will have to be done in a three dimensional coordinate system. If the third (Z) axis is perpendicular to the plane of the image and oriented in the observer's direction, we will get the following results:

1. Rotation around Z-axis:

$$x' = x \cos\alpha - y \sin\alpha,$$

$$y' = x \sin\alpha + y \cos\alpha,$$

$$z' = z.$$

2. Rotation around X-axis:

$$x' = x,$$

$$y' = y \cos \alpha - z \sin \alpha,$$

$$z' = y \sin \alpha + z \cos \alpha.$$

3. Rotation around Y-axis:

$$x' = x \cos \alpha + z \sin \alpha,$$

$$y' = y,$$

$$z' = -x \sin \alpha + z \cos \alpha.$$

In the IP program image shift is carried out by the SHIFT command. This shift can be set to within a fraction of a pixel. The ROT command produces a rotation of any angle around a center that can be located outside the image. With the EXPAND command, we can enlarge a part of the image by a factor of 2 and the function SCALE produces scaling of any factor that can be different along the X-axis and along the Y-axis. We can also add the MIRROR command that permutes the left and right sides of the image to rectify it after the image has passed through different optical systems (e.g. suffered an odd number or reflections on mirrors).

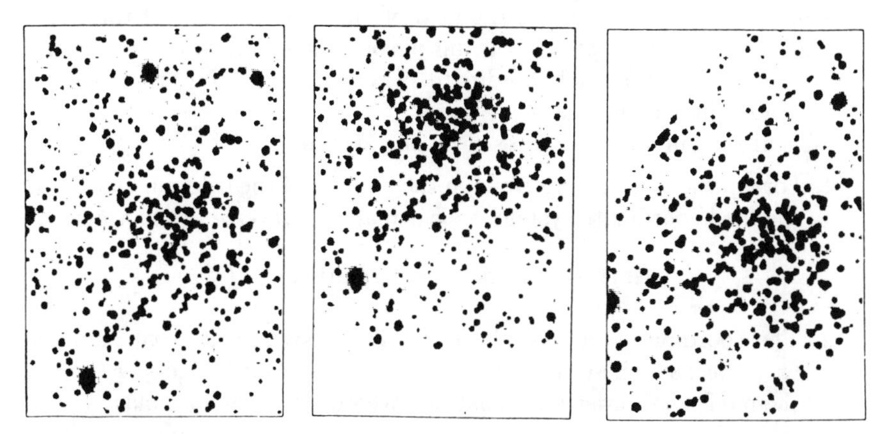

Figure 4.28 *Example of shift and rotation applied to the open cluster NGC 6939. Image taken with a 20 cm telescope and a TH7852 CCD with a 10 minute exposure.*

Resampling

If the position of a pixel on the original image is directly calculated in the final coordinate system, the result is usually disappointing because there is no exact pixel-to-pixel correspondence between the two coordinate systems (except for coincidences, when, for example, we carry out a shift of an integer number of pixels).

In most cases, therefore, a pixel of coordinate `(23,52)` in the initial coordinate system will find itself with the coordinates `(108.26, 14.90)`. Should these values be rounded off to position the pixel in the new coordinate system? Then `(108, 15)` will be the pixel's new coordinates. Let us process the neighboring pixel with the coordinates `(24,52)`. The transformation gives `(108.48,14.72)`. By rounding off we get `(108,15)`, i.e., the same coordinates as the preceding pixel! Two different pixels in the original coordinate system get the same position in the final coordinate system—there is some conflict here. Conversely, we can see that certain positions in the final coordinate system will never be given an intensity. There will then be holes in the final image.

Obviously, we cannot proceed in this way. The problem should be approached from the opposite direction. For each discrete integer position in the final coordinate system, we calculate an inverse transformation that supplies real (i.e., not exact integer) coordinates in the initial coordinate system. These coordinates are rounded off and the intensity of the point located at these coordinates in the initial coordinate system is given to the pixel processed in the final coordinate system. Duplication and holes are thus avoided in the final image.

The algorithm can be expressed in this manner:

1. For a position (i, j) expressed in the number of the column and line in the final coordinate system, we calculate the inverse transformation

$$(i, j) \longrightarrow \text{inverse transformation} \longrightarrow (x, y)$$

2. The coordinates in the initial coordinate system are real numbers (x, y) because these coordinates are unlikely to be integers, we round them off (zero order interpolation) according to the formulas:

$$x' = \texttt{CINT}(x)$$

and

$$y' = \texttt{CINT}(y);$$

3. We read the intensity of the pixel at coordinates (x', y') in the initial image and assign this intensity to the point (i, j) in the final image.

4. If the point of coordinates (x', y') is located outside the limits of the initial image, the corresponding point in the final coordinate system will be given the value 0.

This method is very simple. The fact that some adjacent pixels in the final image can be assigned identical amplitudes or that some intensities in the initial image will not be the final coordinate system is usually not a problem.

The final result can be considerably improved by using interpolation techniques that are less primitive than simple rounding off to determine the amplitude in the initial coordinate system. Bi-linear interpolation gives very satisfactory results and is easy to program:

1. We calculate the integer parts (x', y') of the coordinates (x, y) obtained as previously by an inverse transformation of (i, j):

$$x' = \mathtt{INT}(x)$$

and

$$y' = \mathtt{INT}(y).$$

2. Let A, B, C, D be the intensities of the pixels surrounding the point of coordinates (x, y). The respective integer coordinates of these pixels are as follows:

$$(x', y') \longrightarrow \text{intensity } A$$

$$(x', y'+1) \longrightarrow \text{intensity } B$$

$$(x'+1, y') \longrightarrow \text{intensity } C$$

$$(x'+1, y'+1) \longrightarrow \text{intensity } D.$$

If one of these coordinates falls outside the image, we give the intensity 0 to the points in question.

Let α and β be the fractional parts of x and y:

$$\alpha = x - x'$$

$$\beta = y - y'$$

The coordinate point (i, j) in the final image is then assigned the intensity calculated by

$$(1-\alpha)(1-\beta)A + (1-\alpha)\beta B + \alpha(1-\beta)C + \alpha\beta D.$$

Some Application Examples

Registering several images is done almost daily by the user of a CCD camera either for compositing or for comparing images. Usually the camera is not taken apart between each exposure so that the only transformations to be carried out are the two shifts in the image plane to compensate for tracking defects. For perfect registration of several images, we have to determine precisely the shift that exists between each of them. In the case of deep sky images, stars are extremely helpful.

We should not be satisfied with measuring the position of a star by taking the position of its maximum intensity. At best the precision of the measurement would be the pixel. It is highly preferable to calculate the center of gravity inside a small window enclosing the star (command `CENTRO`). The center of gravity of a discrete function $F(i)$ in the interval n, m is

$$C_{\mathrm{i}} = \frac{\sum_{i=n}^{m} F(i) \cdot i}{\sum_{i=n}^{m} F(i)}.$$

In the case of a two-dimensional image, the center of gravity is calculated independently and successively on the two axes. Let's translate all these mathematics into a small program. If `px%` and `py%` are the coordinates of the corner of the window isolating the star and `la%` the width of the window, the center of gravity `cx` along the X-axis will be programmed as follows:

```
s1 = 0
s2 = 0
FOR i% = px% TO px% + la%
   s3 = 0
   FOR j% = py% TO py% + la%
       IF a% (i%,j%) > threshold % THEN s3 = s3 +
       a% (i%,j%)
   NEXT
   s1 = s1 + s3 * j%
   s2 = s2 + s3
NEXT
cx = s1/s2
```

The calculation along the Y-axis is similar. Note that the calculation takes into account only the pixels whose intensity is higher than the contents of the variable `threshold%`. The latter is adjusted so that the sky background does not intervene. Typically `threshold%` has a value 10 to 30 quantification units higher than the level of the sky background in the neighborhood of the star. The calculation window should be large enough to include the whole stellar image to be measured, but we have to be careful not to include another star. A further basic precaution is not to use a saturated star image to calculate the center of gravity.

Recentering planetary images is more difficult. The ideal would be to use intercorrelation techniques, but they are too hard to set up with a small computer system. We will have to be satisfied with finding the position to the nearest pixel of several characteristic details. We locate these details with a cursor that we move on the image. The relative final shift between two images is determined by calculating the average value of the shift of each detail.

The `IP` program has a command that allows a rapid comparison of two images by displaying them on the screen successively at high speed. Therefore two images can be compared, in real time, to measure the shift between them. This is the `BLINK` command (see also section 5.6).

Geometric transformations can be used to correct an image acquired with a CCD whose pixels are not square (scaling). Also a simple scaling operation can create the illusion of seeing a galaxy face on. Of course, if this galaxy is seen edge-on, there is nothing to be done. Remember the golden rule of image processing: do not invent information (in the edge-on case it's really impossible, even if the operator wants to!)

More powerful techniques enable simultaneous rotation, translation and scaling. This is accomplished by introducing a correspondence between a reference grid, and a grid which is defined on the image by two polynomials having a variable degree (1 polynomial for each axis of the image). The correspondence is calculated by a least square algorithm. For example, if the grids are defined from star positions in two fields of a given object, it is possible to calculate a polynomial which allows the two images to be superimposed after resampling. This technique may be used in supernova survey programs where new CCD images are compared to digitized reference photographs.

Construction of planispheres from planetary images is an important aspect of geometric transformation techniques. The "spreading" of the disk, among other things, facilitates the measurement of details or drawing maps that simultaneously present the whole planet (360° longitude) by placing maps made from elementary images side by side.

To make a planisphere we have to proceed in the usual way with geometric transformations. We start with a representation on the screen to find the corresponding points in the image to be transformed. In the present case, we define a network of rectangular coordinates on the screen: the X-axis (traditionally the horizontal axis) represents the longitudes (l), usually equidistributed; the Y-axis (vertical) represents the latitudes (L). There are many possibilities for projecting the globe on this latter axis:

1. Equally distant parallels;
2. Distribution in tangent L. This is the Mercator projection that sailors know well. It has the advantage of conserving the angles between two points but at the price of a considerable expansion in the polar zones

(the poles themselves are impossible to draw);

3. Distribution in sine L, producing a compression of the high latitudes. This compression probably gives the most reliable representation without resorting to more complex methods.

The first operation to be carried out is the transformation from the screen coordinates (i, j) to planetocentric coordinates (l, L). For example, in the case of a sine-type projection, we will write something like

$$l = K_1 \cdot i + C_1$$

$$\arcsin L = K_2 \cdot j + C_2$$

where K_1 and K_2 scale are factors and C_1, and C_2 are constants.

Starting with the coordinates (l, L) we determine the Cartesian planetocentric coordinates (x_1, y_1, z_1), with the Y_1-axis being oriented toward the north pole and the axes $X_1 - Z_1$ defining the equatorial plane. Then, taking into account the orientation of the polar axis with respect to the line of sight of the observer, we calculate the apparent Cartesian planetocentric coordinates (x, y, z). The difficulty here is to identify the system of planetocentric coordinates (the $X - Y$ axes defining the image plane) with the image that is really observed. The origin of the system of axes (X, Y, Z) has to be identified with the center of the image and a scale factor has to be adjusted. All of this geometry is not always obvious if the planet has a visible flattening or if there is a phase.

If λp is the longitude of the line of the poles projected in the plane X, Y (see Figure 4.29) and if ϕp is the inclination of the axis of the poles, the transformation equations will be

$$x = (x_1 \cdot \sin \phi p + z_1 \cdot \cos \phi p) \cdot \sin \lambda p + y_1 \cdot \cos \phi p,$$

$$y = -x_1 \cdot \cos \phi p + z_1 \cdot \sin \phi p,$$

$$z = (x_1 \cdot \sin \phi p + z_1 \cdot \cos \phi p) \cdot \cos \lambda p - y1 \cdot \sin \lambda p,$$

with

$$x_1 = \sin l \cdot \cos L,$$

$$y_1 = \sin L,$$

and

$$z_1 = \cos l \cdot \cos L.$$

The coordinates in the image plane will be

$$i_m = i_c + x \cdot r,$$

and

$$j_m = j_c + y \cdot r$$

where (i_c, j_c) are the coordinates of the center of the disk, and r is the radius of the disk (in pixels).

Note that these equations allow us to superimpose a network of coordinates on the planet's globe. Figure 4.29, at the bottom, shows this kind of network, which is helpful for locating the position of planetary features. We should of course draw only half a hemisphere (z positive). In the case of the flattening of a globe (Jupiter), one should vary the value of the radius as a function of latitude ϕ to have a more precise superposition of the image and the network.

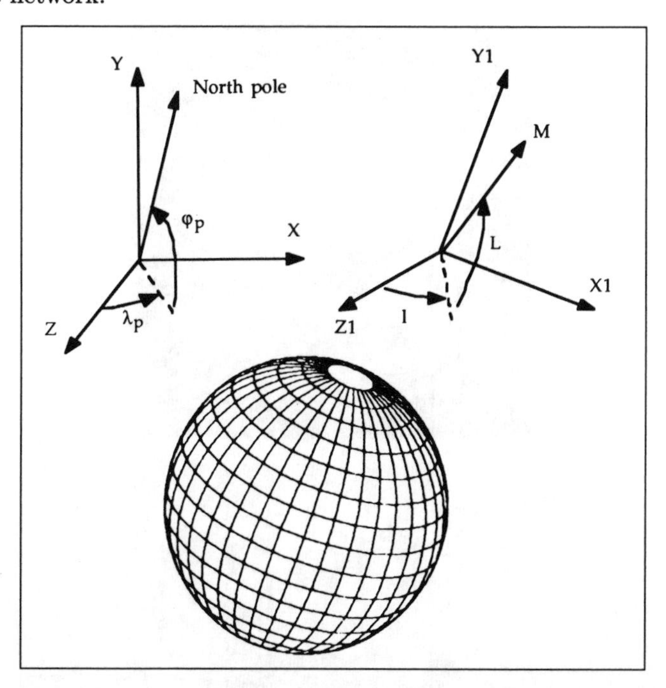

Figure 4.29 *At the upper left, the system of apparent planetocentric axes (linked to the screen). The plane defined by $X - Y$ represents the image plane. Axis Z is oriented toward the observer. At the upper right, the system of real planetocentric axes (OY_1 represents the north pole, and the plane defined by X_1 0 Z_1, the equator). At the bottom, example of the tracing of a coordinate network. Here $\lambda p = 20°$ and $\phi p = 70°$.*

It is not likely that the calculated coordinates (i_m, j_m) will be integers. The intensity of the point on the map will then be calculated by bilinear interpolation. Obviously the interpolation will become larger as we approach the planetary limb.

Figure 4.30 shows an example of planisphere calculation. At the top of that figure we have the initial image, processed by unsharp masking (see section 4.5.7). Note the strong inclination of the axis of the poles. At the bottom is the planisphere built from this image with a sine projection. The parameters are $\lambda p = 85^\circ$ and $\phi p = 54^\circ$ for a disk radius of 64 pixels. The map extends over 140° in longitude and over ±30° with respect to the equator. Since the TH7852 CCD does not have completely square pixels ($28 \times 30\ \mu$m), it was necessary before plotting the planisphere to scale it by a factor $^{28}/_{30}$ in the vertical axis. This correction, which may at first seem too small to be significant, is not negligible with this kind of processing. The bands are noticeably twisted if the correction isn't made.

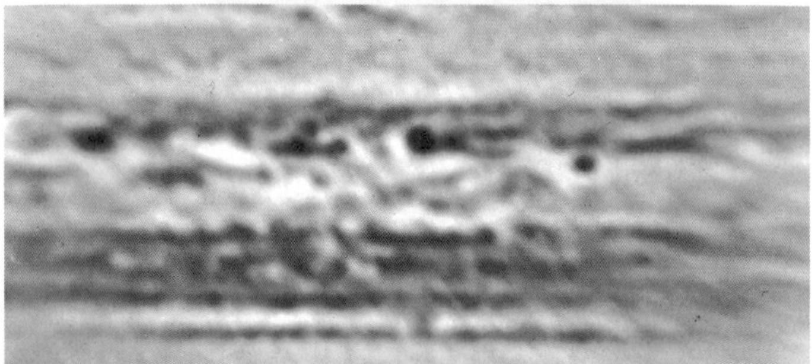

Figure 4.30 *The planet Jupiter observed in the near infrared with the 1 meter telescope at the Pic du Midi and a TH7852 CCD.*

A problem arises because of the center-edge darkening. This phenomenon does not allow a perfect radiometric connection between the sections of the image—far from it. It is therefore necessary to model the

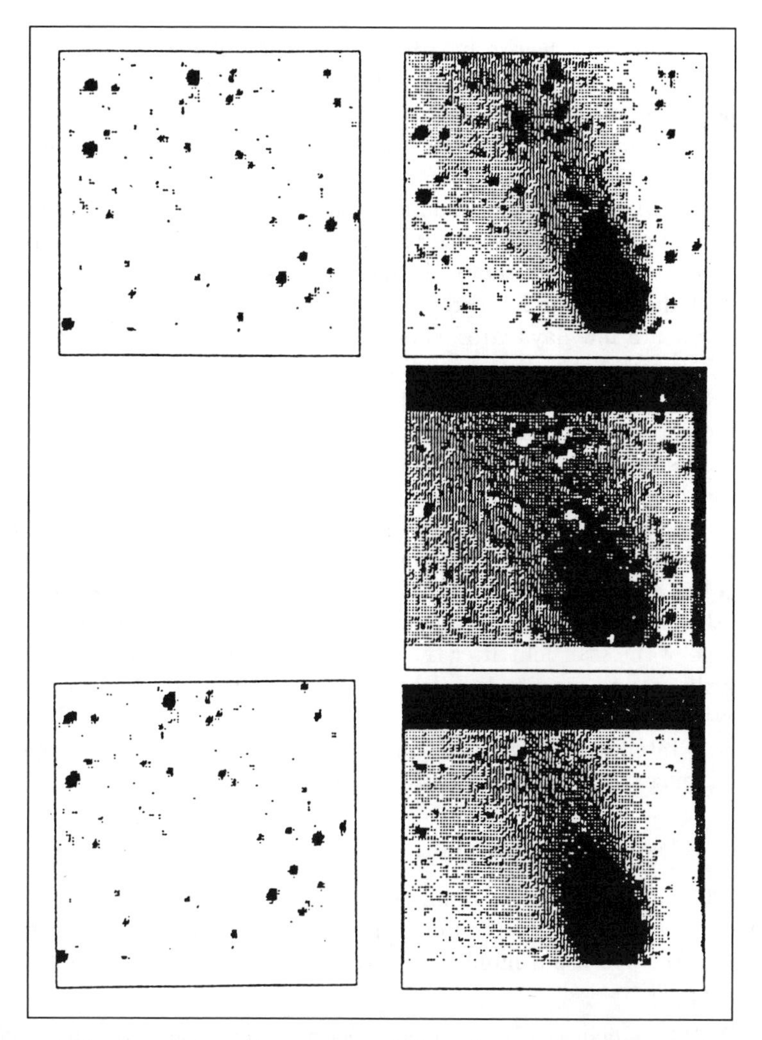

Figure 4.31 *Processing of an image of Halley's comet to remove the stellar background. Taken at Réunion Island on April 2, 1986, at 0H45 UT. The image is obtained by placing a 135 mm lens open at f/2.8 in front of the CCD.*

increase as one moves away from the middle of the disk. If θ is the central angle measuring the distance with respect to the center of the disk, a law approaching the center-edge intensity variation will be

$$I_\theta = \cos^k \theta.$$

But, $\cos\theta = \cos\lambda \cos\phi = z$. Therefore

$$I_\theta = z^k.$$

The parameter k has to be adjusted for each planet, especially in the case of gaseous planets, according to the filter used in obtaining the image. This darkening model will be terribly complicated when the planet presents a phase, i.e., when it is observed too far from opposition. In the example in Figure 4.30 the darkening law used is $\cos^{0.5}\theta$.

Figure 4.31 shows a final example of a geometric image transformation technique. The processing concerns an image of Halley's comet. The original image (upper right) shows the comet in the middle of a dense field of stars that disturbs the interpretation of the comet. The image at upper left, obtained a few days later, shows the same star field, but without the comet. The purpose of the processing is to remove the star trails by determining the difference between the two images. The central image shows a first attempt at subtraction, after recentering with respect to a star marked with a cross (upper left). We see that superposition does not occur over the whole field because the camera was removed from the telescope after taking the first image and was not re-mounted on the telescope exactly the same way the second time. We easily determine that the re-positioning of the camera has caused a field rotation of 3.5°, and we therefore have to rotate one of the two images by this amount. The figure at the bottom left shows the image of the star field after this transformation (the rotation was made arbitrarily around a point located in the lower left corner). The image at the lower right is the result of the subtraction of the image of the comet and the rotated star field. This time most of the stars have disappeared, and the image becomes much more legible.

4.5.10 Cosmetic Corrections

A CCD image is rarely completely free from artifacts such as parasitic electronic noise, defective columns, hot spots, cosmic rays, passage of a satellite, etc. We use a number of tools to clean the image including the following:

1. `CICAT`, which attributes to a parasitic column the average value of the two columns surrounding it;
2. `CCURSOR`, which is a graphic cursor that we move over the screen to correct, more or less manually, the amplitude of deviant pixels;
3. `COSMIC`, which automatically detects cosmic rays by their character (isolated bright points);
4. `FILL`, which makes a parabolic interpolation inside a window of dimensions chosen by the user from points located on the edges of this window. This command is useful for removing large undesirable spots (a star, for example).

Photograph I–1. *Globular cluster M13 assembled from 6 images each covering only a portion of the object. The limited field of the CCD (TH7863) can be expanded with this mosaic technique. The streak at the bottom of the image is due to the passage of a satellite during the acquisition of one of the images. Each 1 minute exposure was taken with a 60 cm telescope at Pic du Midi Observatory.*

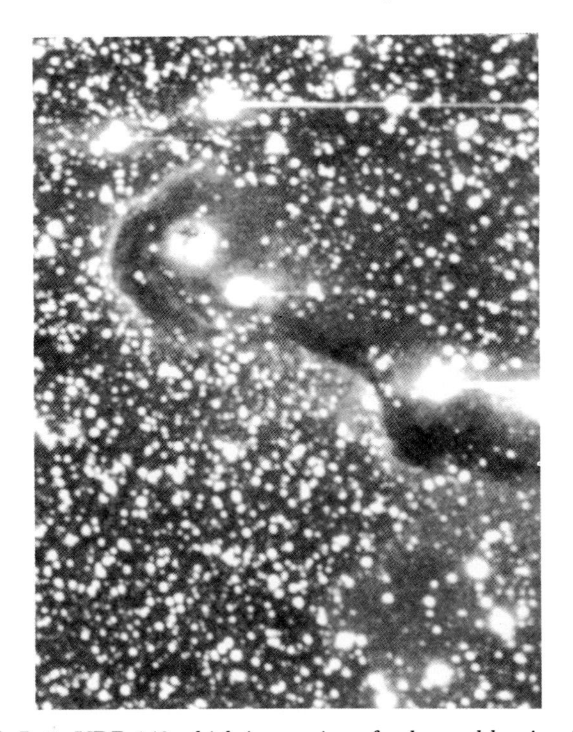

Photograph I–2. *VDB 142 which is a region of enhanced luminosity within the vast nebula group IC 1396. The range of dark and bright nebulosities is remarkable. 15 minute exposure with the T280 telescope at Pic du Midi Observatory.*

Photograph I–3. *The diffuse nebula VDB 152. A pale area extends towards the north and intermixes with a dark nebula. Ten minute exposure with the 280 mm telescope and a TH7863 CCD.*

Photograph I–4. *The Cocoon nebula (IC 5146). Twenty minute exposure with a TH7863 CCD and a 280 mm telescope of the ALCYONE Association.*

Photographs I–5 and 6. *The planetary nebula NGC 7293 (Helix). The negative view (bottom) shows a second external ring, much fainter than the main one. The streak at the top of the negative image is caused by a satellite. Twenty minute exposures with a TH7863 CCD and the 60 cm telescope at the Pic du Midi Observatory.*

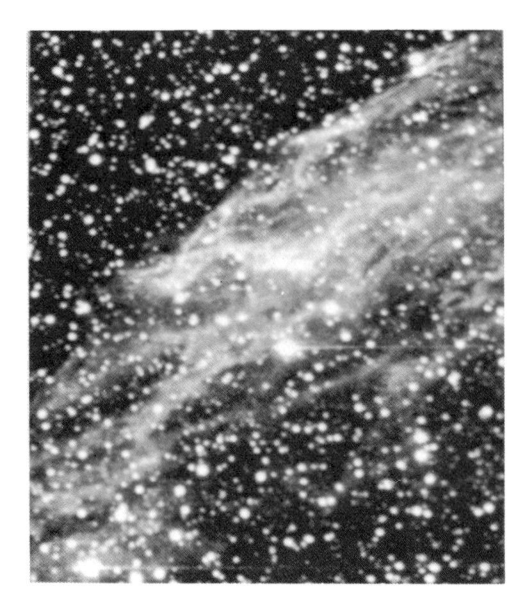

Photograph I–7. *A portion of NGC 6992, the Veil or Cirrus nebula. This image is the result of compositing two images each exposed for 10 minutes with a TH7863 and a 280 mm telescope.*

Photograph I–8. *The nebula NGC 7538. Ten minute exposure, TH7863 CCD and a 280 mm telescope.*

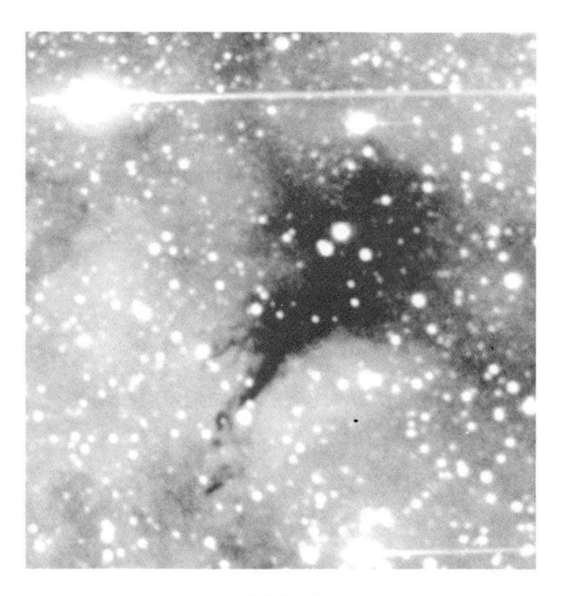

Photograph I–9. *Because current CCD's have small sensitive surfaces extended objects need not be avoided—"close ups" can be interesting as in this image of the dark nebula LDN 889 in IC 1318. Twenty minute exposure with a TH7863 CCD and a 280 mm telescope.*

Photograph I–10. *NGC 7635 is a hybrid object, half-way between a diffuse nebula and a planetary nebula (note the ring structure). Composite of two 10 minute exposures with a TH7863 CCD and a 280 mm telescope.*

Photograph I–11. *These images of the planet Jupiter were exposed through blue, green and infra-red filters. The images were acquired very close together (in time) to minimize the effects of rotation and atmosphere. Notice the difference in the planet's appearance depending on the color (wavelength). Direct focus of the 1 meter telescope of the Pic du Midi Observatory (16 meter focal length) and a TH7852 CCD.*

Photograph I–12. *This illustration was created by taking the three monochromatic images in Photograph I-11 (above) and "stacking" them (with computer software) to create a "true" color image of Jupiter. The contrast of this image was intentionally enhanced. The satellite Io projects its shadow on the planet's disk.*

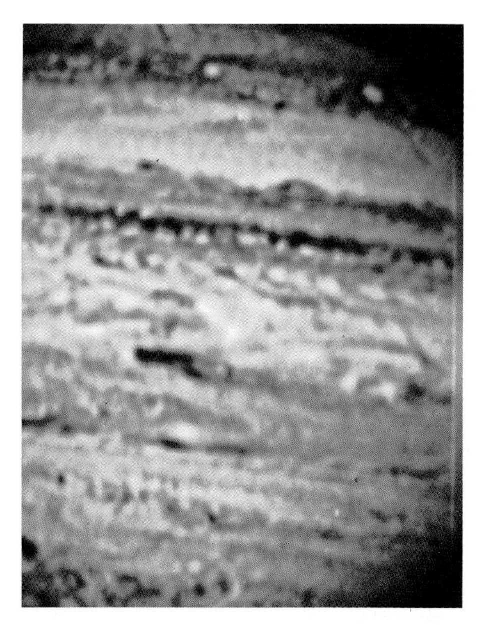

Photograph I–13. *Close up of Jupiter shows details of 0.3 arcseconds. One meter telescope at the Pic du Midi operating at a 30 meter focal length, a TH7852 CCD, 0.7 second exposure and careful image processing.*

Photograph I–14. *The planet Jupiter taken in the infra-red with a wide-band filter (red image on the right) and a narrow filter centered on the methane band at 8800 Å (blue image on the left). These two images, obtained almost at the same instant, do not have much in common. Note particularly the presence of bright arcs indicating the poles in the image taken in the methane band. Also note that the red spot in the blue image (located on the edge of the limb) presents an astonishing darkening on its center. Images taken with the 1 meter telescope at Pic du Midi and a TH7852 CCD. The right image exposure took 2 minutes under difficult tracking conditions while the left image took only 0.4 second.*

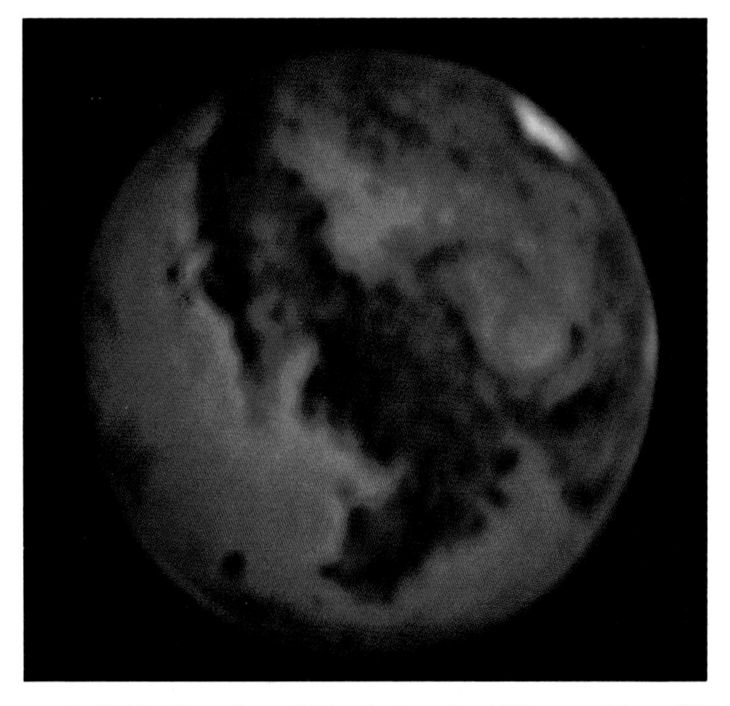

Photograph I–15. *The planet Mars during the 1988 opposition. This color image is a composite of three images each taken through a color filter—blue, green or infra-red). Notice the fog at the edge of the limb, corresponding to the limit between day and night at the moment of opposition. The bluish clouds at the bottom of the image show the presence of the Martian winter which was then raging in the boreal hemisphere. The circular formations visible in the southern hemisphere are huge impact craters. Taken with the one meter telescope at the Pic du Midi with a TH7852 CCD.*

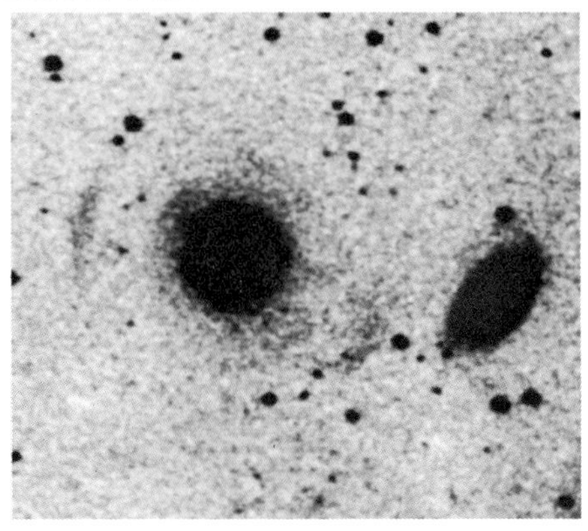

Photograph I–16. *The galaxy NGC 474 is a very peculiar object. It is surrounded by pieces of rings or shells. These details are extremely faint and very difficult to observe without a CCD. 280 mm telescope, 900 second exposure, TH7863 CCD.*

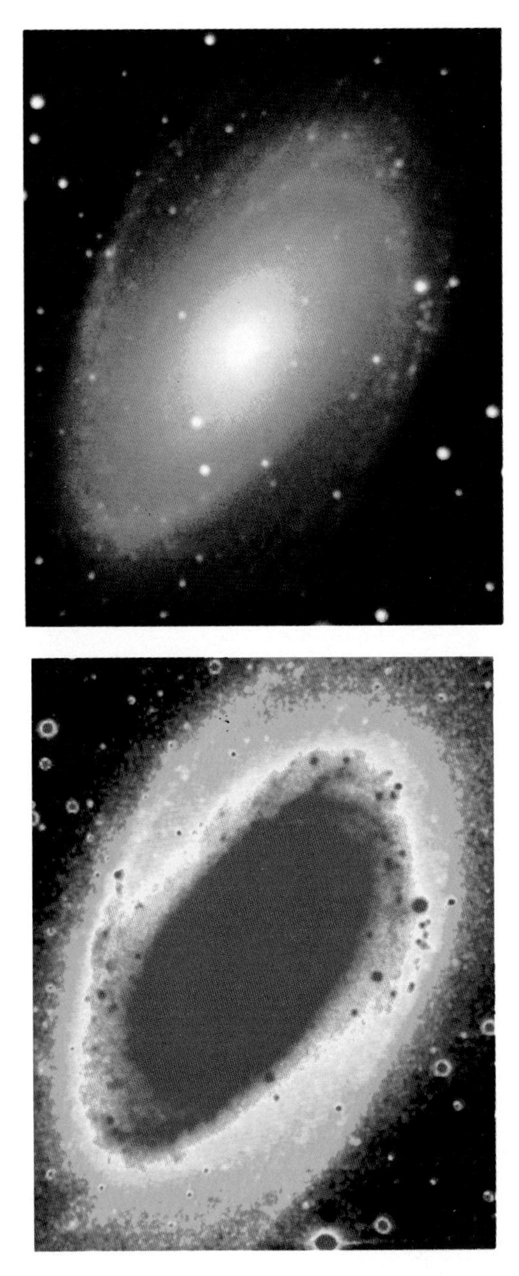

Photographs I–17 and 18. *The galaxy M81 visualized in levels of gray (top) and in false colors (bottom). The latter is useful for determining faint extensions of the object that are not normally apparent. 900 second exposure with a 280 mm telescope and a TH7863 CCD.*

Photograph I–19. *The Sombrero galaxy (M104) visualized in false colors. The part that usually shows in photographs appears as blue. CCD imaging reveals a very pale halo, shown in yellow, extending over 15 arc minutes of diameter.*

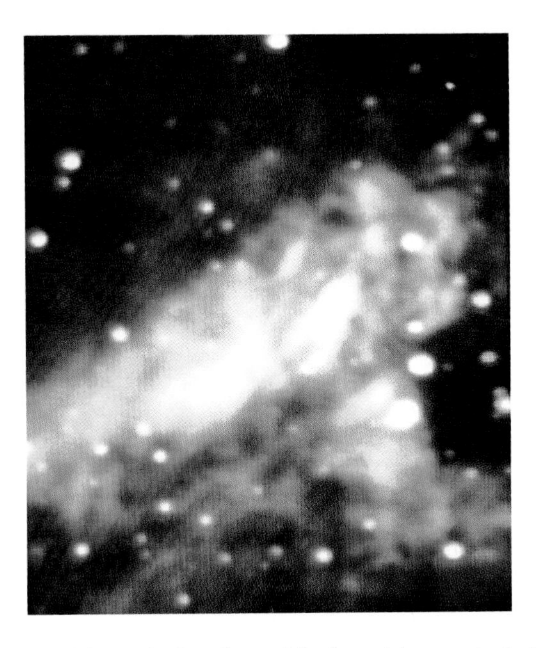

Photograph I–20. *M17 nebula taken with the trichrome technique. The main information comes from the Hα emission line. The reddish zone towards the right indicates the presence of an opaque dust cloud. T60 telescope at Pic du Midi Observatory and a TH7852 CCD.*

Photograph I–21. *The galaxy M82. The external regions were intentionally enhanced in this representation. In these external regions the object has unusual features. Note the "jets" escaping perpendicularly to the principal axis of the spindle. These details are very difficult to photograph using small telescopes. This image is a composite of two 10 minute exposures taken with a 280 mm telescope and a TH7863 CCD.*

Photograph I–22. *The irregular galaxy NGC 6822 of the Local Group. This image easily resolves this object into stars (the limiting magnitude is on the order of 21.5). T60 telescope at Pic du Midi, 900 second exposure and a TH7863 CCD.*

Photograph I–23. *Another dwarf galaxy much more difficult than NGC 6822: IC 1613. Observation made with the 280 mm telescope, 900 second exposure and a TH7863 CCD. Here again the galaxy is resolved into stars.*

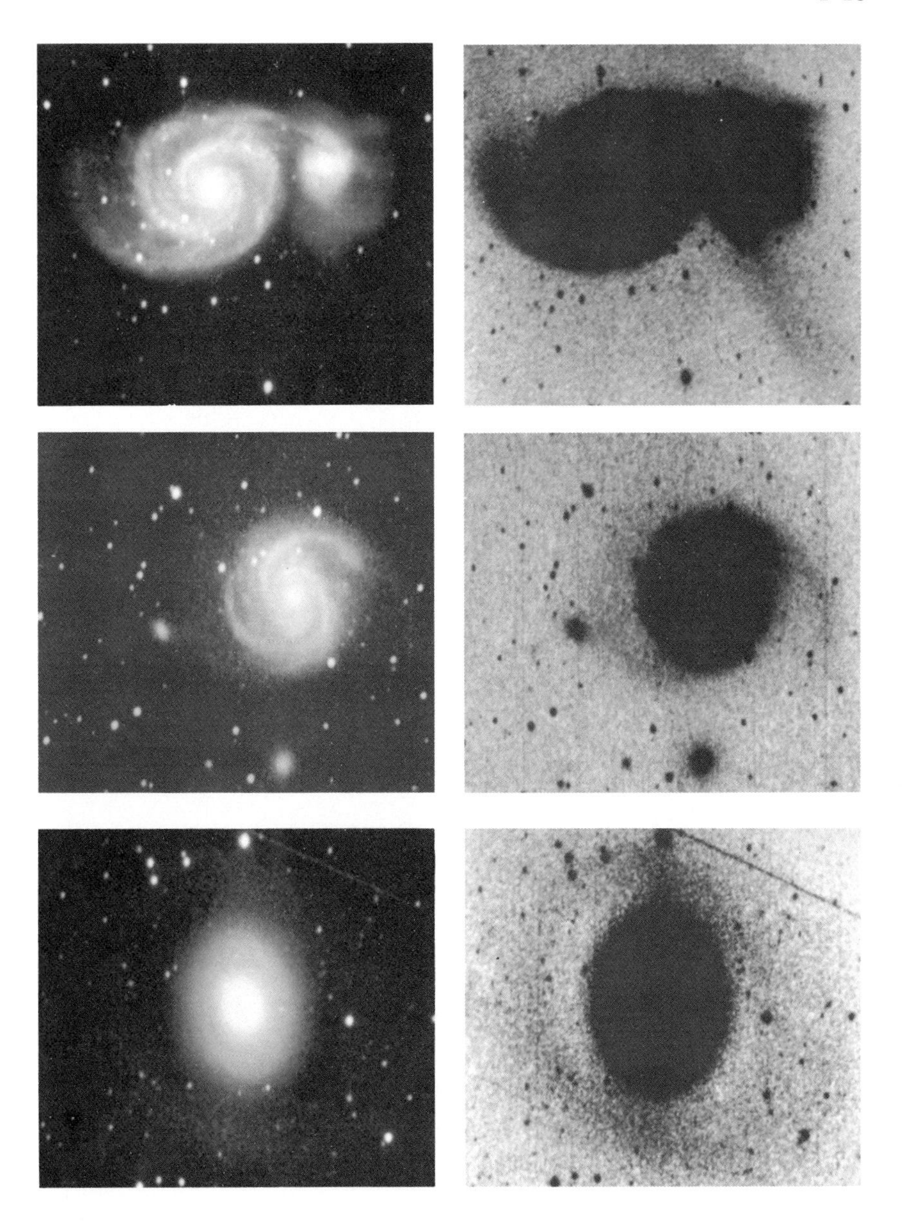

Photographs I–24 through 29. *Weak extensions of some galaxies. From top to bottom: M51, M100, M94. Opposite each positive image is a negative image which shows very faint structures. Note M100's satellite galaxies. M94 shows a huge ring that is hidden in the positive image.*

Photograph I–30. *The galaxy M90 and its companion IC 3583. Composite of two 5 minute exposures taken with the T60 telescope at Pic du Midi.*

Photograph I–31. *This 20 minute exposure of the galaxy NGC 7331 also reveals its procession of satellite galaxies. 280 mm telescope and a TH7863 CCD.*

Photograph I–32. *ARP 331 which is a vast chain of elliptical galaxies (NGC 375-6-7...). Ten minute exposure with a 280 mm telescope and a TH7863 CCD.*

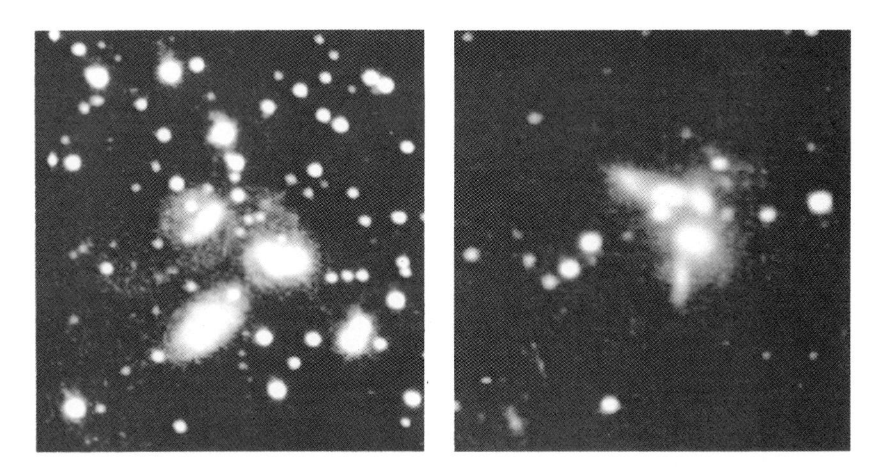

Photographs I–33 and 34. *Stephan's quintet (Left). This group of galaxies is located in the constellation of Pegasus. Seyfert's sextet (right) in the Hercules constellation. Both images taken with the T60 telescope at Pic du Midi, diaphragmed to 40 cm, 15 minute exposure and a TH7863 CCD.*

Photograph I–35 and 36. *The image on the left is a composite of two 10 minute exposures made with a 280 mm telescope of NGC 6946, a well-known spiral galaxy. The image on the right is the pair ARP 94 made up of two interacting galaxies (NGC 3227-8) . Twenty minute exposure with a 280 mm telescope. Both images were made with a TH7863 CCD.*

4.5.11 Image Processing in the Frequency Domain

Up to this point in our discussion we have acted upon the image in the spatial domain, i.e., the image points were addressed by their coordinates (x, y). This is not necessarily the most efficient method since we have seen that certain operations, in particular the filters, were expressed in terms of filtering spatial frequencies. It is often better to describe the image in the frequency domain where the coordinates (x, y) are replaced by frequencies $(1/x, 1/y)$ or (u, v). This representation then translates the presence of a given frequency in the image.

Convolution in the spatial domain, a fairly difficult operation that requires manipulating many pixels, is reduced to a simple multiplication in the frequency domain. Deconvolution is another operation that is practically impossible to do in the spatial domain but is possible in the frequency domain.

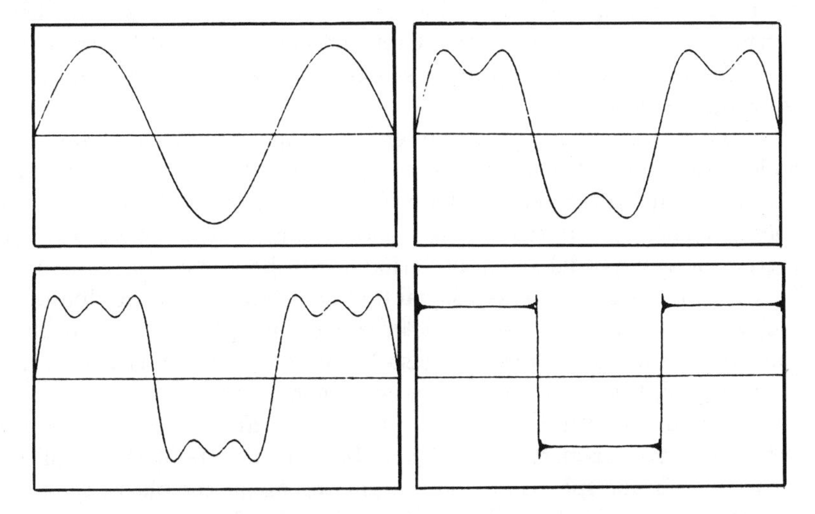

Figure 4.32 *Synthesis of a square signal with Fourier series.*

4.5.12 The Fourier Transform

We know that any signal can be decomposed into an infinite number of sinusoidal components defined by their frequency, amplitude, and phase. The image of a test pattern, made up of a succession of white and black patterns, is comparable to a periodic square-wave signal. Figure 4.32 shows how a sum of sinusoidal functions (Fourier series) gives an approximation of this signal, despite the presence of discontinuities. Going from left to right and from top to bottom in this figure, the rectangular signal is restored from these equations:

$$\sin \pi x$$

$$\sin \pi x + \frac{1}{3} \sin 3\pi x$$

$$\sin \pi x + \frac{1}{3} \sin 3\pi x + \frac{1}{5} \sin 5\pi x$$

$$\vdots$$

(about a hundred terms in this series).

The square signal is approached that much better when we take a larger number of terms. The higher order terms have a very short period, or a large frequency. The rebound effect, when the signal leaves at a right angle, is difficult to remove (we have to calculate more than a thousand terms to make it disappear at the scale of the figure). These oscillations at the levels of abrupt variations of the signal (Gibbs effect) are caused by truncating toward the high frequencies during signal reconstitution. We will discuss this problem below.

The Fourier transform is a mathematical operation that allows us to calculate the spectral density of the signal constituting the image. The word "spectrum" is taken by analogy with the problem of decomposition of a polychromatic light beam into its monochromatic components. Thus the spectrum of a sinusoidal signal of period T will be a line at the frequency $1/T$. Under certain conditions a signal that is the mixture of two sinusoidals of different periods, can show a spectrum containing two strong lines, and so on (lower level harmonics may also be present). In the case of a non-periodic signal, the spectrum becomes continuous.

A spectrum is entirely defined by the amplitude and phase of the sinusoid waves approximating the signal. Information about the amplitude gives the amplitude spectrum, and information about the phase, the phase spectrum. By starting from the spectrum, it is possible to reconstitute the signal by applying an inverse Fourier transform.

Figure 4.33 shows two continuous, one-dimensional signals and their respective amplitude spectra. The signal at the top is smoothed, i.e., it does not contain any high frequencies. This is translated by a spectrum amplitude that is prematurely zeroed toward the high frequencies. The bottom signal is disturbed by noise, i.e., high frequency fluctuations, that show up in the frequency domain in the form of an extension of the spectrum toward the high frequencies. However, beyond at a certain limit, the spectrum stops. This is the cutoff frequency, usually set by the characteristics of the electronics that were used to record the signal. The amplitude of the zero frequency is linked to the average value of the intensity of the points on the spatial image (continuum level). Note that the spectrum is symmetric with respect to the zero frequency. This is a fundamental characteristic of the Fourier transform of real signals.

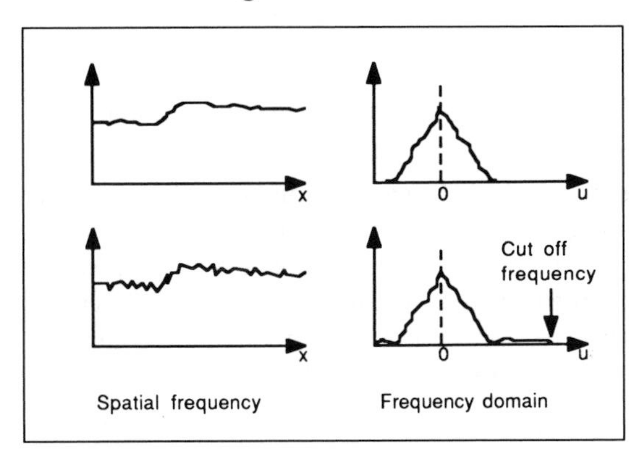

Figure 4.33 *On the left, the signal in the spatial domain; on the right, the corresponding frequency spectrum.*

In practice, the discrete Fourier transform of a series of *real* points $f(x)$, $x = 0, \ldots, N-1$, in the frequency domain gives a real component $R(\nu)$ and an imaginary component $I(\nu)$ that can be calculated by the formulas

$$R(\nu) = \Delta t \sum_{x=0}^{N-1} f(x) \cos(2\pi\nu\, x\, \Delta t)$$

$$I(\nu) = -\Delta t \sum_{x=0}^{N-1} f(x) \sin(2\pi\nu\, x\, \Delta t)$$

with

x = the coordinate value of a point of the sample,

Δt = the sampling step, and

ν = the frequency calculated in the spectrum.

The spectrum's amplitude is the modulus

$$A(\nu) = (R(\nu)^2 + I(\nu)^2)^{1/2}$$

and the power spectrum:

$$A^2(\nu) = R(\nu)^2 + I(\nu)^2.$$

Usually the calculations are carried out for the frequencies given by:

$$\nu = \frac{u}{T} = \frac{u}{N\Delta t}$$

where

$T = 1$ TV screen (the extent of the sample), and

$N = 512, 640, 1024$, etc. pixels in a line.

It is useless to continue the calculation beyond $u = N/2$ because one can show that starting at this limit the spectrum is reproduced symmetrically. This frequency limit, called the Nyquist frequency, is $1/(2 \cdot \Delta t)$. It is therefore determined by the sampling step of the signal. If the sampling is too sparse with respect to the high frequencies contained in the signal, there is distortion or aliasing (Figure 4.34).

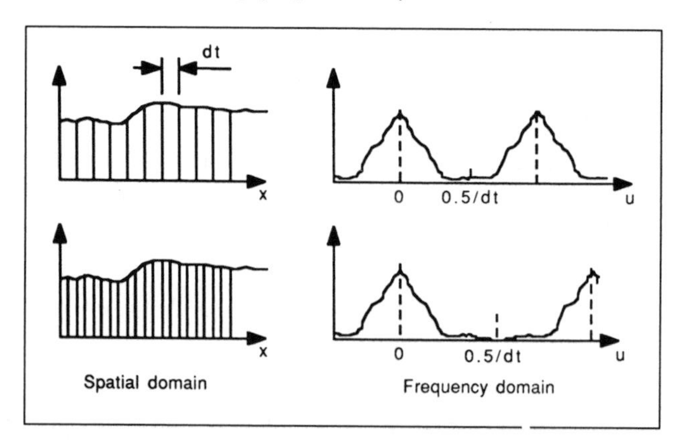

Figure 4.34 *The replica of the spectral pattern beyond the Nyquist frequency. The upper signal is incorrectly sampled, producing distortions in the spectrum (mixing of orders). Overlapping of spectra is avoided by placing the sampling steps closer together.*

By making a few changes in the variable, the Fourier transform of a series of real numbers can be rewritten

$$R(u) = \frac{1}{N} \sum_{x=0}^{N-1} f(x) \cos\left(2\pi \frac{xu}{N}\right)$$

$$I(u) = -\frac{1}{N} \sum_{x=0}^{N-1} f(x) \sin\left(2\pi \frac{xu}{N}\right).$$

The factor $1/N$ is introduced to obtain an equivalence between the direct transform and the inverse transform.

The inverse Fourier transform is given by

$$f(x) = \sum_{u=0}^{N-1} R(u)\cos\left(2\pi\frac{xu}{N}\right) - I(u)\sin\left(2\pi\frac{xu}{N}\right).$$

The small program below applies these equations. We calculate the spectrum of a sinusoidal signal, then reconstitute the signal by an inverse transform. By modifying the number of points in the sample (variable `NP%`) and the unit of the sinusoidal (variable `PERIOD`) we can see how the fineness and the position of the line produced in the spectrum evolves and its position.

```
REM **********************
REM * FOURIER TRANSFORM *
REM **********************
DECLARE SUB direcft (np%, xr(), xi())
DECLARE SUB inverft (np%, xr(), xi())
SCREEN 9
DIM SHARED xr(255), xi(255)
pi = 4 * ATN(1)
n% = 8
period = 10
np% = 2 ^ n%
REM ****************************
REM * Compute signal and draw *
REM ****************************
PSET (0, 50), 11
FOR i% = 0 TO np% - 1
   xr(i%) = COS(2 * pi * i% / period)
   xi(i%) = 0
   LINE -(2 * i%, 100 - xr(i%) * 50), 11
NEXT
CALL direcft(np%, xr(), xi())
REM *****************
REM * Draw spectrum *
REM *****************
PSET (0, 300), 9
FOR i% = 0 TO np% - 1
   modulus = SQR(xr(i%) * xr(i%) + xi(i%) * xi(i%))
   LINE -(2 * i%, 300 - modulus * 250), 9
NEXT
REM ******************
REM * Rebuild signal *
REM ******************
CALL inverft(np%, xr(), xi())
PSET (0, 50), 14
FOR i% = 0 TO np% - 1
   LINE -(2 * i%, 100 - xr(i%) * 50), 14
NEXT

REM ****************************
REM * Direct Fourier Transform *
REM ****************************
SUB direcft (np%, xr(), xi()) STATIC
DIM xr2(255), xi2(255)
pi = 4 * ATN(1)
nb% = np% - 1
FOR j% = 0 TO nb%
   sr = 0
   si = 0
   alpha = 2 * pi * j% / np%
```

```
   FOR i% = 0 TO nb%
      sr = sr + xr(i%) * COS(alpha * i%)
      si = si + xr(i%) * SIN(alpha * i%)
   NEXT
   xr2(j%) = sr / np%
   xi2(j%) = -si / np%
NEXT
FOR i% = 0 TO nb%
   xr(i%) = xr2(i%)
   xi(i%) = xi2(i%)
NEXT
END SUB

REM ******************************
REM * Inverse Fourier Transform *
REM ******************************
SUB inverft (np%, xr(), xi()) STATIC
DIM xr2(255), xi2(255)
pi = 4 * ATN(1)
nb% = np% - 1
FOR j% = 0 TO nb%
   s = 0
   alpha = 2 * pi * j% / np%
   FOR i% = 0 TO nb%
      sr = xr(i%) * COS(alpha * i%)
      si = xi(i%) * SIN(alpha * i%)
      s = s + sr - si
   NEXT
   xr2(j%) = s
   xi2(j%) = 0
NEXT
FOR i% = 0 TO nb%
   xr(i%) = xr2(i%)
   xi(i%) = xi2(i%)
NEXT
END SUB
```

The execution of the preceding program will demonstrate the speed of operations during a Fourier transform. Here we are working with a one-dimensional signal. Just imagine what a whole image will be like! By programming the transform directly, the calculation time is proportional to N^2. The number N reaches several thousand very rapidly.

Fortunately an efficient algorithm to evaluate the FT was found in the 1960's by the mathematician-computer scientists Cooley and Tuckey. It is known as the Fast Fourier Transform or FFT. We will not give all the details of this algorithm (many derived forms exist today), but will give its performance. Calculation time with the FFT is proportional to $N \cdot \log_2 N$. The FFT allows an evaluation 10 times faster than direct calculation when $N = 64$, 30 times faster when $N = 256$, etc. The difference between these two methods increases with N. To reach this efficiency, the FFT algorithm uses to a maximum the calculations' redundancies and replaces many multiplications by additions. The only constraint of the base algorithm is that N must be a power of two ($N = 16, 32, 64, 128, 256, \cdots$).

The Fourier Transform of a Two-dimensional Image

Noting that

$$\cos(n\theta) + i\sin(n\theta) = \exp(in\theta)$$

it is more convenient to write the Fourier transform in exponential form. Thus

$$F(u) = \frac{1}{N}\sum_{x=0}^{N-1} f(x)\exp\left(-2\pi i\frac{xu}{N}\right)$$

with $i^2 = -1$.

The transform into the spatial domain is written

$$f(x) = \sum_{u=0}^{N-1} F(u)\exp\left(2\pi i\frac{xu}{N}\right).$$

In the case of a two-dimensional image we will have to calculate the transform of a function of two variables:

$$F(u,v) = \frac{1}{MN}\sum_{x=0}^{M-1}\sum_{y=0}^{N-1} f(x,y)\exp\left(-2\pi i\left(\left(\frac{xu}{M}\right)+\left(\frac{yv}{N}\right)\right)\right),$$

$$f(x,y) = \sum_{u=0}^{M-1}\sum_{v=0}^{N-1} F(u,v)\exp\left(2\pi i\left(\left(\frac{xu}{M}\right)+\left(\frac{yv}{N}\right)\right)\right).$$

If the image has a square shape of $N \times N$ pixels, which is the most common with processing based on FFT, the relations above become

$$F(u,v) = \frac{1}{N^2}\sum_{x=0}^{N-1}\sum_{y=0}^{N-1} f(x,y)\exp\left(-2\pi i\frac{xu+yv}{N}\right)$$

and

$$f(x,y) = \sum_{u=0}^{N-1}\sum_{v=0}^{N-1} F(u,v)\exp\left(2\pi i\frac{xu+yv}{N}\right).$$

The separability property of the function $F(u,v)$ allows us to express the transform in the form

$$F(u,v) = \frac{1}{N}\sum_{x=0}^{N-1} G(x,y)\exp\left(-2\pi i\frac{xu}{N}\right)$$

with

$$G(x,y) = \frac{1}{N}\sum_{y=0}^{N-1} f(x,y)\exp\left(-2\pi i\frac{yv}{N}\right).$$

In other words, the Fourier transform of a two-dimensional image is obtained by calculating the FT according to the lines, and then the FT of the result according to the columns.

A Program for Calculating an FFT

We have written a module called `FTI` which allows very complete manipulation of images in the frequency domain (filtering, deconvolution, correlation, etc.). This module treats images of varying sizes up to 128×128, which represents a data segment of 64K if the real and imaginary parts are coded using two four-byte floating point numbers (single precision format in MS Basic 7.0) in two arrays (`x()` and `y()`).

Just as for processing in the spatial domain, we need at least one `BUFFER`. In fact this `BUFFER` is divided into two parts for storing the real and imaginary parts (`x1()` and `y1()`). The user interface is identical to that of `IP` and `MAT` and we find some of the functions of these programs in `FTI` (visualization, management of mass memory, etc.).

The heart of `FTI` is obviously the calculation routine of the FFT. This routine is called `FFT`. Arrays `x()` and `y()` contain respectively the real and imaginary parts of the image. The contents of the variable `di%` defines the direction of the FFT. If `di%=1`, the transform is carried out in the spatial-frequency (direct) direction, if `di%=-1` the transform is carried out in the frequency-spatial (inverse) direction. The variable `ft%` contains the dimension of the image to be processed (always a power of two). Printing an '`*`' on the screen marks the end of the FFT in one direction. The first part of the routine transposes the data in the vectors `xl()` and `yl()`. The subprogram `FFTL` carries out the one-dimensional FFT itself from these vectors. The execution of this routine takes about one minute for a 128×128 point image using an 8MHz 80286 CPU and 80287 math coprocessor.

```
SUB fft (x(), y(), di%) STATIC
  ft2% = ft% + 1
  DIM co(129), si(129), xl(128), yl(128)
  pi = 3.14159265#
  np% = ft%
  nm% = CINT(LOG(np%) / LOG(2))
  FOR i% = 0 TO ft%
    par = 2 * pi * i% / ft%
    co(i%) = COS(par)
    si(i%) = -SIN(par) * di%
  NEXT
  FOR il% = 1 TO np%
    FOR ipl% = 1 TO np%
      xl(ipl%) = x(il% - 1, ipl% - 1)
      yl(ipl%) = y(il% - 1, ipl% - 1)
    NEXT
    GOSUB fftl
    FOR ipl% = 1 TO np%
      x(il% - 1, ipl% - 1) = xl(ipl%)
```

```
      y(il% - 1, ipl% - 1) = yl(ipl%)
    NEXT
  NEXT
  PRINT "* ";
  FOR ipl% = 1 TO np%
    FOR il% = 1 TO np%
      xl(il%) = x(il% - 1, ipl% - 1)
      yl(il%) = y(il% - 1, ipl% - 1)
    NEXT
    GOSUB fftl
    FOR il% = 1 TO np%
      x(il% - 1, ipl% - 1) = xl(il%)
      y(il% - 1, ipl% - 1) = yl(il%)
    NEXT
  NEXT
  PRINT "* ";
EXIT SUB

fftl:
  FOR lo% = 1 TO nm%
    lmx% = 2 ^ (nm% - lo%)
    lix% = 2 * lmx%
    scl% = 2 ^ (lo% - 1)
    FOR lm% = 1 TO lmx%
      arg% = (lm% - 1) * scl%
      cc = co(arg%)
      ss = si(arg%)
      j0% = lix% - lm%
      FOR li% = lix% TO np% STEP lix%
        j1% = li% - j0%
        j2% = j1% + lmx%
        t1 = xl(j1%) - xl(j2%)
        t2 = yl(j1%) - yl(j2%)
        xl(j1%) = xl(j1%) + xl(j2%)
        yl(j1%) = yl(j1%) + yl(j2%)
        xl(j2%) = cc * t1 + ss * t2
        yl(j2%) = cc * t2 - ss * t1
      NEXT
    NEXT
  NEXT
  nv2% = np% \ 2
  j% = 1
  nml% = np% - 1
  FOR i% = 1 TO nml%
    IF i% < j% THEN
      SWAP xl(j%), xl(i%)
      SWAP yl(j%), yl(i%)
    END IF
    k% = nv2%
bif:
    IF k% < j% THEN
      j% = j% - k%
      k% = k% \ 2
      GOTO bif
    END IF
    j% = j% + k%
  NEXT
  IF di% = 1 THEN
    FOR i% = 1 TO ft%
       xl(i%) = xl(i%) / ft%
       yl(i%) = yl(i%) / ft%
    NEXT
  END IF
  RETURN
END SUB
```

The FFTD routine shapes the data for the execution of a direct FFT, calls the FFT routine, and permutes the final data to have a 0 frequency in the center of the image. The FFTD routine follows:

```
SUB fftd STATIC
   FOR i% =1 TO ft%
     FOR j% =1 TO ft%
       x(i% -1,j% -1)=a% (i% ,j% )
       y(i% -1,j% -1)=0
     NEXT
   NEXT
   di% =1
   CALL fft(x(),y(),di% )
   CALL permut
END SUB
```

Note that array y() contains zeros since only real data are manipulated when starting from the spatial domain. The first valid element of arrays x() and y() is located at the coordinates (0,0), a fact which explains the shift of one unit with respect to the array a%().

After calculating the FFT, the routine permut positions the symmetry point of the complex image in the center of the arrays x() and y() (coordinates (65,65) if TF%=128). This point contains the zero frequency. Figure 4.35 shows the difference in appearance of a spectrum before and after calling the permut routine. In the first case the symmetry point is found in a corner of the image, which makes interpreting the spectrum more difficult. In addition, the fact of having the point of zero frequency in the center of the arrays eases the programming of most processing algorithms.

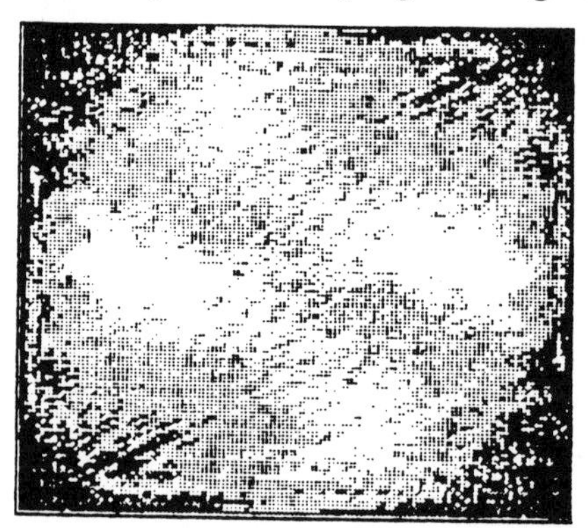
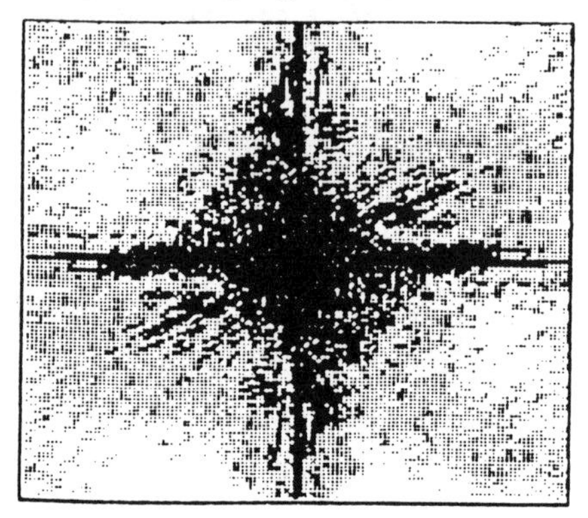

Figure 4.35 *Appearance of the two-dimensional spectrum before (at left) and after (at right) calling the* permut *routine. On the figure at left, the zero frequency is in the lower left corner. These two pictures contain the same information, but the one on the right is easier to understand.*

There are two ways of permuting the quadrants of the spectral image:

1. If the image has N columns and N lines, we carry out a circular permutation according to the columns of $N/2$ pixels, then a permutation of $N/2$ pixels according to the lines.

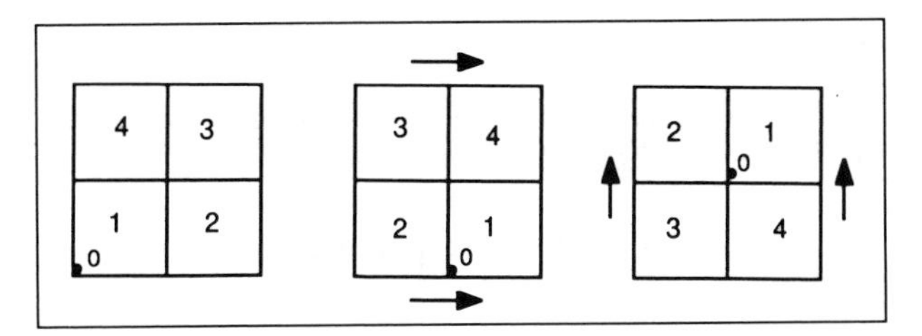

Figure 4.36 *Method for permuting the 4 quadrants in place (it doesn't need a memory zone other than the one containing the spectrum).*

2. We use the buffer as an intermediate storage zone for the quadrants.

It is this latter method that we chose.

```
SUB permut STATIC
   ft2% =ft% /2
   np% =ft% -1
   FOR i% =0 TO np%
     FOR j% =0 TO np%
     IF i% +1>ft2%  THEN
        IF j% +1>ft2%  THEN
               x1(i% ,j% ) = x(i% -ft2% ,j% -ft2% )
               y1(i% ,j% ) = y(i% -ft2% ,j% -ft2% )
        ELSE
               x1(i% ,j% ) = x(i% -ft2% ,j% +ft2% )
               y1(i% ,j% ) = y(i% -ft2% ,j% +ft2% )
       END IF
     ELSE
       IF j% +1>ft2%  THEN
              x1(i% ,j% ) = x(i% +ft2% ,j% -ft2% )
              y1(i% ,j% ) = y(i% +ft2% ,j% -ft2% )
       ELSE
              x1(i% ,j% ) = x(i% +ft2% ,j% +ft2% )
              y1(i% ,j% ) = y(i% +ft2% ,j% +ft2% )
       END IF
     END IF
     NEXT
   NEXT
   FOR i% =0 TO np%
      FOR j% =0 TO np%
          x(i% ,j% )=x1(i% ,j% )
          y(i% ,j% )=y1(i% ,j% )
      NEXT
   NEXT
END SUB
```

After the various processings are carried out in the frequency domain, we return to the spatial domain by calling the routine **FFTI**:

```
SUB ffti STATIC
  di% =-1
  CALL permut
  CALL fft(x(),y(),di% )
  FOR i% =1 TO ft%
     FOR j% =1 TO ft%
        a% (i% ,j% )=CINT(x(i% -1,j% -1))
     NEXT
  NEXT
END
```

Convolution

If $G(s)$, $I(s)$ and $E(s)$ are the Fourier transforms (i.e., $\Longleftrightarrow$) of $g(f)$, $i(f)$ and $e(f)$, we will have these identities:

$$G(s) = I(s) * E(s) \Longleftrightarrow g(f) = i(f) \cdot e(f)$$

and

$$G(s) = I(s) \cdot E(s) \Longleftrightarrow g(f) = i(f) * e(f).$$

In other words, a convolution of two signals in the spatial domain is translated by a simple multiplication of the transforms of the two signals in the frequency domain and vice versa. These remarkable properties of the FT allow us to use particularly efficient filtering techniques. The amplitude spectrum is simply multiplied by a weighting law that is a function of the frequency. This function can have as complicated a form as one likes. The processing will not be any more difficult.

Usually in the frequency domain, the product of a complex number by a real one (the weighting law) is easier if the complex number is expressed in modulus and argument. The `FFTI` program has a command allowing the transformation of the complex image of a Cartesian shape into a polar shape:

```
SUB recpol STATIC
   pi=3.141592654
   nb%=ft%-1
   FOR i%=0 TO nb%
     FOR j%=0 TO nb%
       xx=x(i%,j%)
       yy=y(i%,j%)
       zz=SQR(xx*xx + yy*yy)
       x(i%,j%)=zz
       IF xx=0 THEN
       zz=SGN(yy)*pi/2
       ELSE
       zz=ATN(yy/xx)+(1-SGN(xx)*pi/2
       END IF
       y(i%,j%)=zz
     NEXT
   NEXT
END SUB
```

After passing through `recpol`, array `x()` contains the moduli, and array `y()` contains the arguments. In the opposite direction, the passage from polar coordinates to rectangular coordinates is obtained by the following routine:

```
SUB polrec STATIC
   nb% =ft% -1
   FOR i% =0 TO nb%
      FOR j% =0 TO nb%
         xx=x(i% ,j% )
         yy=y(i% ,j% )
```

```
        x(i% ,j% )=xx*COS(yy)
        y(i% ,j% )=xx*SIN(yy)
      NEXT
   NEXT
END SUB
```

The weighting law, or filter transfer function, is applied to the modulus (the amplitude spectrum). As an example, here is how a low-pass filter with a Gaussian shape will be carried out (the FT of a Gaussian is also a Gaussian):

```
ft2% =ft% /2
nb% =ft% -1
FOR i% =0 TO nb%
   i1% =i% -ft2%
   FOR j% =0 TO nb%
      j1% =j% -ft2%
      k% =i1% *i1%  + j1% *j1%
      x(i% ,j% )=x(i% ,j% )*EXP(-k% /cut)
   NEXT
NEXT
```

A change of coordinates (`i1%,j1%`) is made so that the Gaussian surface is centered on the 0 frequency. The parameter `cut` is linked to the cutoff frequency of the filter. Its value can of course be modified by the user.

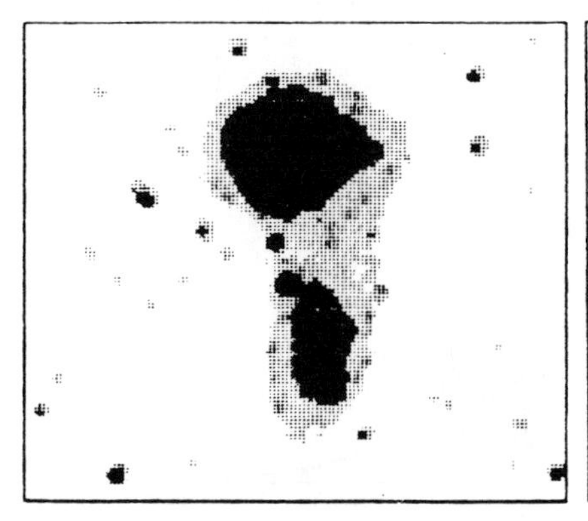

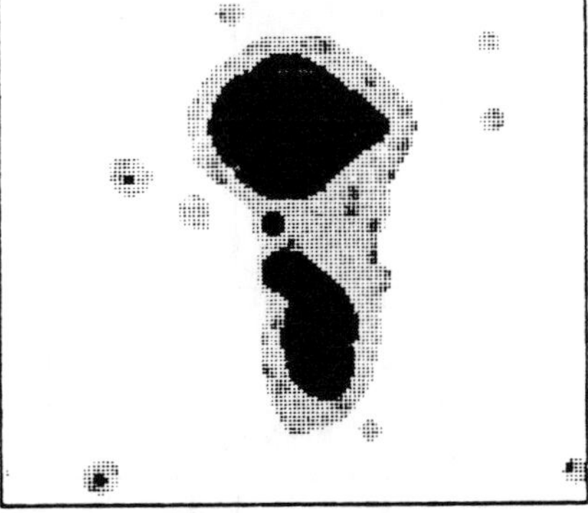

Figure 4.37 *Low-pass type filtering applied to the galaxy pair Arp271. On the right, value of the variable* `CUT` *is 1500; on the left, 160.*

The Butterworth filter is also in the family of low-pass filters, in the form

$$\frac{1}{1+(F/k_1)^{k_2}}$$

where k_1 and k_2 are parameters that can be adjusted according to requirements ($k_2 = 2$ or 4) and F is the frequency.

To carry out unsharp masking we not only have to remove the low frequencies but also the high frequencies to reduce the noise. This kind of operation is almost impossible to carry out at one time in the spatial domain. In the frequency domain, we merely have to create a band-pass filter whose transfer function is

$$\exp(-k_1 \cdot F) * (1 - \exp P(-k_2 \cdot F)).$$

By acting upon the coefficients k_1 and k_2 we can modify the shape and the width of the filter.

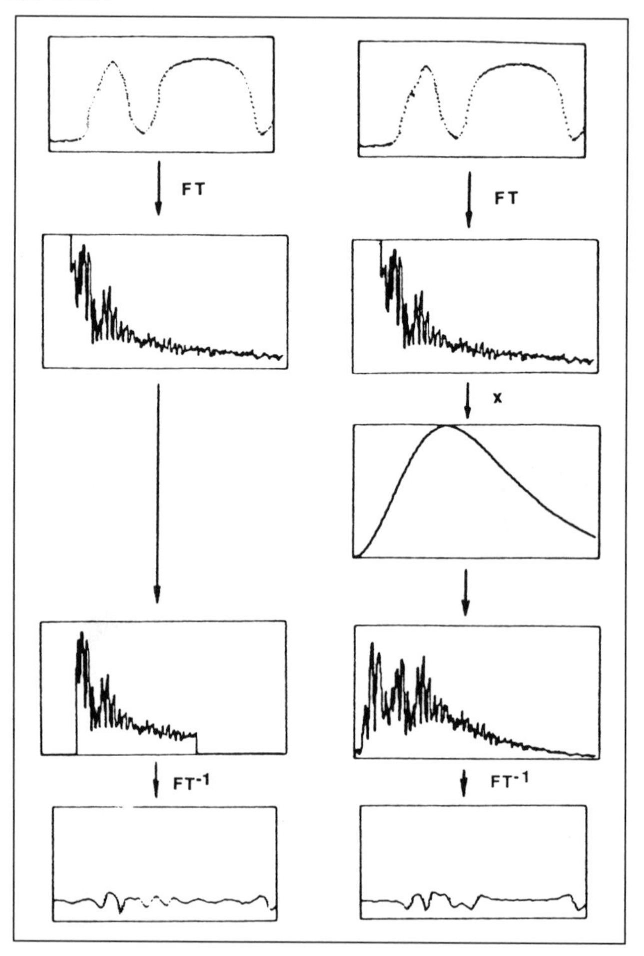

Figure 4.38 *Application of a band-pass filter to a photometric cut of the planet Saturn to emphasize the Cassini division.*

Figure 4.38 clearly shows the disastrous effect of truncating the filter and the need to apodize, that is, use filters with gentle slopes. The processing on the left is summary, to say the least: the low and high frequencies are abruptly truncated, removing the low frequency variations in the image as well as the noise. Unfortunately a secondary effect appears in the form of oscillations in the high contrast zones. To carry out the filtering we multiplied the spectrum by a rectangular transfer function with unit amplitude. But the Fourier transform of this window is a $(\sin X)/X$ type function, and the signal in the spatial domain is therefore convolved by this function, causing the appearance of oscillations in the contrasted parts. For the processing on the right we used a filter with much gentler slopes. The transfer function is built from the product of two Gaussians, whose effect is to convolve the image with another Gaussian in the spatial domain and thus to remove most of the oscillations. In this case we say that the image is apodized.

We may act very locally in the frequency spectrum by "unscreening", which eliminates periodic-shaped signals from the image. These signals usually have an artificial origin (radio frequency parasitic, interference from the 60 Hertz line current, etc.). The pattern is seen in the frequency domain as two symmetric peaks on either side of the origin. By assigning the amplitude of neighboring frequencies to the pixels containing these peaks, we can remove the pattern when we transform back to the spatial domain.

We could multiply the number of examples of filtering that can be applied in the frequency domain. All the different shapes of transfer curves can be experimented with equal simplicity.

To let us choose the filters more easily, the `FTI` program allows us to visualize the spectrum and to observe the modifications that are made. This visualization is obtained by transferring the modulus of the frequency image into array `a%()`. The spectrum can then be examined with the program's tools (cross-sections, isophotes, etc.).

Usually the amplitudes in the neighborhood of frequency 0 are much higher than the amplitudes of the other frequencies. Correct visualization of the modulus requires adopting a logarithmic scale on the amplitude axis and a normalization at the `maxdyn%` level (4095 by default). For this we use the transformation

$$\log\bigl(1 + F(u,v)\bigr).$$

This equation has the advantage of eliminating the problem posed by the zero value of the modulus during visualization.

Convolution of two images is straightforward. After calculating the FT, the product of the moduli and the sum of the arguments are determined; then the inverse FT is applied to this result. The product of the two images in the frequency domain causes the displacement of one component with respect to the other in the spatial domain. For example, Figure 4.39 presents the result of convolving the image of galaxy Arp271 and the image of a Gaussian surface.

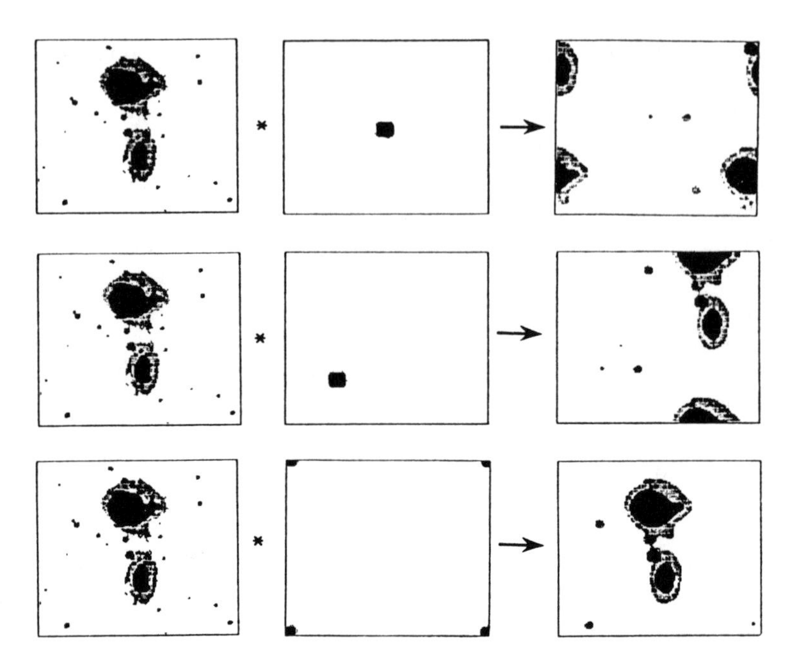

Figure 4.39 *Convolution of the image of the galaxy Arp271 with an image containing a "big" star with a Gaussian profile.*

The origin of the galaxy's image (located at lower left) is shifted after convolution to the level of the centroid of the Gaussian. To get an image that is not sliced into pieces, we either have to bring the center of the Gaussian onto the origin of the image to be processed or place the Gaussian in the center of its image (coordinates (65,65) with an image of 128×128 points) and then permute the quadrants of the result of the convolution. The permutations are obtained with a procedure identical to that already used for the `permut` command, but this time acting on array `a%()`.

Correlation

In image processing, it is sometimes desirable to compare images two by two. Shape recognition is one of these applications.

The simplest way to compare two images is to shift one with respect to the other and to measure the degree of similarity according to this shift. If $x()$ and $y()$ are the two functions to be correlated, this is mathematically translated in the one-dimensional case by the formula:

$$a(k) = \sum_{l=-\infty}^{l=+\infty} x(l)\, y\,(l+k).$$

The signal $a(k)$ is called the cross correlation function. The higher the value of this function, the more similar are the two compared images.

In the frequency domain, the cross correlation function is obtained by the product of the transforms of the two images, one of which is a complex conjugate:

$$A(u) = X(u)Y^*(u).$$

Recall that the conjugate of the complex number $Z = x + iy = \rho \cdot e^{i\theta}$ is $Z^* = x - iy = \rho \cdot e^{-i\theta}$. Therefore if we work in polar coordinates, the correlation of the two images will be programmed as follows:

```
nb% =ft% -1
FOR i% =0 TO nb%
   FOR j% =0 TO nb%
      x(i% ,j% )=x(i% ,j% )*x1(i% ,j% )
      y(i% ,j% )=y(i% ,j% )*y1(i% ,j% )
   NEXT
NEXT
```

One application of correlation to astronomy is the automatic extraction and classification of faint galaxies lost in the middle of a stellar field. The difference in appearance between stars and galaxies allows us to make this distinction. To understand these mechanisms, we will use a simpler example.

Let us take the text in Figure 4.40. Our purpose is to extract the letter "O" from the first three lines.

Figure 4.40 *The test image.*

The first operation is to extract an "O" from the text, then to calculate its spectrum and that of the whole text (Figure 4.41).

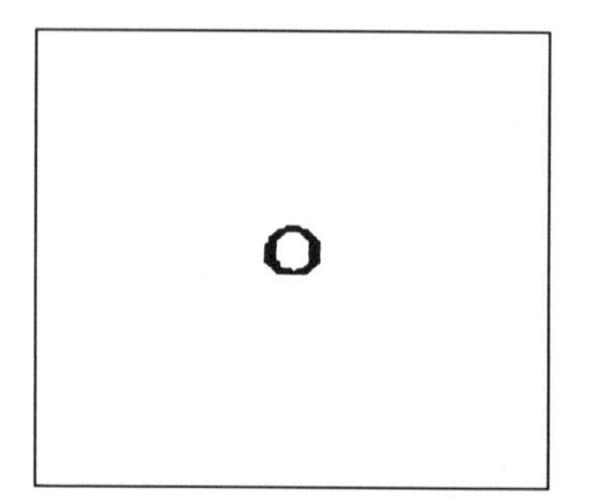

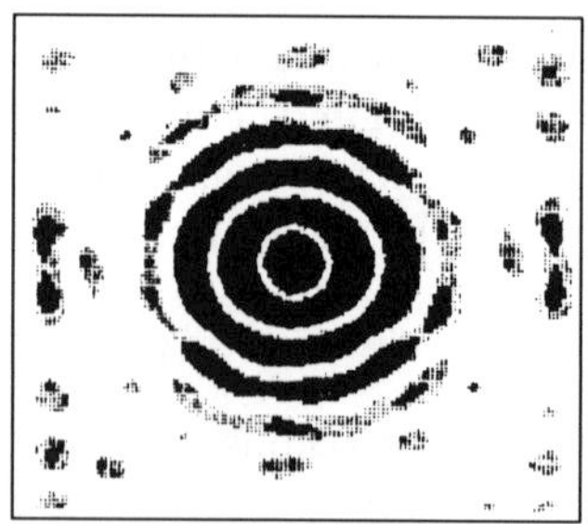

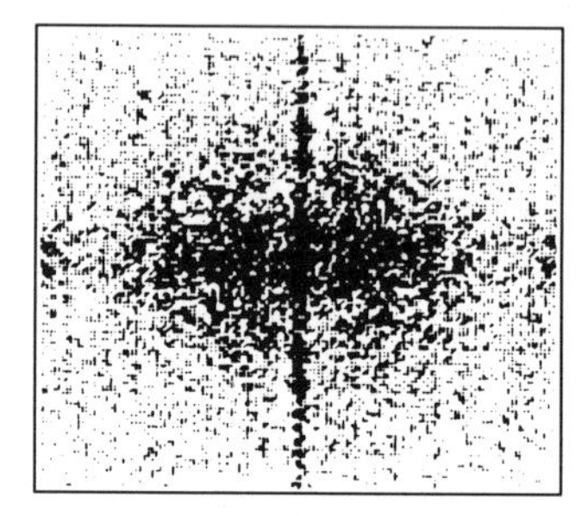

Figure 4.41 *At the top left, the "O" of the word Symphony has been isolated by the* `WINDOW` *function of the* `TI` *program. Its spectrum is shown at the top right. At the bottom center we see the spectrum of the text to be processed.*

The result of the cross correlation of the text and the "O" is visible on Figure 4.42. We find several peaks that mark the strongest occurrences between the text and the "O." By thresholding, we isolate two peaks that rise high above the others. These two are located at the extreme right and we easily note that their position corresponds to the center of the "O's."

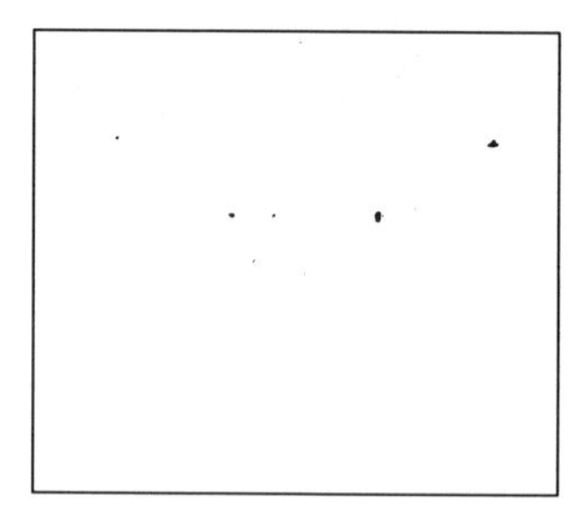

Figure 4.42 *Visualization of the cross correlation in the spatial domain. The positions of the "O" are marked by high amplitude peaks. The third spot left of center is localized on the loop of the "P" which looks like an "O," giving a relatively high correlation.*

When we do the thresholding, we make sure that the two correlation peaks are represented by a single strong amplitude point. We thus approximate what we call a Dirac impulse or Dirac peak. The convolution of the latter by a function $f(x)$ reproduces this function at the position of the impulse.

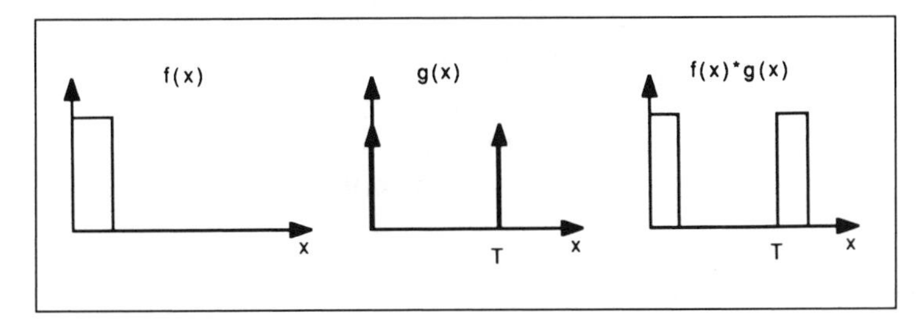

Figure 4.43 *On the left, the signal $f(x)$; in the center, the signal $g(x)$ containing two Dirac peaks (represented by the vertical arrows); on the right, the result of the convolution of $f(x)$ and $g(x)$.*

We carry out the convolution of our two Dirac pulses with the "O" (Figure 4.44).

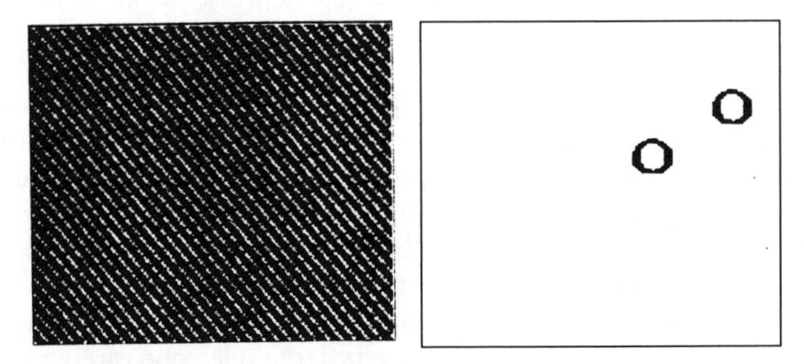

Figure 4.44 *The Fourier transform of the two Dirac functions gives a network of two fringe systems in the frequency plane (on the left). By multiplying this spectrum by that of the "O" and by going back to the spatial domain (on the right), we see that the two "O's" are centered on the position of the correlation peaks.*

All we have to do now is subtract the image of the two "O's" from the image of the complete text and it's finished (Figure 4.45).

The residuals found in the "O" in "Kullervo" are caused by the fact that "O" is not identical to the "O" in "symphony."

Now, let's take the "O" in symphony and return to the high resolution observation of double stars. Here we will see how the association of a very good photon detector (the CCD) and an image processing technique (correlation) modifies the way we perceive an image that at first glance contains no technical surprises.

Figure 4.45 *After subtracting the O's from the original text (but subtracting in the spatial domain is really simpler) we get the final result.*

We know the effects of atmospheric turbulence on focus: the images are spread and move erratically around an average position. In fact, the spreading itself is induced by a motion at a very small scale and very high frequency of the various parts of the diffraction spot. If the telescope is larger than 1 m, the Airy spot is usually completely destroyed. This will also be the case with smaller amateur telescopes as soon as the seeing degrades to a point where arcsecond features cannot be resolved, which is after all fairly usual.

We can distinguish between two types of turbulence. The first produces a very rapid trembling of the star (period less than 0.1 second). This type of turbulence is responsible for the spreading. The second is responsible for a large amplitude movement of the object around its average position with a low frequency. This movement can be in the order of one second or more.

The overall consequence of this turbulence is a blurring of the image when we take a long exposure because all these motions are integrated in time. Thus, in a long exposure even a very large telescope rarely reaches a resolution that is less than $0.8''$. If it were not for this turbulence the laws of optics would place the theoretical resolution at less than $0.1''$.

One way to reduce the effects of turbulence is to shorten the exposure time to "freeze" movement. This method is routinely used in professional observatories. For telescopes larger than one meter, a 5 to 50 millisecond exposure will completely freeze the low frequency motion. But unfortunately we still observe a spreading that is always larger than the size of the diffraction spot. However, a careful examination of the enlarged image shows that the star's image is covered with tiny points. These points are the expression of diffraction spots distributed throughout the interior of the stellar image. They are called *speckles.*

If it is a double star, each speckle splits in two if the separation of the components is higher than the dimension of the diffraction spot. Under certain conditions, this division can be seen by the naked eye, allowing a skillful observer to measure the separations. But the speckles' fugitive and constantly moving character requires the use of a more efficient recorder than the eye to work correctly. In addition, substantial calculation methods are needed to dissect the contents of an image filled with speckles. Astronomers know this technique well, it is called speckle interferometry, and was developed by the French astronomer Antoine Labeyrie.

It is nearly impossible to record the speckle of very faint stars because of the short exposure time. This problem is solved by using image intensifiers and by accumulating a large number of images that, taken alone, do not contain enough information, but when summed, clearly reveal the speckles. In this way objects down to magnitude 18 could be observed (the separation, then the determination, of the orbit of the Pluto-Charon pair is a good example of speckle interferometry).

If we do not have a large enough telescope or a detector allowing very short integration (the case of the camera described in this book), the speckles cannot appear, or when they exist, they are irreparably blurred (it is exceptional for a speckle structure to remain stable longer than 0.1 second). However, if the length of the exposure can be less than 1 second, it is still possible to freeze to a certain extent the large amplitude motions. When we observe a star, we generally notice a deformed image with a granular structure ("super speckles" in a way). A basic point is that there is a certain degree of coherence in a small angular field (of the order of $10''$ of arc). This coherence means that when we observe a double star, the components will be deformed identically. Therefore, by measuring the positions of the corresponding granules in the two spots, we can determine the separation and the angular extent of the system.

Here again the problem of sensitivity arises, since an integration of several tenths of seconds does not allow for the collection of many photons. To increase the signal-to-noise ratio, we acquire many images (5 to 100) that we register and "stack" to generate an image that is bright enough. To increase the efficiency of the system in terms of resolution, we only composite those images that are sharp enough. The good images can be selected manually by visual inspection, or automatically (you then have to establish a selection model to allow the computer to work correctly). This technique that consists in acquiring many "instant" images and then compositing them later with selection is called "CCD movies". The gain in resolution can be very significant with respect to a single image exposure for an equivalent length of time. Values of 2 to 4 have been reported depending on the strictness of the selection criteria.

With a 300 mm telescope and a CCD we can record 8th magnitude stars in $1/2$ sec., but we have to accumulate several dozen of these images to obtain a usable final image.

In practice, many difficulties arise. We will see in section 5.5.2 that the degradation of the signal-to-noise ratio becomes intolerable if we divide a long exposure up into too many shorter ones. It is furthermore tiresome to re-center dozens of images (long calculation time). Above all, we have to define a method to re-center each individual image. Such a method is not straightforward since stars are far from being figures of revolution.

These problems caused us to abandon the traditional spatial domain for CCD movies and move to the frequency domain. The key point is a fundamental property of the Fourier transform: the invariation of shift in the spatial plane into the frequency plane. Figure 4.46 illustrates this point. We numerically simulate a double star. In the frequency domain we obtain a system of fringes whose orientation is perpendicular to the line joining the pair in the spatial domain. The distance P between the fringes is given by

$$P = \alpha d^{-1},$$

where d is the separation of the components in pixels, and α a constant whose value is the dimension of the image, also in pixels (here $\alpha = 128$). When the double star moves in the observation field, the fringe system remains exactly the same.

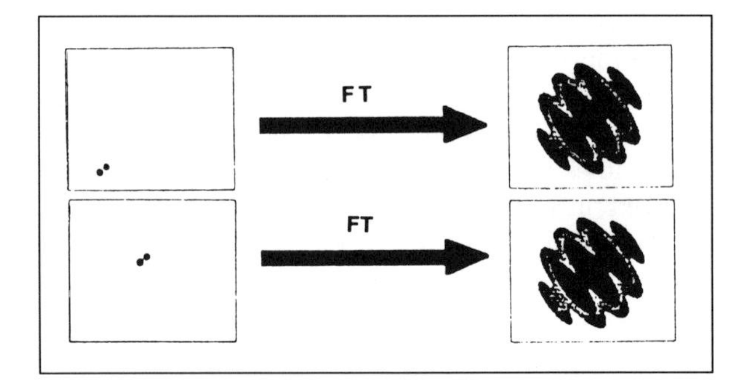

Figure 4.46 *On the left, two positions of a double star in the field of observation. On the right, for each of these positions, the Fourier transform (power spectrum).*

To improve the contrast of the fringes and thus facilitate the star's measurement, we simply sum the power spectra calculated from the "instant" images. The fringe systems are superimposed perfectly whatever the object's position on the CCD. This is a decisive advantage compared to processing in the spatial domain.

Image selection is still possible and even desirable. We will choose images that produce a maximum spectral density at the high frequencies, i.e., a high contrast of the fringes, a sign that the image contains significant details. We will apply all this knowledge to a well known double star: ϵ^1

Lyr. The observation was made with a 280 mm telescope and the image was enlarged by projection onto the CCD with an ocular. The resulting focal length was 15.4 meters, giving a scale of 0.306″ per pixel (TH7863 CCD). A long focal length is necessary to over-sample the star so that the components are separated by at least 3 or 4 pixels. The determination of the exact scale and the orientation of the camera are accomplished by measuring a wide and stable binary (β Cyg, for example). The orientation can also be measured by making a "trailed" image of a relatively bright star, which of course gives the East-West axis. In this way we free ourselves from a systematic error that can be caused by the imprecision of the elements of the reference double star. The trailing technique also gives us access to the scale. If L is the length of the segment in pixels and T the exposure time in sidereal seconds, we will get

$$\text{Scale} = \frac{15 \cdot T \cdot \cos\delta}{L}.$$

We should note however, that it is not always easy to estimate the exact length of the track left by the star because of the turbulence that randomly spreads the image.

Figure 4.47 *The double star ϵ^1 Lyr. On this individual exposure we see how the components are disturbed by the turbulence.*

Summing nine power spectra of the star ϵ^1 Lyr produces the fringe system in Figure 4.48. The energy contained in each of these images and the fringes are superimposed perfectly. The observation was made under fairly poor conditions, a strong wind and turbulence produced an average stellar profile width at half-maximum of 1.3″. Even though the individual images were exposed for only 0.2 of a second the stars were so deformed that it would have been impossible to composite them in the spatial domain.

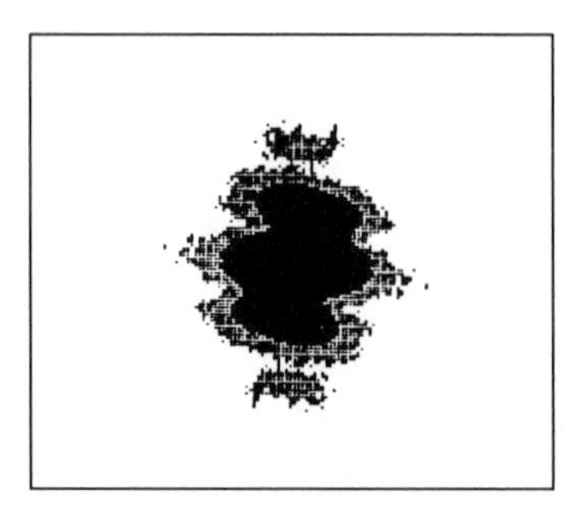

Figure 4.48 *Spectrum of the double system ϵ^1 Lyr.*

Measuring the separation of the fringes and their orientation enables us to determine the parameters of the double star. However, these measurements are not easy; we have to be able to determine precisely the place where the fringes are maximum (or minimum) and to make them rotate by an exact angle so that their axes are parallel with the lines of the matrix. All of these criteria are quite subjective. It is much easier to return to the spatial domain by calculating the *autocorrelation* function.

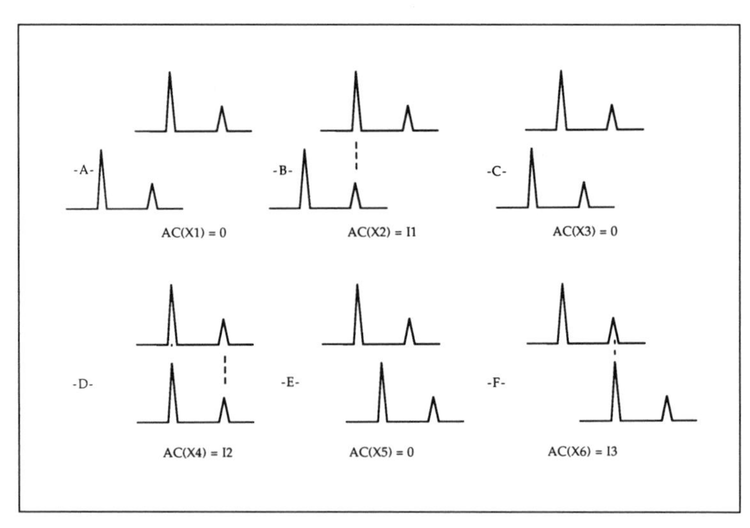

Figure 4.49 *The autocorrelation principle.*

The autocorrelation function (AC) is obtained by correlating the image with itself. Figure 4.49 shows what to expect when working with a typical double star. This figure shows a cross-section passing through the center of the two components. This cross-section was divided in two, then the two resulting profiles were shifted with respect to each other. Finally to determine the correlation we calculated the point-by-point product of these two profiles according to their shift (X).

There is no correlation in 4.49A (the product of the two profiles gives a zero result). Let us shift the lower cross-section with respect to the upper at position X_2 (Figure 4.49B). This time the principal component on the upper cross-section coincides with the secondary component of the lower cross-section. The autocorrelation function then takes the value I_1. When we shift as far as position X_3 (Figure 4.49C), the components no longer coincide, and the autocorrelation function is 0. In 4.49D the correlation is very strong (I_2) since the components are exactly superimposed. In 4.49E the autocorrelation function is 0 again; then it takes the value I_3 in 4.49F. Note that $I_1 = I_3$.

Finally the autocorrelation function will look like Figure 4.50.

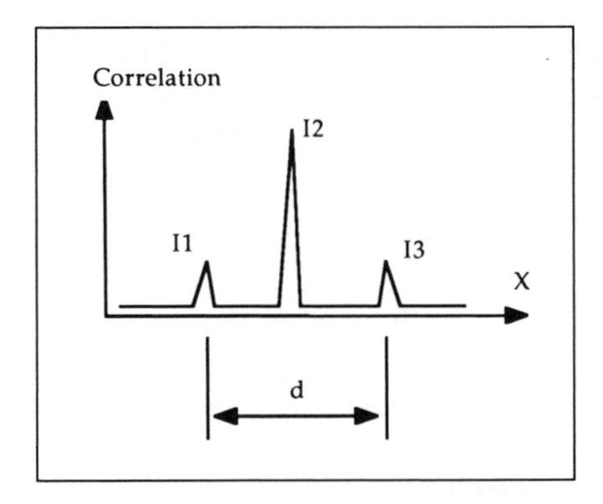

Figure 4.50 *The autocorrelation function of a double star.*

The distance d between the two secondary peaks of the correlation function is twice the separation of the components. The orientation of the line joining these two peaks gives the angle of the binary system.

We saw in this chapter that the cross correlation of two functions is obtained by multiplying one of the functions by the conjugate of the other in the frequency domain. In an autocorrelation, the function is multiplied by its own conjugate, giving the squared modulus (the phase term is 0). From then on the autocorrelation function is simply obtained by calculating the Fourier transform of the squared modulus (the power spectrum) of the fringe system. Figure 4.51 shows the result for ϵ^1 Lyr. Note the strong correlation point in the center of the "image" and the secondary component on each side. Here, before determining the Fourier transform of the power spectrum, we calculated its logarithm which allowed us to increase the contrast between the secondary correlation peaks by enhancing the fringes (high frequencies).

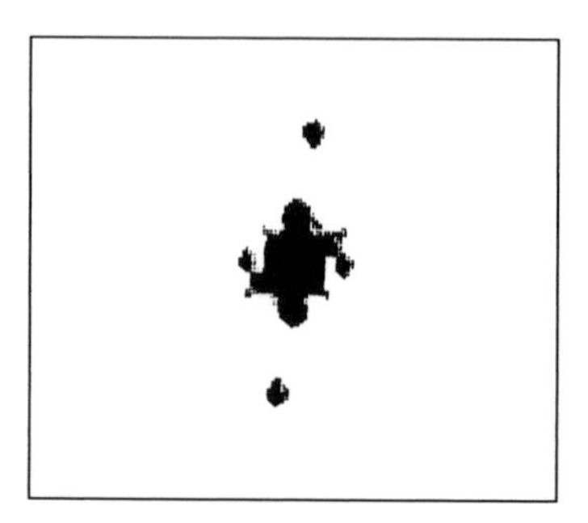

Figure 4.51 *The autocorrelation function of the star ϵ^1 Lyr.*

All we have to do now is measure the coordinates of the two secondary peaks, (X_1, Y_1) and (X_2, Y_2), using tools like the `CENTRO` function of the `IP` program. The separation of the double star will of course be

$$\rho = \frac{\sqrt{(X_1 - X_2)^2 + (Y_1 - Y_2)^2}}{2}.$$

And the position angle to within π will be

$$\theta = \arctan \frac{X_1 - X_2}{Y_1 - Y_2}.$$

We compared our result to a measurement given by Paul Couteau in his book *Observing Visual Double Stars*. For the observation date (1989.6) we obtained: $\rho = 2.39''$ and $\theta = 352°$.

Taking into account the orbital motion of the pair, we calculated the following positions for 1975, the date of Couteau's measurements: $\rho = 2.46''$ and $\theta = 356°$. Couteau gives $\rho = 2.54''$ and $\theta = 358°$. Agreement is to within $0.1''$ in separation. The position angle is always more difficult to determine and the agreement is within 2°. These measurements were made without a filter, which is not ideal given the large spectral range covered by the CCD and the strong chromatic aberration resulting from our optical combination. Filtering is advisable, but the wavelength value and the width of the optimum wave band are very hard to determine. Therefore, working in the blue part of the spectrum allows us to reduce the size of the diffraction spot, but turbulence is lower in the red. Working in a narrow wave band increases the duration of the image's coherence and thus its stability, but to gain flux we have to expose longer, blurring the image. Therefore, there is no clear-cut rule. We have had to experiment, and the result depends on the aperture of the telescope, the quality of the observing site, the detector's sensitivity, etc.

While the theoretical resolution of our instrument is 0.4″ we obtained a position measurement at 0.1″. This phenomenon is well known to astrometrists who routinely position stars with respect to each other with a precision much tighter than the size of the image spot. But beware—precision does not mean resolution. With our instrument we cannot separate stars closer than 0.4″.

We also measured the star ϵ^2 Lyr. Figure 4.52 shows the power spectrum and the autocorrelation function of this pair. Notice the change in orientation of the fringes.

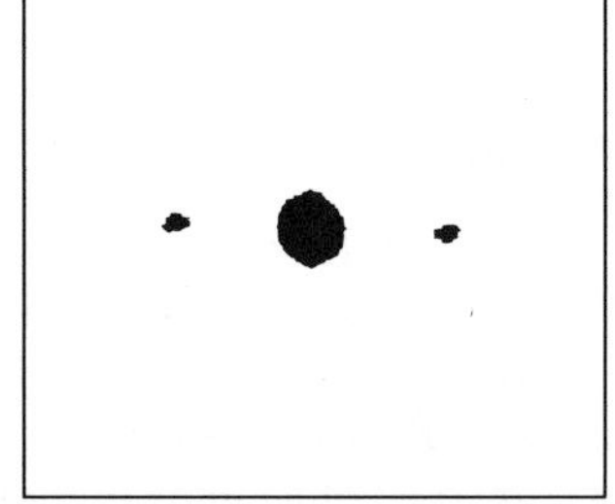

Figure 4.52 *On the left, the power spectrum of ϵ^2 Lyr (composite of 5 images exposed 0.2 seconds each). On the right, the autocorrelation function.*

We can again compare our measurements with those of Couteau (still taking the motion into account):

Buil	Couteau
$\rho = 2.27''$	$\rho = 2.20''$
$\theta = 90°$	$\theta = 92°$

Finally, we present a case that is much more difficult (STF 2438, Figure 4.53) since it is much fainter (magnitude 7) and closer (0.89″). Notice the wide separation of the fringes, a sign that the components are very close to each other.

Here are the results which are essentially the same, $\pm\pi$, and which do not take into account possible orbital motion:

Buil	Couteau
$\rho = 0.91''$	$\rho = 0.89''$
$\theta = 2°$	$\theta = 359°$

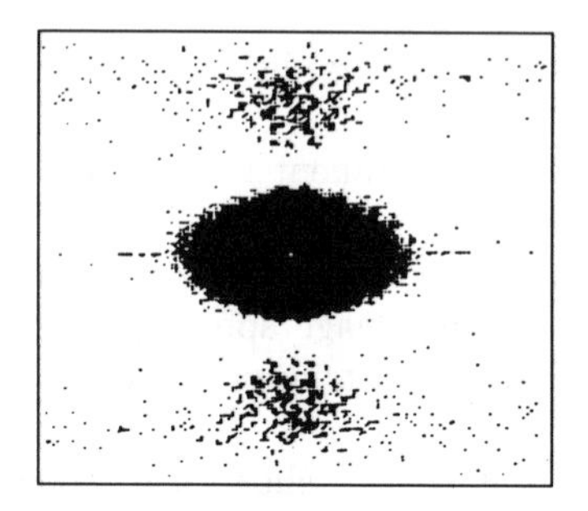

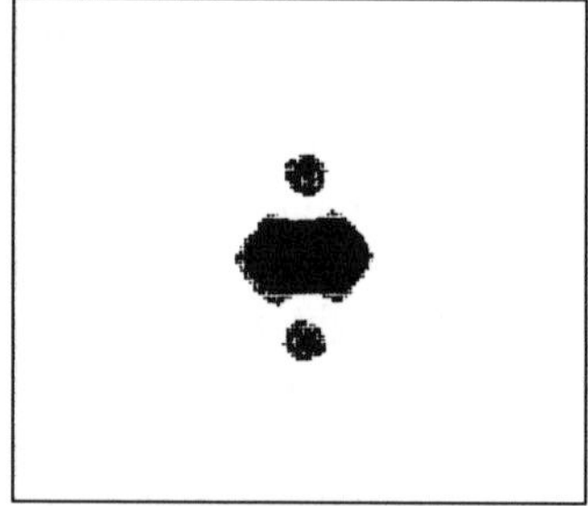

Figure 4.53 *System of fringes produced by the system STF2438 (on the left). We composited 20 spectra calculated from images integrated over 0.8 seconds. The components are perfectly separated in the autocorrelation function (on the right).*

Deconvolution

Deconvolution techniques are used to compensate for signal deterioration and to restore, as well as possible, its original appearance. We know that the sharpness of astronomical images is deteriorated by a variety of causes (atmospheric turbulence, optical aberration, etc.). This deterioration is shown by convolution of the initial image $g(x, y)$ by the point spread function $h(x, y)$ The latter representing the point of light from a star after it passes through the atmosphere and the telescope optics. Thus

$$f(x, y) = \text{ observed image } = g(x, y) * h(x, y).$$

After a Fourier transform we get

$$F(u, v) = G(u, v) \cdot H(u, v).$$

$H(u, v)$ is the transfer function of the acquisition system, also called the Modulation Transfer Function (MTF).

Once we know the point spread function—we will shortly describe how this is done—we can restore the image by applying the operation;

$$G(u, v) = \frac{F(u, v)}{H(u, v)}.$$

This technique is known as inverse filtering. In the frequency domain, the image $F(u, v)$ is multiplied by a filter whose transfer function is $1/H(u, v)$. Despite its apparent simplicity, inverse filtering raises several difficulties. In the frequency domain, $H(u, v)$ can have zero values. $F(u, v)$ and $H(u, v)$ can also have simultaneous zero values, a situation which leads to indeterminacy.

In practice, an additive noise $N(u, v)$ is present in the images. Taking it into account, the above equations are written

$$F(u, v) = G(u, v) \cdot H(u, v) + N(u, v)$$

and

$$G(u, v) = \frac{F(u, v)}{H(u, v)} - \frac{N(u, v)}{H(u, v)}.$$

When $H(u, v)$ is small (at high frequencies), the ratio $N(u, v)/H(u, v)$ can become non-negligible compared to the first term. Noise is then amplified, often dramatically.

To remedy these problems we have to intervene. Processing can be stopped at a certain frequency to avoid values of $H(u, v)$ that are too low, or a constant can be added to $H(u, v)$. This means the initial image can only be approximately restored.

So that the intensity of the restored image is not modified significantly with respect to the final image, $H(u, v)$ is normalized to one.

If σ is a frequency beyond which filtering poses too many problems, the transfer function will be

$$H'(u, v) = \frac{1}{H(u, v)} \text{ for } u^2 + v^2 \leq \sigma^2$$

$$H'(u, v) = 1 \text{ for } u^2 + v^2 > \sigma^2.$$

If we suppose the noise to be negligible, we will apply

$$G(u, v) = F(u, v) \cdot H'(u, v).$$

The preceding approximations are completely arbitrary, and more elaborate techniques should be used to approach the initial image. One such technique is called restoration by least squares or Wiener filtering; it consists in writing the restored image in the form,

$$G(u, v) = \frac{F(u, v) \cdot H^*(u, v)}{H(u, v) \cdot H^*(u, v) + \eta},$$

with $H^*(u, v)$ the complex conjugate of $H(u, v)$ and η a parameter equal to the ratio of the spectrum of noise power to the spectrum of the signal power. Noise always being difficult to quantify in practice, the parameter η will be chosen by trial and error to reach the most satisfactory result (typical starting value: 0.0001). If the noise is nonexistent, then $\eta = 0$, and the Wiener filter becomes identical to the inverse filter. The Wiener filter can be programmed in the following way (polar coordinates):

```
nb% =ft% -1
FOR i% =1 TO nb%
   FOR j% =1 TO nb%
      x(i% ,j% )=x(i% ,j% )*x1(i% ,j% )/(x1(i% ,j% )*x1(i% ,j% )+dzeta)
      y(i% ,j% )=y(i% ,j% )-y1(i% ,j% )
   NEXT
NEXT
```

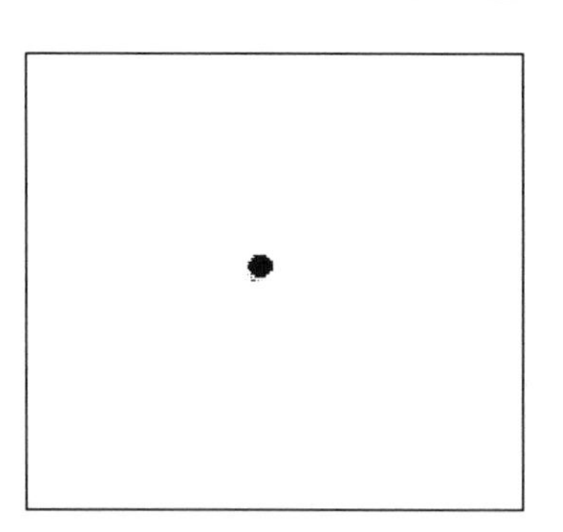

Figure 4.54 *At top left, original image. Top right, fuzziness produced by the passage of a Gaussian filter. Bottom left, the impulsive response. Bottom right, enhancing high frequencies by a high-pass filtering in the spatial domain.*

Figure 4.54 presents an example of artificial deterioration of an image by a Gaussian low-pass filter. The impulse response $h(x, y)$ of the Gaussian filter is obtained as follows:

1. A white point of amplitude 4095 is created in the center of a completely black image (image produced by the `ZAP` command of the `IP` program);
2. This image is convolved by a Gaussian identical to that used to deteriorate the image to be processed (`GAUSS` command).

The first thing to try to improve a deteriorated image is increasing the contrast. For this process, energetic high-pass type filtering is carried out in the spatial domain. The result (Figure 4.54) is fairly interesting, since the large title becomes easily legible.

Figure 4.55 shows the stages of restoration by inverse filtering. The division of the Fourier transform of the image by that of the impulse response is only carried out as far as the radial frequency 0.5 (conventionally, the half-side of the frequency plane extends to a frequency of 1.0).

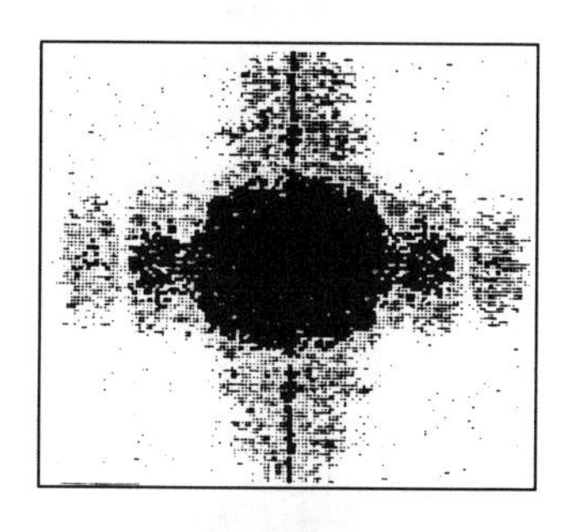
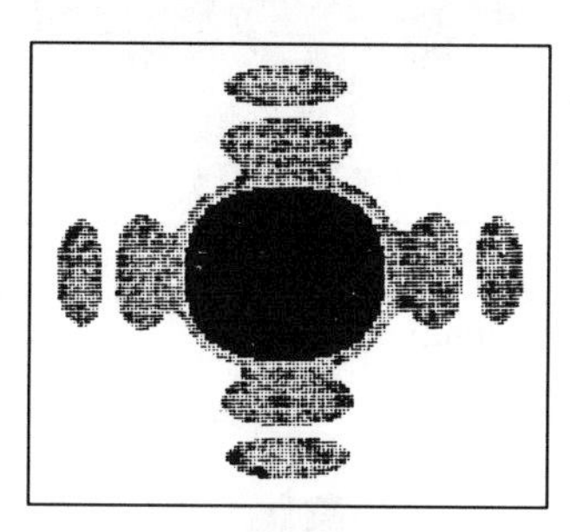
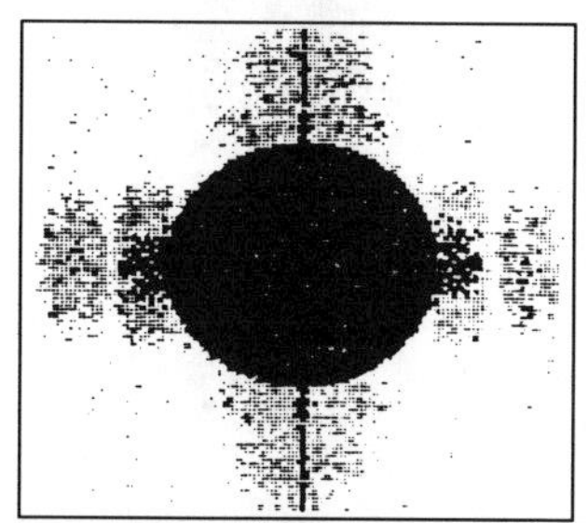

Figure 4.55 *From top left to right, spectrum of the image to be processed, spectrum of the impulsive response (this one has no symmetry of revolution because the Gaussian filter used has a slight directional effect). Bottom left, result of the division. To limit noise, the spectrum of the image is not modified beyond frequency 0.5. Bottom right, the result.*

The image is partially deconvolved. We will note in passing that the essential information is contained in the first half of the spectrum starting from 0 frequency. The result of a Wiener filtering is presented in Figure 4.56.

Figure 4.57 shows the result of deconvolving our usual image of Arp271. The point spread function was obtained by isolating a field star. This method is quite rigorous. Obviously it cannot be used to treat a planetary image. In that case, the simplest thing to do is to synthesize an artificial star, as in the preceding example. Its spreading will be adjusted by trial and error, by judging the result of the deconvolution.

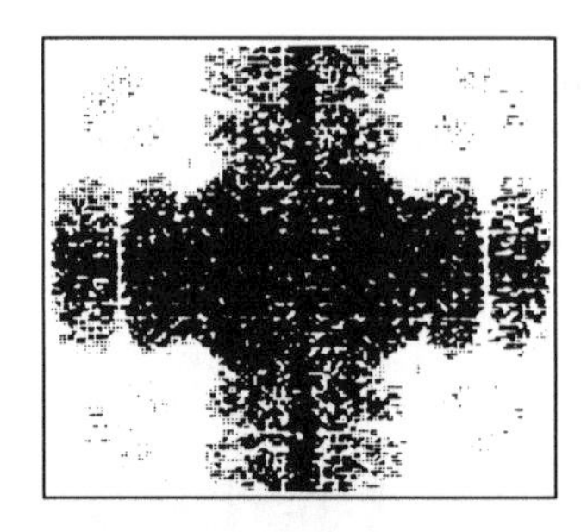

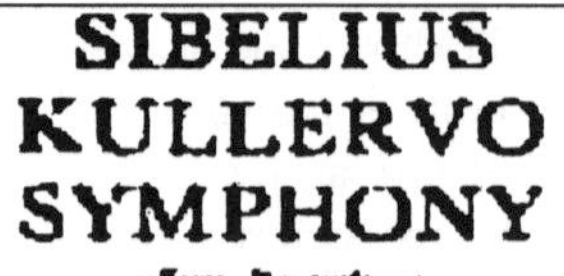

Figure 4.56 *At left, the spectrum after Wiener filtering. At right, the result in the spatial domain.*

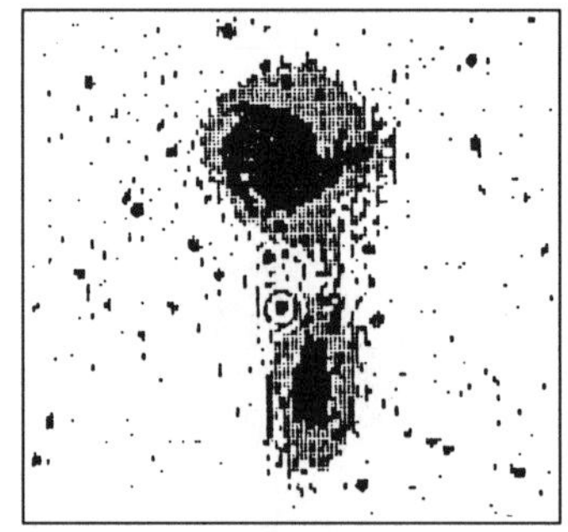

Figure 4.57 *At left, the original image. At right, deconvolution by a point spread function obtained from an isolated field star.*

In its most simple form, the Wiener filter perceptibly increases the noise and produces a fringe effect close to high contrast details (this effect is obvious in Figures 4.56 and 4.57). However, a class of image restoration algorithms appreciably limits these two problems. Essentially this is accomplished with an iterative process:

$$q(x,y)_{k+1} = q(x,y)_k + [f(x,y) - q(x,y)_k * h(x,y)]$$

where

$k = 1, 2, \cdots, n$ is the iteration number

$q(x,y)_k$ is the estimate of the image at the kth iteration

$f(x,y)$ is the observed image

$h(x,y)$ is the Point Spread Function (PSF)

$*$ is the convolution symbol.

When k increases, the $[f(x,y) - q(x,y)_k * h(x,y)]$ terms tend to 0, and $q(x,y)$ converges to the ideal image.

An important point is that the convolution can be done in the spatial domain because only five to ten iterations are necessary. Since a Fourier transform is not necessary, the programming will be simpler and larger images can be processed with a smaller machine.

This method uses the following steps:

Step 1: Subtract an offset to $f(x,y)$ to get a zero mean sky background

Step 2: Initialize the iteration by making $q(x,y)_k = f(x,y)$

Step 3: Make the convolution of $q(x,y)_k$ by $h(x,y)$. It is generally sufficient to approximate the PSF by a Gaussian function:

$$h(x,y) = \exp\left(-\frac{x^2+y^2}{2\sigma^2}\right).$$

We use the cross convolution method (section 4.3.3). The value of σ is estimated by comparing the analytical profile with the star's real profile (typically, $\sigma = 1$ to 1.5 pixel with a 2-meter focal length telescope and a 20-micron pixel.) To correct asymmetric blur a different σ may be used for each axis.

Another, more rigorous approach uses the sum of normalized profiles from several stars found in the field to build $h(x,y)$.

Step 4: Calculate $q(x,y)_{k+1}$. Suppress the negative values which have no physical meaning (positive constraint). Return to step 3 until $k = n$.

Another form of the interactive equation is

$$q(x,y)_{k+1} = q(x,y)_k + w(x,y)_k \left[f(x,y) - q(x,y)_k * h(x,y)\right],$$

where w_k is a relaxation function which depends on the intensity of q_k.

The w_k function rejects noise and suppresses negative values. Then, $w = 0$ for negative values and w is small for small intensities (e.g. the background). A typical form of the w function is

$$w(x,y)_k = 1 - 2 \cdot \left|\frac{q(x,y)_k}{q_0} - 0.5\right|$$

where q_0 is a normalization parameter which allows us to adjust the w_k function. One can take for example $q_0 = \mathrm{MAX}[q(x,y)]$.

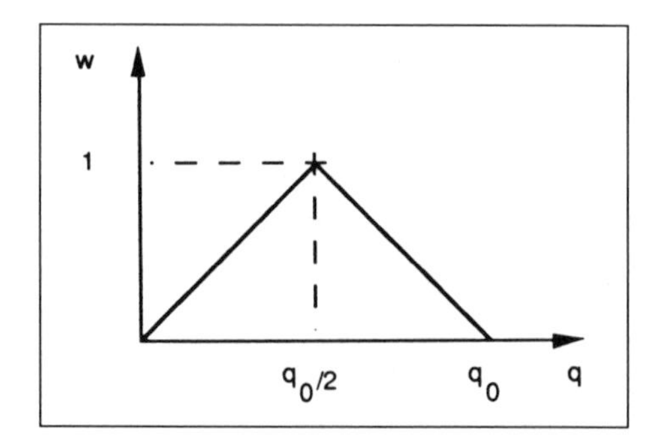

Figure 4.58 *The relaxation function w as a function of q.*

Other w functions can be used. They give different results according to the histogram of the image and to the kind of details that one wants to restore

$$w(x,y)_k = 1 - \left[0.5 + 0.5\cos\frac{2\pi q(x,y)_k}{q_0}\right]^4$$

$$w(x,y)_k = \left[0.5 - 0.5\cos\frac{2\pi q(x,y)_k}{q_0}\right]^4$$

and $w = 0 => w = 0$.

When compared to most other image restoration methods (e.g., the maximum entropy method), the iterating restoration algorithm is very simple to program and does not require large computers. One can say that the performance/complexity ratio of this method is exceptionally beneficial.[2]

Figures 4.59 and 4.60 show an example of the iterating algorithm applied to an image of the edge-on galaxy NGC 4517. The PSF is determined with the mean stellar profile of several stars (Figure 4.59). Note that the PSF is slightly different between the two axes to correct for poor guiding during the image acquisition. With this technique the convolution can be accomplished with either the real profile, or with a synthesized Gaussian profile. The relation between the FWHM (Full Width at Half Maximum) and the σ for a Gaussian profile is

$$\text{FWHM} = 2 \cdot \sigma \cdot \sqrt{2 \cdot ln(2)}$$

where $ln(x)$ means natural logarithm of x.

[2]For a good review of interactive restoration methods see E.S. Meinerl in *Instrumentation in Astronomy* Vol. 627, p. 715 (1986), P.A. Jansson *et al.* in *Journal of the Optical Society of America* Vol. 60, **5**, p. 596 (1970), and H. Maitre in *Computer Graphics and Image Processing*, Vol. 16, p. 96.

Figure 4.60 shows the galaxy before and after processing. Note in the processed image the increased details in the "dust lanes" of the galaxy. This was accomplished without an increase in the noise. The fringing effect around the saturated star at the edge of the galaxy is negligible (in spite of its extreme contrast). Another consequence of the restoration is the increased sharpness of the stars, which in turn increases the detectivity of the image. In fact, the blurred images in the original picture are now concentrated in a small area in the final image.

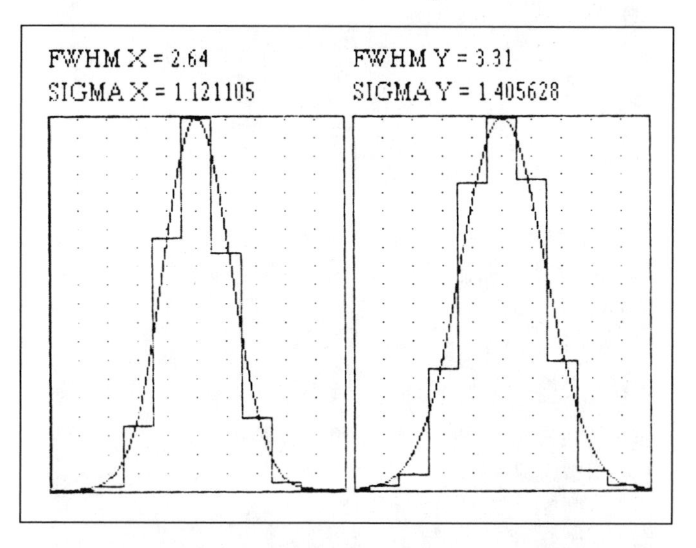

Figure 4.59 *Mean profile along the two axes of four stars in the field of the image. Each step is a pixel. The curve is the Gaussian that fits the PSF.*

Figure 4.60 *The top picture is a blurred image of the galaxy NGC 4517 (turbulence + telescope). This is a print of a 10-minute exposure at the 60-cm telescope of the Pic du Midi Observatory. The CCD is a TH7863. The bottom image is the product of 8 iterations using a triangular relaxation function and a real stellar PSF for convolution.*

Chapter 5

Operating a CCD Camera

Anyone can buy a photographic camera and attach it to a telescope. But this is not to say that anyone can take exceptional photographs. Even assuming that one is willing to master the technical aspects of astrophotography there will always be that individual who gets better results than his colleagues. Small jealously kept secrets, natural gifts, hard work—these make the difference. Despite all the technology surrounding the CCD, the electronic image responds like silver on film. Give the same camera, telescope and object to two different astronomers, and they will produce two different images. Beyond the technique, the CCD brings out the qualities of an observer. In short, the human element remains important.

Today, the CCD is the Formula 1 car of amateur astronomy. Here, we will try to provide a road map of the traps and blunders that await the unwary. First, we will explore the process of photometric image reduction. This part is essential for getting useful scientific information from a CCD camera. Next we will discuss taking images of stars and planets.

5.1 Radiometric Correction

Optimum operation of a CCD requires systematic radiometric correction of the image, a process which entails a great deal of work. In this regard there are two types of bias which especially interest us here:

1. Additive biases are parasitic signals that merge with the useful signal. The most visible is the dark current. Even if charges of thermal origin have been nearly eliminated by deeply cooling the CCD, a continuous level of intrinsic origin (electroluminescence of the amplifier, interference fringes due to night sky lines, etc.) is superimposed on the image.

2. Multiplicative biases result from variations in sensitivity from one pixel to another. (It corresponds to a variation of the detector's gain along its surface.) The quantity of signal detected can also depend on the regularity of the thinning of the substrate, on the optical vignetting, etc.

These biases introduce spatial noise in the image (that should not be confused with the temporal noise, which is a random variation of a signal in time). The spatial noise fundamentally limits the CCD's detectivity. Thus inter-pixel variations of the dark current produce a high-frequency spatial noise (shot noise) and a fine, faint star could be easily "lost in the noise". Also when observing an extended object, the actual structure of the object can be lost in the slow signal variations caused by vignetting (low frequency noise).

To eliminate the effects of these two types of bias, it is necessary to subtract an approximation of the additive bias from the rough image and to divide the result by an image which compensates for the multiplicative bias.

5.1.1 Offset Correction

An amateur CCD will often be operated warmer than it should be for optimum results. The result is a thermal signal which is added to the signal produced by the observed scene. For example, a TH7852 CCD cooled to $-40°$C still gives, in total darkness, a continuous background of 5 mV during a 300 second exposure. This is far from negligible, since the interpixel variation of the dark signal can reach 10% peak-to-peak of the parasitic signal, or 0.5 mV in our example. Because the readout noise is usually 5 times weaker, we see that the dominant spatial noise comes from dark current.

A simple and efficient method of suppressing the effect of the dark signal's spatial variation is to obtain an image in complete darkness with the same operating temperature and the same exposure time as the image to be corrected. From the latter we then subtract the map thus obtained. If (x, y) represent the coordinates of a pixel and if $R(x, y)$ is the rough image, $N(x, y)$ the dark map (often called "dark") and $I(x, y)$ the treated image, we will get

$$I(x, y) = R(x, y) - N(x, y).$$

The dark map is made by covering the telescope's opening with an *opaque* mask. Be especially careful in designing this opaque mask—low level light can originate through the mesh of "black" cloth or the material can be transparent to infrared light. Any light "leaks" must be hunted down.

Light may reach the CCD from unlikely places, for example, from the area between the camera and the instrument tube. Test the effectiveness of your light blocking under the most severe condition—in broad daylight—and compare that result to the result you obtain at night.

The correction procedure described above is only valid if the conditions under which the "dark map" is made duplicate those prevailing during actual image acquisition. Although the length of the exposure is assured by the computer's internal clock, special precautions should be taken to keep the CCD's operating temperature constant. Three cases may arise:

1. The CCD's temperature is very low, below $-80°$C. In this case there is no real reason to worry; the thermal charges are so few that their effect can be considered as negligible. From time to time, short exposures should be taken in the dark to determine the electronic offset (see below);

2. The CCD's temperature is known to within a tenth of a degree. This allows us to subsequently recreate the conditions under which the rough images were acquired to make post-recorded dark maps.

3. The CCD's temperature is not known exactly.

The last case is the most common one for amateurs. There are three possible ways of correcting it:

1. Bracket each exposure with two dark frames;

2. Take regular dark exposures throughout the night;

3. Model the dark frame to create the lowest noise signal.

First Method

It consists in following a deep sky image with a dark exposure, hoping that the temperature does not change too much between the two. Ideally, the exposure of the object should be framed by two "darks," a natural consequence when many objects are observed one after the other. Basically, one spends as much time taking dark maps as deep sky images. Imagine observing an object with an integration time of 10 minutes. If the temperature is not precisely regulated or is unknown, a dark exposure should be made after this image. That means 10 minutes of inactivity. If we take another dark exposure to limit the noise by averaging the "darks" a simple 10 minute exposure is stretched out to 30 minutes! This is an intolerable situation.

Second Method

This method is applicable only if the CCD's temperature does not fluctuate more than 5 or 6° during the night:

1. A dark map is made by summing many darks acquired during preceding observations. Let $N_{\mathrm{t}}(x, y)$ be this map, M_{t} its average amplitude, and n the number of darks summed;
2. From time to time while observing, we take dark exposures (3 or 4 spaced regularly throughout the night). Let M_{o} be the average amplitude of one of these maps made as closely as possible to the image $R(x, y)$ to be corrected. Finally, let T_{n} and T_{r} be the respective exposure times used to acquire the map and the image to be corrected;
3. Therefore, the following treatment is applied:

$$I(x, y) = \big(R(x, y) - O(x, y)\big) - \big(N_{\mathrm{t}}(x, y) - n \cdot O(x, y)\big) \frac{M_{\mathrm{o}}}{M_{\mathrm{t}}} \frac{T_{\mathrm{r}}}{T_{\mathrm{n}}}.$$

The ratio $M_{\mathrm{o}}/M_{\mathrm{t}}$ allows us to adjust the level of the cumulative dark map to that of the acquired map closest to the image to be corrected. The ratio $T_{\mathrm{r}}/T_{\mathrm{n}}$ brings the dark map to the integration time of the image to be corrected.

The image $O(x, y)$ gathers the offsets of non-thermal origin. It is made by using the shortest possible exposure at low temperature (e.g., about a hundred milliseconds) to render the thermal signal completely negligible. The image $O(x, y)$ does not have nil intensity because we register the slight difference that exists between the level of the signal and the reference level in the video signal. This offset, intrinsic to the CCD, is of the order of 15 to 20 mV. A level of 0V in darkness is the sign of a malfunction. Generally, it means that the offset voltage of the DC component of the video signal is incorrectly set. (See Section 2.3.8 and Figure 4.1.) With double sampling, external offsets are totally absent from the image (immediate elimination of all noise originating in the drift of characteristics of the electronic components).

Third Method

As before, we use an image $N_{\mathrm{t}}(x, y)$ synthesized from many darks. As we sum a higher number of darks to produce $N_{\mathrm{t}}(x, y)$, the thermal noise in this total map decreases. The appearance of the image $N_{\mathrm{t}}(x, y)$ changes very little during the lifetime of a CCD and can thus be considered as an instrumental constant. The corrected image is:

$$I(x, y) = \big(R(x, y) - O(x, y)\big) - \big(N_{\mathrm{t}}(x, y) - n \cdot O(x, y)\big) \cdot \rho.$$

The parameter ρ is adjusted so that the variance of the intensity of the pixels on a small area of the image will be minimal. Typically this is a star free region about 30 pixels square. The amount of adjustment can be determined graphically by plotting the curve of the variance according to ρ. The curve should pass through a minimum at which point the value of the parameter will be taken. This is a powerful technique for correcting images because the intensity of the dark is calculated according to the best possible criterion: minimizing noise in the treated image.

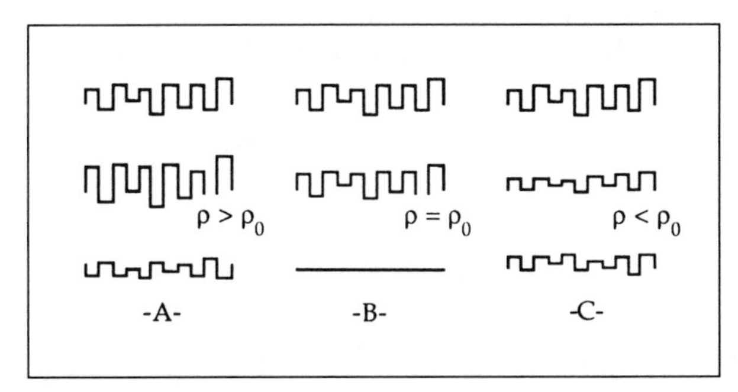

Figure 5.1 *Illustration of the procedure that minimizes thermal noise by optimizing the level of the dark map. In A at the top, we have a break in the image to be processed. The difference in level from one pixel to another is caused by non-uniform generation of the dark signal. Still in A, but in the middle, a break in the same place in a badly adjusted dark map (multiplicative coefficient ρ too strong compared to the optimal value). At the bottom, the result of the subtraction of the rough image and the dark map. The residual noise is important. In B, we have the same kind of break in the image and in the dark map, but this time the coefficient ρ has the right value. The noise is thus almost null in the processed image. In C, the coefficient is again poorly adjusted (its value is too low) and we get noise in the final image. We therefore passed through an optimum that is $\rho = \rho_0$. The coefficient should have a typical value less than 0.1. This means that the dark map has a level ten times higher than the image to be processed. The map's temporal noise is then virtually absent (division by ten), and only the spatial noise remains, allowing the correction.*

The third correction method sometimes fails when there are so many stars in the image that the sky background is no longer visible. This situation occurs in some areas of the Milky Way. The algorithm can no longer converge or converges very badly. We then use the second method.

Averaging several dark maps is essential to limiting the contribution of thermal noise. A median dark map can be created by stacking a number of individual maps. The median map has the advantage of being less noisy than the average one since many percussional parasitics disappear in the operation (cosmic rays, signal points, 60 Hz interference, etc.). The difficulty here is the requirement that all the individual maps be memory resident to determine the median value of each pixel. Substituting disk memory for RAM memory is not a viable solution because disk access times make this procedure prohibitively long. Using a standard PC (640 KB) and a compact program, it is possible to make a median map of 145×218 points from 7 memory resident images.

In fact, professional astronomers, although they rarely have problems with the dark current, regularly take offset maps, called biases, because whatever the offset's amplitude, it has to be subtracted from the rough image. A good way to judge the importance of correcting for thermal effects is to subtract one's dark maps from others and to compare these darks with the result of the subtraction. This is what we did in the example in Figure 5.2.

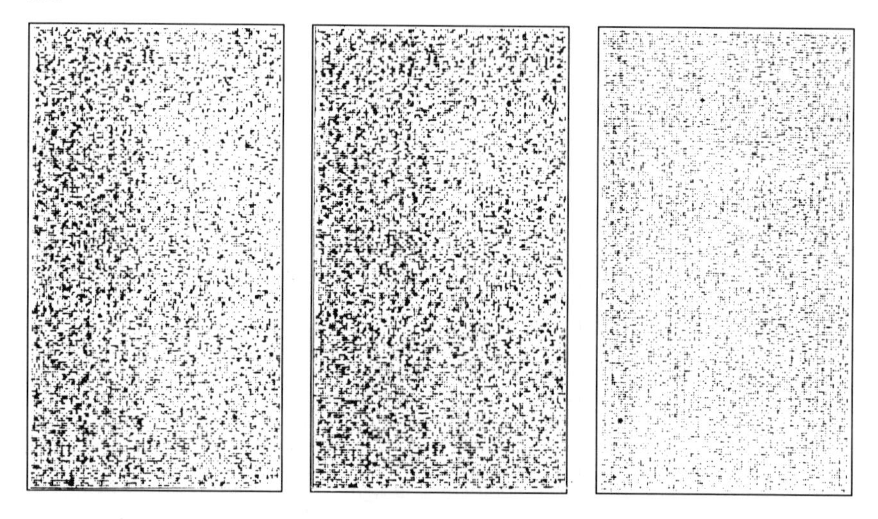

Figure 5.2 *From left to right, two dark maps made within a short interval, then the result of the subtraction of these two maps. After subtraction, the noise has noticeably decreased. The visualization thresholds are equal for these three images (a constant equal to the average value of the maps is added to the result of the subtraction).*

5.1.2 Correcting Differences in Sensitivity

The flat-field is the bane of any CCD camera user. But let's not get ahead of ourselves.

The CCD camera user routinely observes objects whose brightness is equal to or less than 1% of the sky background. There is little room for error when one works with a number this small. Attention to small details is critical. For example, the relative sensitivity of the individual photosites that form the mosaic is not constant, as uniformly illuminating the detector shows. The resulting image presents spatial noise because adjacent pixels did not react in the same way to identical incident flux. Local variations in quantum output are due to problems in manufacturing the CCD. Silicon's physical characteristics change according to the part of the chip we consider. In the case of a thinned CCD, the quantum output is strongly linked to the silicon's thickness, and this parameter is difficult to control during the manufacturing process. Variations in sensitivity can reach 1 to 2% over the array. This variation is far from negligible when observing extremely faint objects or in photometry.

To all this we have to add other causes which are external to the non-uniformity of the sensitivity, such as optical vignetting or dust on the camera's window and its filters. Normally when a focal reducer is used the illumination of the focal plane will vary between the center and edge by 10%.

Correction of non-uniform sensitivity is obtained by dividing the acquired image (after subtracting the dark map) by another image made on a bright uniform sky background: a flat-field. If $I(x, y)$ is the image to be corrected for non-uniformity and if $F(x, y)$ is the flat-field, then the processed image will be $E(x, y)$:

$$E(x, y) = K \frac{I(x, y)}{F(x, y)}.$$

In this equation the multiplicative coefficient K allows us to find, approximately, the initial level of the processed image. The value of K is equal to the average intensity of the flat-field.

It is absolutely essential that the uniform image be acquired under conditions identical to those of the image to be processed, i.e., the same CCD, the same telescope, the same filter, etc. The simplest and safest method of obtaining a flat-field is to point the telescope at the sky background at dusk. The sky is then bright enough to give a dense image even with an integration time shorter than 30 seconds. Under these conditions the thermal noise will be negligible, and the sky background signal such that it drowns out the stars. The ideal time period within which a flat-field can be made is very short (usually no longer than half an hour). If there is too much daylight, the CCD becomes saturated; if it is too dark, the exposures are long and stars appear. Acquiring flat-fields within this window of opportunity is always done in a feverish atmosphere. As an added precaution, it is wise to make flat-fields at dawn *and* at dusk.

There are quite a few pitfalls to be avoided while making flat-fields. Here is a list of some common problems:

1. The camera should not be removed from the telescope between the exposures for the objects and the flat-fields. You must accept the

reality that there is no mechanical way to reposition the camera in *exactly* the same position once it is removed. The slightest movement of the detector invalidates the flat-field correction because various obstructions modify the light distribution on the sensitive surface. This is even more true when the optical system vignettes;

2. The corollary of the preceding point is that mechanical flexure must be ruthlessly eliminated. On a large telescope the problem is so severe that flat-fields have to be made with the telescope pointing in a direction identical to the one in which it was pointing when the image to be corrected was made. (This requires as many flat-fields as there are images to process);

3. When one is working in a dome, it is essential to ensure that the opening through which the telescope aims does not block part of the mirror. Since the obstruction takes place far from the instrument's entry pupil (the main mirror), there will be a variation in illumination in the focal plane that is very difficult to correct;

4. Pay attention to the uniformity of the sky background. Do not aim toward cumulus-type clouds. At dusk, even when the sky is clear, it presents a brightness gradient centered on the place where the sun just set or will rise. The gradient has almost no effect with a focal length of about one meter or more given the smallness of the fields typical with these instruments but the problem becomes acute with short focal lengths (telephoto lenses). To achieve consistency when making flat-fields it is better to aim the instrument toward the zenith, thereby avoiding variations in luminosity near the sun and the effect of the Earth's shadow in the direction opposite to the sun. When it is not possible to do otherwise, a sky uniformly covered with cirrus or alto-stratus can act as a substitute flat-field;

5. At dusk, the telescope receives indirect light much stronger than at night, even when working under a dome. Light rays coming from all directions, but mostly from the place where the Sun rose or set, hit the interior of the tube and cause a deceptive diffusion. No instrument, whether it is a Newtonian or Cassegrain type, can be shielded completely from this phenomenon. Well-planned baffling can help and the tube of Schmidt-Cassegrains can be lengthened but make sure this addition does not move and modify the vignetting during the night. Of course, the assembly connecting the camera to the telescope must be completely lightproof;

6. When using colored filters, flat-fields must be made for each one. First, the relative inter-pixel sensitivity is modified according to the spectral domain. A blue flat-field will not look like a red one. Using

filters can present problems for the unwary. For example the filter's effective wavelength depends on the spectral distribution of the observed source. But the sky at dusk and at night are not the same color, creating a bias, which is fortunately very weak in most cases. Second, a filter always produces reflections (multiple reflections on its surfaces, reflections between the surface of the filter and the window of the camera, etc.), and putting it in place while making a flat-field does not always get rid of the parasitic light that was induced at the time when the images were taken. In fact, the nature of the observed objects is different: on one hand the point-source stars, on the other, the uniform background. This situation poses great difficulties because of the different angular distribution of rays. If we use several filters, the repeatability of positioning and orienting must be as close as possible;

7. Do not change focus for flat-fields. This requirement may seem strange because we observe a continuous background without details for making flat-fields. The explanation is that focusing changes the geometry of the CCD camera-telescope assembly. The result is a modification in the vignetting. Furthermore, experience shows that we inevitably accumulate dust along the optical path. This dust creates shadows on the sensitive surface whose shape is linked to the angle of incidence of the rays and the distance of the dust from the sensitive surface. All these parameters will change with focusing changes. With fairly long focus telescopes ($f/10$), dust located a few millimeters from the sensitive surface will cast shadows with very clear edges onto the CCD. If the dust is further away from the sensitive surface, the shadow has a characteristic ring shape—a negative of the telescope's pupil with the central obstruction;

8. The sky background is slightly polarized and the optical configuration of some telescopes can act as analyzers, causing a change in the light distribution on the CCD's surface. Although the overall effect of this polarization is generally small, flat-fields should be made with the telescope pointed away from the sun, where the polarization rate is low;

9. If it is already dark, a few stars will appear on the flat-field. We will see later that this is not a major problem if we make several flat-fields, each of which were made after changing the pointing of the telescope to change the stellar field. This will help us to better eliminate the stars during reduction. The telescope should track during these exposures to avoid star trails which are always more difficult to process out of the image;

10. As with dark maps, it is important to make several flat-fields which will be summed later. Here too we have to be sure not to introduce too much noise into the image to be treated. Three or four flat-fields through each filter is a minimum. Since all of these flat-fields have to be done within half an hour, we can better understand the feverish atmosphere in which this is accomplished. To decrease the noise, the flat-fields' amplitude must be consistent to get a good signal-to-noise ratio. With a camera functioning on 12 bits, the exposure time will typically be adjusted so that the average amplitude of an individual flat-field will be around a level of 2000. The CCD should not be too strongly illuminated to avoid possible problems of linearity;
11. Generally, exposure times for flat-fields are relatively short—usually less than a minute. The dark signal will therefore be negligible. Despite this, it is necessary to subtract a dark map to determine the zero of the flat-field and thereby make a correct division;
12. A golden rule: make flat-fields whenever possible. Never delude yourself into believing that the flat-fields made the week before can be used to correct today's observations. The reality is that a flat-field image evolves with time. Sometimes after a long and tiring night of observing, common sense tells us to go to bed as soon as the first rays of the sun appear, but those flat-fields must be made.

The ideal flat-field does not exist—as the above list indicates. But in following these principles, we can at least try to get close.

Sometimes there are no flat-fields available to process an observation. Take, for instance, the following hypothetical case. The night is horrible; the sky is full of clouds. Then suddenly at midnight, the sky clears. Action! The telescope is uncovered, the computer is booted, the camera is turned on, the CCD is already at $-40°$C. After taking about ten images, it is 5 a.m., and the sky is getting cloudy. Worse, it begins to snow! It is impossible to make the flat-fields. The situation is grim, but do not despair. There are two remedies:

1. The flat-fields are made subsequently by pointing the telescope toward a white diffusing screen located two or three meters in front of the tube. The screen is lit by carefully adjusted lamps that distribute the light uniformly on the screen. If necessary the interior of a white dome can act as a screen. The lamps are equipped with filters to get a color close to that of the night sky background. In professional observatories, flat-fields are sometimes made with the telescope pointing inside the dome. But as professionals will tell you, this is resorted to when all else fails. Indeed, the screen only roughly reproduces

the conditions of a night sky background (geometric extent, diffusion indicator, color, shadow effects, etc.);

2. The flat-fields are directly extracted from the observed images to be corrected. As the sky is not perfectly black even in the middle of the night there is a continuous background in the image. The idea is to remove all the objects from the image and keep only the background, which will act as the flat-field. If there are two or three stars in the field, they can easily be eliminated by sky background interpolation techniques (`FILL` command on the `IT` program). If there are many stars and diffuse objects, this type of cleaning will not work. The solution then is to calculate a median image from images acquired during the night. Since there is little probability that the objects will be present in the same places on the observational images, the median picture will only contain the sky background. If the images were obtained with different exposure times, they have to be multiplied by a constant to bring them to an average level of the common sky background before the median is calculated. The flat-field thus obtained is of high photometric quality because it is made at exactly the same time and in the same way as the images to be corrected. We cannot be more precise. The only problem is that the level of the flat-field is very low and when it is divided by the image to be processed, much noise might be added. This method is frequently used with thinned CCD's to remove interference fringes produced by the spectral lines present in the light of the night sky. In the last case, it is better to subtract the flat-field from the image because the bias to be corrected is additive.

A situation in which there wouldn't be any flat-field to correct one or two images taken one night is perfectly possible. In this case, for want of anything better, we calculate a mathematical surface which passes through points chosen in the image, between stars and galaxies (`POINTING` function of the `CURSOR` command in `IT`). We usually use a polynomial surface of the type:

$$\begin{aligned} I(X,Y) &= A + B\cdot X + C\cdot Y + DX\cdot Y + E\cdot X^2 + F\cdot Y^2 + GX^2\cdot Y \\ &\quad + H\cdot X\cdot Y^2 + I\cdot X^3 + J\cdot Y^3 + \cdots \end{aligned}$$

where $I(X,Y)$ is the intensity of the coordinate point (X,Y). We usually limit ourselves to the third or fourth degree. The coefficients of this equation are calculated by the method of least squares (command `POLY`). Then an artificial image is produced (command `SYNTHE`) which is subtracted from the image to be treated. This method is not very accurate. In particular the photometric quality of the resulting image is very poor. However, this technique can be useful, if only to visualize the image properly. Another application of polynomial adjustment of sky backgrounds is the elimination of non-uniform residue in images that are badly corrected by the flat-field.

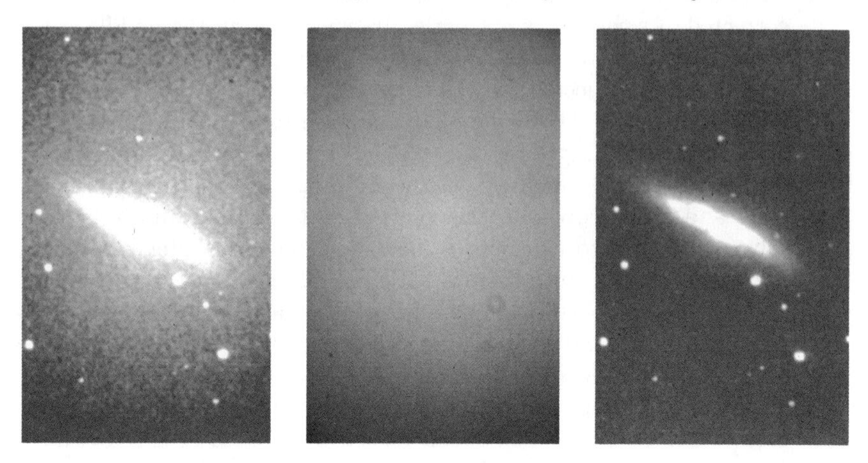

Figure 5.3 *On the left, an image of the galaxy M82 made with a 5 minute exposure of a TH7852A CCD mounted on a 200 mm Schmidt-Cassegrain telescope in an urban environment. A focal reducer converts the relative aperture to 6 but introduces significant vignetting. The sky background, caused by city lights, is very strong, making it difficult to observe faint details of the galaxy. In the center, the average of 3 flat-fields, obtained in 10 second exposures at dusk. The vignetting is clearly visible. A more careful examination shows diagonal zoning, a section with low quantum efficiency at the upper right of the image (2% lower than the rest of the array), a variation in sensitivity on one line out of two (this is characteristic of this type CCD which has anti-blooming drains) and some dust. On the right, the result of the processing.*

5.1.3 Summary

Extreme care must be used in radiometric correction of images if we want to observe low-contrast objects. These objects are barely visible—typically 1% above the sky background. If the image is not processed, it is likely that these objects will be lost in the background, although the information does exist. With a precise acquisition procedure, we were able to observe stars down to magnitude 19 with a 200 mm telescope located in an urban environment (with street lights only a few dozen meters away and a naked eye limiting magnitude of three!) Needless to say, the sky background was especially bright, but by summing many images and with good quality flat-field correction, much is possible.

The equation below summarizes the corrections to be made on any worthwhile CCD image before it is analyzed:

$$E(x,y,t) = K\frac{R(x,y,t) - N(x,y,t)}{F(x,y,t') - N'(x,y,t')}$$

where

(x, y) are the coordinates of a pixel;
t, the integration time of the image to be treated;
t', integration time of the flat-field;
$R(x, y, t)$, the rough image;
$N(x, y, t)$, the dark map of the rough image;
$F'(x, y, t')$, the flat-field;
$N'(x, y, t')$, the dark map of the flat-field;
K, the average intensity of the flat-field; and
$E(x, y, t)$, the processed image.

All this might appear very complicated compared to what is done in taking a photograph. But let's examine it more closely. The photographic film base is the equivalent of the dark current in CCD's. The film grain is a local variation of the sensitivity, identical to that produced by the photosites of the CCD. Optical vignetting affects a film as well as a CCD. There is however an enormous difference between these two types of detectors. With photographs we can see all these, but that is all we can do. With a CCD we can correct them. Correcting these problems is the fundamental reason for the superiority of the CCD for the detection of faint objects and the precision of photometric analysis.

5.2 The Measurement of Flux

Some of the CCD's qualities (linearity, large dynamic range, reproducibility, extended spectral response) are much appreciated in photometry. CCD's are widely used today for both point source and extended object photometry requiring a precision of 0.01 magnitude. Since astrophysical study requires photometric measurement we will briefly explore this topic.

On one hand we have a star with its magnitude and color; on the other hand we have the stimulus that it produces in the image: a spot of greater or lesser dimension and a certain intensity.

If the flux F coming from the star is expressed in an adequate unit, we calculate the instrumental magnitude (M) by Pogson's classic equation:

$$M = -2.5 \log F.$$

The problem is determining the flux F from the image spot. First, the atmosphere absorbs the signal coming from stars in a complex way and second, the CCD's response to the luminous flux is also subtle. All these factors mean that it is almost impossible to measure the absolute value of the flux. We will always proceed by comparing the signals received from several objects. The flux measurements will therefore be relative. If F_1 and F_2 are the fluxes received from the two objects, the difference in magnitude between the two objects will be:

$$M_1 - M_2 = -2.5 \log \frac{F_1}{F_2}.$$

Atmospheric turbulence, optical defects, and tracking errors are some of the causes that spread the star's image in the focal plane. The CCD itself also contributes through the charge diffusion phenomenon. The flux information is therefore most often dispersed over several pixels. There are therefore two measuring methods: aperture photometry, and mathematical adjustment of the image spot's profile.

5.2.1 Aperture Photometry

The signal produced by the star is integrated over an area that includes the star's zone of irridation. Usually the measurement surface has a circular contour. We then talk of synthetic aperture photometry by analogy with monochannel photomultiplier photometry in which the star is isolated from its environment by a circular diaphragm. This integration is simply carried out by summing the intensity of the pixels found inside the measurement area.

Of course the sky background contributes to the measurement since the star is superimposed on it. To take this into account, the measurement zone is moved to one side of the star, and the sky background alone is taken. If F_e is the signal measured on the star and F_c the signal recorded on the sky background, the contribution of the star (N) is obviously $N = F_e - F_c$.

This number is expressed in quantification units. The signal value corresponds to a certain integration time. To enable comparison between exposures taken with different integration times, we adjust the value of the signal to a standard unit of time, a second for example. If the result of integrating a stellar image gives 1245 ADU (Analog Digital Units) for a 300 second exposure, we will say that the signal is $1245/300 = 4.15$ ADU/s.

The `IP` program has a command (`PHOT`) that allows us to carry out aperture photometry easily. After asking for the diameter of the measurement zone in pixels (an option gives the choice of using a rectangular aperture, for measuring a spectrum for example), a circle of the chosen dimension appears superimposed on the image. The circle can be moved with the mouse and positioned around the star to be measured. Just pressing on a key supplies the intensity integrated inside the circle in a fraction of a second. The measurement zone should include the star's entire signal. We can check this by visualizing the star with tight thresholds to determine the image spot's outer limits. The sky level is obtained by measuring the background at several points around the star with the same sized circle. During this operation we have to avoid faint stars in the analysis zone. Doing so is not always easy when the stellar field is crowded. After determining the average sky background the contribution of the star alone is calculated. A variation of this method measures the sky in a zone concentric to the star, but of course using a more extended measurement area (we will take the difference in size of the measurement zone into account to calculate the contribution of the star alone). Warning: before any measurement of a flare star, it is critical to do the dark frame subtraction and flat-field correction! Note that certain processings preclude any further photometric measurement in the image (for example after an unsharp masking).

Aperture photometry with a CCD is similar to that done with a monochannel detector (usually a photomultiplier). In both cases the star is measured by being isolated from its environment; the sky background is taken into account; the flux is integrated in time, etc.

Just as in traditional photoelectric photometry, local variation in sensitivity on the CCD's surface is a source of concern. With a photomultiplier, the problem is solved by forming the image of the telescope's entrance pupil on the target with a Fabry lens. This configuration is stable in time and the distribution of energy on the photocathode does not depend on the star's position in the measuring diaphragm. This technique cannot be used with a CCD operating in the classic imaging mode: the star moves on the CCD's surface according to the pointing. Correction of sensitivity differences with the flat-field and offset compensation with dark maps render the harmful effects of the star's movement in the field negligible. However, biases will be found with some CCD's because of the variation in sensitivity inside the pixel itself. This is especially true for CCD's that have an anti-blooming device, such as the TH7852, because the drains implanted between the pixels produce zones that are not photosensitive. Imagine the image of a star that is so small that it occupies only a fraction of the pixel. If the star is positioned on a sensitive zone of the pixel, there is no problem. On the other hand, if the star is on a drain, the amplitude will decrease significantly. There is then an important photometric error. The inter-pixel sensitivity variation is much less obvious if the star takes up many pixels. In some cases it will be necessary to defocus to produce sufficient spreading (this will be the case with instruments having a short focal length, typically less than 2 meters). By taking this precaution, we were able to measure stars with a precision of 0.02 magnitude with a TH7852. Dead zones are nearly absent in CCD's without an anti-blooming device, like the TH7863 CCD.

5.2.2 Mathematical Fitting of a Stellar Profile

This technique consists in approximating the surface of the stellar image by a mathematical equation, usually in a Gaussian form:

$$I(r) = I(0)\exp\left(-\frac{r^2}{\sigma^2}\right)$$

with:

r the radius with respect to the center of the star;

σ, a parameter characteristic of the star's spreading; and

$I(0)$, the central intensity.

The signal contained in a circle of radius r is given by

$$E(r) = \pi\sigma^2\big(I(0) - I(r)\big).$$

The volume of the image spot in quantification units is obviously

$$N = \pi\sigma^2 I(0).$$

All the stars within the same image produced by optics with no field aberrations have the same Gaussian profile; therefore, σ is a constant of the image. On the other hand, the coefficient $I(0)$ is a function of the brightness of each star. Thus the profiles of the stars on the image are deducted from each other by simple scaling. This property is used to erase stars by subtracting from the mathematical surface. It is very useful for photometry in rich stellar fields (globular clusters), in which the bright stars hide the faint stars. The former are removed by subtraction of a surface previously adjusted to isolated stars; the latter can then be measured. The operation can be repeated with an iterative process to reach even fainter stars. Some highly refined professional programs carry these operations out automatically. The prototype of these programs is `DAOPHOT`,[1] which requires a VAX.

A datum derived from the adjustment by a Gaussian is the full width at half maximum of the star:

$$FWHM = 2\sigma\sqrt{ln2}.$$

[1]P. Stetson, 1987, *Publications of the Astronomical Society of the Pacific*, Vol 99, p. 191

This width is expressed in pixels, but it can be translated into arcseconds once the image scale is known. The full width at half maximum allows us to estimate the resolution of the image.

The first step in adjusting a stellar profile is to remove the sky background from the image. The sky background is estimated by sampling around the star with a graphic cursor. The centroid of the star is then determined, and the star is moved so that its center coincides exactly with a pixel. The adjustment can then be made by the least squares method. For that, the Gaussian equation is linearized and written in the form

$$ln\ I = ln\ a + bR$$

with $a = I(0)$ and $R = r^2$.

On a regular grid we pick up pairs (I, R) inside the stellar image. The grid can have a tighter step than the pixel, the additional points being calculated by interpolation. Afterwards all we have to do is make the adjustment of the star's bidimensional profile. The equation above can be written $Y = A + BX$.

The coefficients (A, B) are calculated by linear regression from the recorded points (X_i, Y_i) with $i = 1 \dots N$:

$$A = \frac{\alpha\delta - \beta\gamma}{\alpha^2 - N\beta},$$

$$B = \frac{\alpha\gamma - N\delta}{\alpha^2 - N\beta},$$

and

$$\alpha = \sum X_i; \quad \beta = \sum X_i^2; \quad \gamma = \sum Y_i; \quad \delta = \sum X_i Y_i.$$

Then

$$I(0) = \exp(A),$$

and

$$\sigma^2 = \frac{-2}{B}.$$

The technique of flux measurement by profile adjustment is fairly clumsy to apply. It only really works if the star's image extends over enough pixels (3 or 4 pixels at half maximum). The measurement precision is then high. Otherwise the star is under-sampled and the precision drops. Sometimes with short focal length instruments, the star only occupies one pixel. This is a case where the adjustment method cannot be used. In conclusion, adjustment is reserved for images taken with instruments having a large focal length. For crowded fields where conventional photometry is impossible, aperture photometry is faster and sufficiently precise most of the time.

5.3 Photometric Reduction

In this section we will transform instrumental magnitudes that are taken directly from flux measurements in ADU's into magnitudes expressed in a standard photometric system.

Instrumental magnitudes simply cannot be used in photometric studies since they are dependent on the telescope's size, the CCD's wavelength sensitivity, the kind of filters used, etc. There are many variables that complicate the quest to find a relationship between measured quantities and the physical reality of the studied objects. In addition the parameters considered belong to a given instrumentation, effectively preventing any comparison with work done in other observatories. Therefore, if we want our own observations to be seriously considered by the scientific community, we have to tie them to a standard photometric system. Calibrating our observations photometrically means going beyond the stage of contemplation and entering into a rigorous scientific procedure.

First we will explore the characteristics of some photometric systems. Next we will examine the problems posed by atmospheric transmission. Then various reduction methods will be proposed.

5.3.1 Photometric Systems

A knowledge of an object's distribution of spectral energy enables astrophysical interpretation. The observations must therefore be made in well defined spectral bands that are isolated with filters. Several photometric systems exist that essentially differ by the spectral characteristics of the filters. The basis of any photometric system is precise observation of standard stars using a set of filters. These stars are obviously known to be non-variable to within a specified level of accuracy.

The most commonly used photometric system is the one defined by Johnson and Morgan,[2] known as the UBV system (for Ultraviolet, Blue, Visible). The equivalent wavelength and the width at half-maximum in the spectral bands are given below:

	R	B	V
$\lambda 0$ (Å)	3600	4400	5500
$d\lambda$ (Å)	700	1000	900

By "equivalent wavelength" we mean the wavelength of the braycenter of the product of the filter's transmission weighted by the CCD's spectral response.

[2]Johnson, H.L., Morgan, W.W., 1953, *Astrophysical Journal*, Vol 117, p. 313.

Subsequently, Johnson and Morgan's system was extended into the red and near infrared:

	R	I
$\lambda 0$ (Å)	7000	9000
$d\lambda$ (Å)	2200	2400

The most commonly used system today for working with CCD's is the Kron-Cousins' UBVRI.[3] The characteristics of this system allow us to take greater advantage of the sensitivity of modern detectors in the near infrared. It therefore differs from the Johnson-Morgan system in the R and I bands:

	R	I
$\lambda 0$ (Å)	6500	8000
$d\lambda$ (Å)	1000	1500

Some observers work in an intermediate system u,v,g,r called Thuan and Gunn's [4] that extends into the infrared (i band).[5] The choice of spectral bands in this system, whose characteristics are given below, allows, among other things, the rejection of parasitic spectral lines of the night sky.

	u	v	g	r	i
$\lambda 0$ (Å)	3530	3980	4930	6550	8200
$d\lambda$ (Å)	400	400	700	900	1300

There are equations that enable the translation of one photometric system into another. Here is the "bridge" between Johnson's (index j) and the Cousins' system (index c):

$$V_c = V_j$$

$$\begin{array}{ll} (V-I)_c = 0.713(V-I)_j & (V-I)_j < 0 \\ (V-I)_c = 0.778(V-I)_j & 0 < (V-I)_j < 2 \\ (V-I)_c = 0.835(V-I)_j & 2 < (V-I)_j < 3 \end{array}$$

$$(R-I)_c = 0.856(R-I)_j + 0.025$$

$$\begin{array}{ll} (V-R)_c = 0.73(V-r)_j - 0.03 & (V-R)_j < 1 \\ (V-R)_c = 0.62(V-r)_j - 0.08 & 1 < (V-R)_j < 1.7 \end{array}$$

---

[3]Bessel M.S. 1979, *Publications of the Astronomical Society of the Pacific*, Vol 91, p. 589; Cousins A.W. 1973, *Mem. R.A.S.* Vol 77, p. 223; Cousins A.W. 1976, *Mem. R.A.S.* Vol 81, p. 25; Cousins A.W. 1978, *Mon. African Astron. Obs. Circ.* Vol 1, p. 234.

[4]Thuan T., Gunn, J.E. 1976, *Publications of the Astronomical Society of the Pacific*, Vol 88, p. 543; Kent, S. 1985, *Publications of the Astronomical Society of the Pacific*, Vol 97, p. 165.

[5]Wade R.A., Hoessel, J.G., Elias J.H., Huchra J.P. 1979, *Publications of the Astronomical Society of the Pacific*, p. 35.

The bands of a photometric system must be isolated by a careful choice of filters that are adapted to the detector's spectral sensitivity. As an example, the U band of the initial UBV system is obtained by observing through a Corning 9863 filter, the B band with the Corning 5030 joined to the Schott GG13, and the V band with a Corning 3384 filter, all of them associated with a type 1P21 photomultiplier. The latter's photocathode has a very different spectral response from a CCD. With a CCD (silicon response), for Cousins' UBVRI system, the following set of filters is chosen:

U : 1 mm UG2 + $CuSO_4$

B : 1 mm BG12 + 2 mm BG18 + 2 mm CG385

V : 2 mm GG495 + 1 mm BG18 + 1 mm WG305

R : 2 mm OG570 + 2 mm KG3

I : 3 mm RG9 + 1 mm WG305

Note that the filters' thicknesses have been included. In this assembly, WG305 (a clear filter), is transparent to UBVRI—its function is to equalize the optical thickness for all spectral bands and, consequently, one does not have to refocus every time a filter is changed. All these filters are referenced in the Schott Company's catalog, except the $CuSO_4$ which is, in fact, a transparent cell containing liquid copper sulphate (sometimes a $CuSO_4$ crystal is substituted). Note that the combinations of filters that isolate a band in the visible are carefully masked for the infrared, i.e., the infrared is not transmitted. This masking is done with "cold" filters such as the BG18 or the KG3. Never trust appearances to decide that a filter is opaque to the infrared because the eye is insensitive to this radiation. We should remember that silicon's spectral response favors the infrared over the visible. If we use a blue filter that is approximately masked, the measurements will actually be made in the infrared. That is not really what we are trying to do! Colored filters used for photography are not masked at all. All the filters with Wratten gelatin (Kodak) are completely transparent from about 700 nm. Even interference filters have surprises in store for us in this domain. To know whether a filter that has to work in the blue or the visible is correctly masked, make an image with an incandescent lamp through the filter in question, then redo it with the same filter combined with a BG18. There should be no observable difference in signal level between the two images.

With thick substrate frontside illuminated CCD's (TH7852, TH7863), the U band is inaccessible because of the low sensitivity in the blue.

If we are using interference filters, we have to make sure that they are crossed by light beams that are parallel or only very slightly divergent because spectral transmission depends on the rays' incident angle. The filters should be placed on a sliding mount or a wheel so they can be chosen quickly.

5.3.2 Atmospheric Extinction

When it crosses the terrestrial atmosphere, the light of celestial objects is reduced because of absorption and diffusion by the medium. This reduction of incident flux is called atmospheric extinction. Its strength depends on the zenith distance of the object, the wavelength, atmospheric conditions during observation, and the observatory's altitude. Taking atmospheric extinction into account is obviously an integral part of photometric reduction.

If M is the instrumental magnitude, k_λ the extinction factor in magnitudes at the zenith and z the zenith distance of the object, the extra-atmospheric magnitude M_0 is given by:

$$M_{0(\lambda)} = M_{(\lambda)} - K_\lambda \sec z \qquad (\sec z = 1/\cos z).$$

This equation is known as Bouguer's law. The coefficient K_λ is a function of the wavelength λ. We thus know that the atmosphere reddens rays that cross it, showing that extinction is not identical for all colors.

The term $\sec z$ is called air mass. It defines the thickness of the atmosphere crossed by light rays. By definition, air mass is 1 at the zenith. This air mass rises to an infinite value at a zenith distance of 90°, at the horizon. In practice this is not a problem because we never point so low. Air mass is often represented by the letter χ:

$$\chi = \frac{1}{\cos z} = \sec z.$$

An equation that is slightly more complicated supplies a more precise value of χ when the air mass is greater than 3:

$$\chi = \sec z(1 - 0.0012 \tan^2 z).$$

The zenith-distance is obtained from the relation:

$$\sec z = 1/(\sin \psi \sin \delta + \cos \psi \cos \delta \cos h)$$

with

ψ = latitude of observing location, and

δ = observed object's declination,

h = object's hour angle.

Figure 5.4 shows an example of a Bouguer straight-line plot. This kind of curve is obtained by measuring the brightness of a star according to its zenith-distance. Its shape depends on the star's color and of course on atmospheric conditions. If these do not change during the night, the plotted curve is a straight-line. The intersection of the straight-line with the ordinates' axis supplies the value of the extra-atmospheric magnitude (M_0). The line's slope is $-K_\lambda$. Obviously the measurement points are not aligned because of measurement errors or slight differences in atmospheric transmission. The best straight-line fit to these points is determined by using a linear regression.

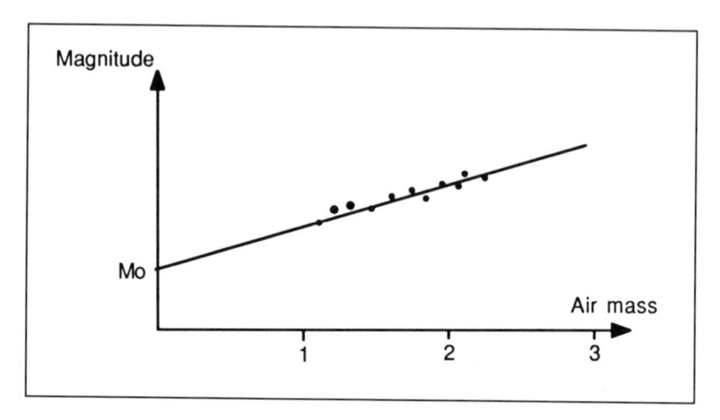

Figure 5.4 *A Bouguer straight-line.*

The fact that the bandpass of the filters covers an appreciable spectral range creates an added complication. Since the spectral distribution of the object's flux varies within the bands of the photometric system, the effective working wavelength will be a function of the object's color. In other words, two stars observed through the same filter and at the same zenith-distance will not have the same extinction if their spectral type is different. In wide band photometry (typical UBVRI system), the expression of the extinction coefficient must be decomposed in the following way:

$$K_\lambda = K'_\lambda + K''_\lambda(B - V).$$

K'_λ designates the main extinction coefficient or the first order coefficient and K''_λ the color correction coefficient or second order coefficient. The value of the first order coefficient is characteristic of the transparency of the atmosphere for the night of observation. The second order coefficient depends on the width of the spectral bands used and the appearance of the curve $K = F(\lambda)$. The index (B-V) is the parameter specifying the color of the observed star.

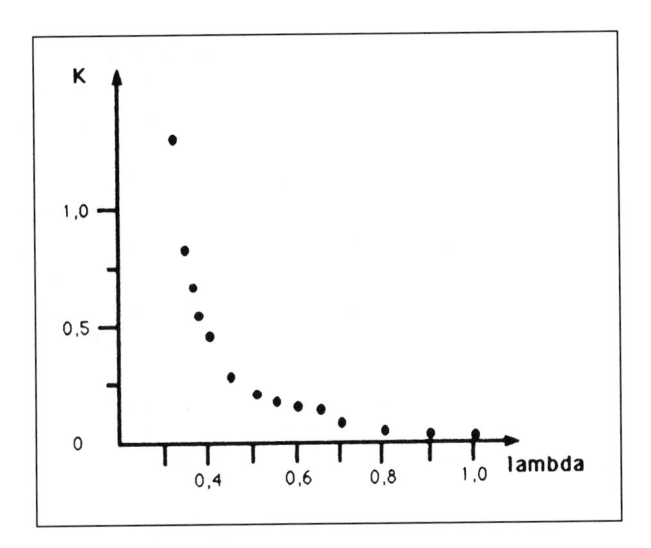

Figure 5.5 *Variation of the extinction coefficient in magnitude as a function of wavelength in microns. (From Allen,* Astrophysical Quantities, *3rd edition).*

Because of the slight slope of the curve $K = F(\lambda)$ at the level of the V band, K''_V oscillates around -0.02 and -0.04 magnitude. The situation is more complicated for K''_U because of the presence of the limit of the Balmer series at 3647Å, which has a very different appearance from one star to another. K''_U is between 0 and -0.04. This illustrates the fact that it is almost impossible to take measurements in the U band with a camera equipped with a frontside illuminated CCD. The second order coefficients can be ignored in the red and infrared bands because of the appearance of the $K = F(\lambda)$ curve at these wavelengths.

Introducing the color coefficient, the expression linking the instrumental magnitude with the extra-atmospheric magnitude is:

$$M_0 = M - XK'_\lambda - XK''_\lambda(B - V).$$

This transformation must be carried out independently for each spectral band. The equations below are commonly used for the reduction, in which V, $(B - V)$, and $(V - R)$ are the elements measured:

$$V_0 = V - X\big(K'_V + K''_V(B - V)\big),$$

$$(B - V)_0 = (B - V) - X\big(K'_{BV} + K''_{BV}(B - V)\big),$$

$$(V - R)_0 = (V - R) - X\big(K'_{VR} + K''_{VR}(V - R)\big).$$

The first order extinction coefficient in each spectral band is determined by experimentally plotting Bouguer's law for each filter. For this we observe the same star throughout the whole night in the bands of the system and we plot the curve of the observed magnitude according to the air mass. Measuring the slope of the straight-line gives the K' coefficient for each band.

Gathering data to precisely determine the slope of a Bouguer line is tedious because it means observing a good part of the night just for that purpose. Moreover, the atmosphere's characteristics can change during the observation due to the appearance of cirrus type clouds or airplane contrails, for example. Should this happen, the Bouguer straight-line method fails.

A more rapid method measures two standard stars of the same spectral type at very different zenith-distances. These spectral types must be identical so that the color terms will be eliminated in the equations. To simplify things, we will suppose that we are doing photometry in the UBV bands. Let Δv, Δ(b-v) and Δ(u-b) be the differences in magnitude and in color observed between star 1, called high, and star 2, called low. The difference in air mass between these stars is ΔX. If V_1, $(B\text{-}V)_1$, $(U\text{-}B)_1$ and V_2, $(B\text{-}V)_2$, $(U\text{-}B)_2$ are the magnitudes and color indices found in the catalog, we calculate the differences thus:

$$\Delta v' = \Delta v - \big((V_1 - V_2)\big),$$

$$\Delta'(b-v) = \Delta\big((b-v) - ((B-V)_1 - (B-V)_2)\big),$$

$$\Delta'(u-b) = \Delta(u-b) - \Big(((U-B)_1 - (U-B)_2)\Big).$$

The extinction coefficients then become

$$K'_v = \frac{\Delta v}{\Delta X}, \quad K'_{bv} = \frac{\Delta'(b-v)}{\Delta X}, \quad K'_{bv} = K'_{ub} = \frac{\Delta'(u-b)}{\Delta X}.$$

This method, called “high star–low star”, allows us to obtain very rapidly instantaneous first order extinction coefficients. By repeating the procedure several times during the night, we can follow any possible change in atmospheric transmission.

To determine the second order coefficients, we measure two stars located at the same zenith distance (the first order terms are simplified). For maximum precision, these two stars should have color indices that are as distant from each other as possible. If ΔV, Δ(B-V) and Δ(U-B) are differences in magnitude and color index given by the catalog (the difference is carried out in the direction blue star–red star) and if Δv; Δ(b-v), and Δ(u-b) are the corresponding values observed at the air mass X, we easily find:

$$K''_v = \frac{v-V}{\Delta X(b-v)},$$

$$K''_{bv} = \frac{\Delta(b-v) - \Delta(B-V)}{\Delta X(b-v)},$$

$$K''_{ub} = \frac{\Delta(u-b) - \Delta(U-B)}{\Delta X(u-b)}.$$

These corrections highlight the importance of not neglecting atmospheric extinction—even though we are using differential photometry on a CCD image, where the objects are of necessity included in a reduced field. Remember that two objects that are very close to each other undergo different extinction and reddening if they are not of exactly the same spectral type.

5.3.3 Transformation to a Photometric System

An instrument's spectral characteristics, set by the combination of the detector and the filters, cannot be exactly the same as the reference photometric system. Consequently, when observing standard stars whose magnitudes and color indices are defined in the reference system, we always find differences between the values in the catalog and those measured. These are "transformation" errors.

Tying the instrumental system to the standard system is done by observing many standard stars whose magnitude and extra-atmospheric color indices (V_0, $(B\text{-}V)_0$, $\cdots$) have been calculated. The transformation equations are in the form:

$$V_{\text{cat}} = V_0 + \epsilon(B-V)_{\text{cat}} + C_v,$$

$$(B-V)_0 = \mu(B-V)_{\text{cat}} + C_{bv},$$

$$(U-B)_0 = \psi(U-B)_{\text{cat}} + C_{ub}.$$

The magnitudes with a "cat" index are taken from the list of stars describing the system ("catalog" data).

ϵ, μ and ψ are the transformation coefficients. C_v, C_{bv}, and C_{ub} are terms that set the 0 of the magnitude scale (magnitude constants). All these elements are calculated by linear regression from the observation of a large enough number of standard stars (about 10).

A rapid method of obtaining an approximate value of the magnitude constant C in the spectral band being used is to take the integrated flux I_s of a standard star with magnitude M_s and the sky background F_s:

$$C = M_s + 2.5\log(I_s - F_s).$$

The magnitude constant then represents the magnitude that a star causing a stimulus of one quantification unit above the sky would have.

The magnitude of any star causing a signal I_e (brought to the same integration time as that used to acquire the standard star) on a sky background F_e will be:

$$M_e = -2.5 \log(I_e - F_e) + C.$$

The magnitude constant is 20.9 in the red for a one minute exposure with a camera equipped with a TH7863 CCD mounted on a 280 mm telescope. However, a magnitude constant of 20.9/min does not mean that we will reach such a high magnitude by integrating for such a short time. To be detected, the star has to emerge four or five times above the noise (in our example, 1.5 ADU RMS). In addition, the star is spread over several pixels, and the sky background's brightness will make the image noisy. Generally, we have to expose at least 15 minutes under a very black sky to detect a star of magnitude 20.9.

Knowing the extinction coefficients, the transformation coefficients, and the magnitude constants allows us to find the magnitude of an unknown star in a particular photometric system.

5.3.4 A Speedy Reduction Method

We can go directly from instrumental magnitudes to photometric system magnitudes without having to calculate extra-atmospheric magnitudes. To do so, many standard stars are measured during the night for various air masses (X). Then by least squares method, the following equations are resolved for the coefficients a_i, b_i, and c_i (example of reduction in the VRI bands, the most accessible to a non-thinned CCD):

$$v = V + a_0 + a_1(V - R) + a_2 X,$$

$$r = R + b_0 + b_1(R - I) + b_2 X,$$

$$i = I + c_0 + c_1(R - I) + c_2 X.$$

The data of the standard star catalog are V, R, and I when the instrumental magnitudes are v, r, and i.

The parameters a_0, b_0 and c_0 set the zero point of the magnitude scale. The coefficients a_1, b_1, and c_1 represent the color terms; a_2, b_2, and c_2 are the extinction coefficients.

For a given telescope, set of filters, detector and site, the color terms are almost constant over time. When they are well established by observations spread over several nights, they can be set during calculation by least squares of the linear equations above. Only the extinction coefficients and the zero points need to be determined for each night.

Conducting observations is not recommended when the air mass is greater than 2 (zenith-distance of about 60°), because beyond 2 the linear model that we have established is not valid. We are then required to introduce terms of order higher than 1, a situation which considerably complicates the method.

To check the exactness of the calculation, the instrumental magnitudes of the measured standard stars should be compared to the calculated magnitudes. A significant differences between these two values is warning of either a poor quality standard (badly calibrated, if not variable!) or an error in identifying the correct star. In this case, the suspect data is rejected and the coefficients recalculated.

By inverting the above equations, a star's instrumental magnitude can be transformed to that of the system by the following relations:

$$Y_1 = v - a_0 - a_2X, \quad Y_2 = r - b_0 - b_2X, \quad Y_3 = i - c_0 - c_2X,$$

$$I = \frac{Y_3 + Y_3b_1 - Y_2c_1}{1 + b_1 - c_1},$$

$$R = \frac{Y_2 - Y_2c_1 + Y_3b_1}{1 + b_1 - c_1}, \quad \text{and}$$

$$V = \frac{Y_1 + a_1R}{1 + a_1}.$$

5.3.5 A List of Standard Stars

Below is a small list of standard stars taken from a study by Landolt that contains more than 200 stars.[6] These data constitute a recognized standard and are widely used professionally to calibrate CCD images photometrically. The system used is that of Kron-Cousins. The stars in our list are bright, usually located around the celestial equator, and there are enough of them so that at least one is well situated when we need to point at a standard. The list also contains the SAO number and the equinox 2000 coordinates for each star.

Another list of standards that is easy to obtain is the "Arizona-Tonantzintla" catalog of 1,325 stars that are visible to the naked eye and measured in Johnson's UBVRI.[7] However, this catalog, is less reliable than Landolt's.

[6]The complete list appeared under the references Landolt, *Astronomical Journal*, Vol. 88, number 3, March 1983, pp. 439–460 and Landolt, *Astronomical Journal*, Vol. 88, number 6, June 1983, pp 853–866.

[7]*Sky and Telescope*, July 1965, pp 21–31.

A List of Selected Standard Photometric Stars

SAO	α	δ	V	B-V	U-B	V-R	R-I	V-I	Sp
074596	1 16 27	+23 35 22	6.693	+0.047	+0.041	+0.058	−0.005	+0.052	A0
130471	3 24 59	− 5 21 50	7.866	+1.150	+1.141	+0.686	+0.558	+1.246	K5
132211	5 31 27	− 3 40 39	7.960	+1.474	+1.231	+0.985	+1.095	+2.076	M1
132291	5 34 56	− 0 07 22	7.169	−0.085	−0.449	−0.031	−0.054	−0.085	B5
132378	5 37 37	− 4 56 03	7.150	−0.160	−0.776	−0.066	−0.085	−0.151	B3
058292	5 39 15	+30 09 02	7.709	+0.115	−0.147	+0.117	+0.086	+0.200	B8
077381	5 40 35	+28 58 38	7.033	+1.154	+0.959	+0.609	+0.534	+1.143	G5
094795	5 42 58	+14 10 43	6.731	+1.096	+1.013	+0.579	+0.511	+1.090	K1
114536	6 52 04	+ 1 15 04	7.861	+1.535	+1.744	+0.826	+0.757	+1.583	K5
134061	7 01 27	− 3 07 04	7.702	−0.088	−0.958	−0.011	−0.030	−0.038	09
116160	7 57 04	+ 2 57 03	7.832	−0.182	−0.786	−0.055	−0.075	−0.130	B3
136651	9 11 51	− 6 58 46	7.601	+1.628	+1.933	+0.990	+1.100	+2.087	M2
137229	9 55 35	− 1 07 35	7.997	+1.108	+1.037	+0.575	+0.515	+1.091	K0
137239	9 56 39	− 0 27 43	7.835	+1.485	+1.787	+0.795	+0.728	+1.524	K5
137898	10 56 40	− 1 10 08	7.916	+0.309	+0.095	+0.173	+0.173	+0.346	A5
081958	11 43 47	+24 00 37	7.411	+0.858	+0.450	+0.475	+0.438	+0.914	G2
081968	11 44 44	+28 40 13	7.024	−0.002	+0.027	−0.004	−0.004	−0.009	A0
138673	12 15 06	− 7 15 27	7.354	+0.814	+0.451	+0.437	+0.389	+0.826	G5
100009	12 15 13	+16 54 27	6.819	+1.184	+1.165	+0.591	+0.528	+1.120	K2
100038	12 19 06	+16 32 54	7.028	+0.602	+0.073	+0.330	+0.313	+0.642	G3
139437	13 36 14	− 0 55 52	7.062	+0.528	−0.010	+0.313	+0.311	+0.624	F8
140659	15 37 29	− 0 53 05	7.779	+1.275	+1.309	+0.663	+0.585	+1.249	K2
140680	15 39 01	− 0 18 42	7.500	+0.543	+0.051	+0.312	+0.297	+0.608	F8
141281	16 37 21	− 0 24 50	7.964	+1.303	+1.413	+0.682	+0.589	+1.269	K2
122374	17 25 45	+ 2 06 40	7.540	+1.356	+1.276	+0.854	+0.768	+1.622	K7
085357	17 43 15	+21 36 32	7.521	+0.745	+0.284	+0.451	+0.415	+0.867	K0
122683	17 44 03	+ 6 03 43	7.435	+0.326	+0.180	+0.216	+0.236	+0.451	A2
085402	17 46 40	+25 45 00	6.982	+0.147	+0.138	+0.123	+0.143	+0.266	A2
141889	17 48 37	− 2 11 46	7.782	+0.224	−0.747	+0.148	+0.158	+0.306	B0
141956	17 54 57	− 7 44 02	6.926	+0.759	+0.415	+0.405	+0.354	+0.759	G5
186037	17 58 57	−22 31 01	6.991	+0.263	−0.754	+0.206	+0.221	+0.427	O8
123809	18 41 26	+ 0 33 51	7.474	+1.449	+1.710	+0.767	+0.681	+1.448	K2
124055	18 55 46	+ 0 15 54	7.395	+0.107	−0.671	+0.074	+0.077	+0.151	B3
124878	19 38 21	+ 0 20 43	7.216	+0.278	+0.113	+0.154	+0.166	+0.319	A2
126040	20 38 16	+ 1 00 59	7.885	+1.641	+2.043	+0.931	+0.955	+1.884	K5
144957	20 56 18	− 3 33 41	6.566	−0.076	−0.306	−0.034	−0.043	−0.078	B9
145050	21 02 59	− 0 55 29	6.498	−0.099	−0.496	−0.037	−0.045	−0.083	B8
126992	21 41 56	+ 0 20 45	7.653	+0.488	+0.013	+0.291	+0.291	+0.582	F5
146261	22 42 49	+ 0 13 55	6.969	+0.311	+0.105	+0.187	+0.189	+0.376	F0
127706	22 42 58	+ 0 24 07	7.737	+0.864	+0.473	+0.480	+0.462	+0.941	G5
108392	23 05 33	+14 57 33	6.783	+0.004	−0.043	−0.014	−0.009	−0.024	A0
128034	23 12 38	+ 2 41 10	7.708	+0.620	+0.116	+0.355	+0.339	+0.694	G5
191760	23 15 19	−24 51 10	7.283	+0.181	+0.100	+0.086	+0.096	+0.181	A5
146962	23 54 03	− 9 17 24	7.200	+1.581	+2.003	+0.906	+0.947	+1.854	M1

Besides calibration lists like Landolt's, astronomers use stellar fields containing a range of standards. This provides a range of stars distributed in magnitude and spectral type on a single image and saves a considerable amount of time. In addition, Landolt's stars are often too bright for straightforward observation with very large telescopes because of the risk of saturation. These "standard" stellar fields often are clusters or globular clusters which are rich in stars. As reference works, there are Christian[8] (stellar fields, generally quite faint, in NGC 2264, NGC 2419, NGC 4147, NGC 7790, M 92) and Landolt[9]. There are also Kunkel[10] (stars included in the Selected Areas, sky zones that are well known to photometrists) for the Northern Hemisphere and Graham[11] for the Southern Hemisphere.

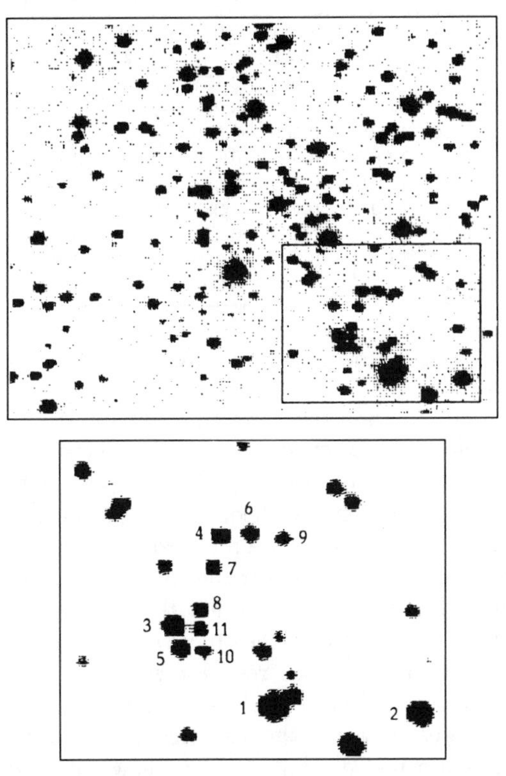

Figure 5.6 *Above, a general view of the open cluster M67. Below, enlargement of the framed area in the top picture containing the photometric sequence. Image acquired with a 4 minute exposure using a TH7852A CCD and a 280 mm f/6.1 telescope.*

As an example we give (Figure 5.6 and the list below) the photometric sequence established by Schild[12] in the open cluster M67. The magnitudes

[8]Christian C.A., 1985, *Publications of the Astronomical Society of the Pacific*, Vol. 97, p. 363.

[9]Landolt A.U., 1973, *Astronomical Journal*, Vol. 78, p. 959.

[10]Kunkel W.E., Rydgren A.E., *Astronomical Journal*, Vol. 84, p. 633.

[11]Graham S.A. 1982, *Publications of the Astronomical Society of the Pacific*, Vol. 94, p. 244.

[12]Schild R.E. 1983, *Publications of the Astronomical Society of the Pacific*, Vol. 95, p. 1021.

are in the Kron-Cousins system.

Photometric Sequence for M67					
Star	V	B-V	V-R	R-I	V-I
1	9.692	1.351	0.683	0.656	1.339
2	10.007	−0.098	−0.047	−0.020	−0.068
3	11.465	1.051	0.558	0.551	1.069
4	12.147	0.466	0.289	0.276	0.565
5	12.287	0.569	0.340	0.344	0.684
6	12.677	0.800	0.470	0.442	0.912
7	12.816	0.553	0.338	0.334	0.672
8	12.932	0.449	0.282	0.293	0.576
9	13.096	0.582	0.344	0.309	0.653
10	13.181	0.577	0.349	0.345	0.694
11	13.218	0.601	0.361	0.375	0.736

5.4 Calculating the Radiometric Flux

It is always useful to know the theoretical possibilities of a CCD camera. This knowledge allows us, for example, to optimize the instrumentation for a given observation (choice of telescope, filtering possibilities, etc.). It also enables comparisons between the theoretical performance and the actual results to help us monitor and maintain the correct functioning of the camera. It is much easier to model performance in spatial resolution (the pixel size is known and the telescope's focal length sets the angular resolution) than photometric characteristics. The object of the radiometric response is to determine the detector's electric response when we observe a given star. This calculation includes factors such as atmospheric extinction, optical transmission, the spectral band, the detector's sensitivity, noise, etc.

We will describe a real example to show how this is determined. We have to determine the radiometric flux with the following data: we observe a star of magnitude 20 (spectral type G2) at 45° above the horizon with a telescope having a diameter of 60 cm, at f/3.5. The instrument is situated at a high mountain site (Pic du Midi, altitude 2861 m). The seeing is 2″ at the zenith (angular spreading at half maximum of the stellar image caused by turbulence). We cool the array to −30° C. The camera is used without a filter. We calculate whether or not the star is detectable with an integration time of 300 seconds. We recall that the difference in magnitude between two stars producing luminous fluxes F_1 and F_2 is calculated by Pogson's equation:

$$M1 - M2 = -2.5 \log \frac{F_1}{F_2}.$$

Inversely, knowing the magnitude of the two objects, we deduce the flux ratio by the equation:

$$\frac{F_1}{F_2} = 10^{(M_2-M_1)/2.5} = 2.512^{M_2-M_1}.$$

The table below gives, according to wavelength, the necessary astrophysical data:

1. The number of incident photons by unit of wavelength, surface and time outside the atmosphere for a magnitude 0 star of spectral type G2 (Nλ, expressed in photons /cm^2/s/Å);
2. Atmospheric transmission for a unit air mass (star at the zenith) at sea level (T_1) and at an altitude of 2861 meters (T_2). For an air mass X, the value of the atmospheric transmission is written:

$$T_X = e^{ln(T_{X=1})X} \quad ;$$

3. The number of photons, N_{sky}, coming from the night sky with a 3 day moon (in photons/m^2/s/arcsec^2, in a spectral band of 500 Å);
4. The quantum output, Q_λ, of the array, here a TH7852 (Qλ).

λ(Å)	N_λ	T_1	T_2	N_{sky}	$Q\lambda$
4000	544	0.54	0.76	8	0.02
4500	873	0.63	0.84	11	0.07
5000	999	0.69	0.88	10	0.17
5500	1069	0.72	0.89	19	0.23
6000	1109	0.74	0.90	20	0.25
6500	1077	0.78	0.94	18	0.29
7000	1050	0.81	0.96	18	0.30
7500	1002	0.62	0.97	25	0.24
8000	958	0.79	0.98	32	0.24
8500	895	0.90	0.98	48	0.20
9000	851	0.56	0.98	55	0.15
9500	811	0.63	0.99	57	0.07
10000	755	0.97	0.99	170	0.04

We first calculate the dimension of the star at the telescope's focus. The seeing depends on the air mass that is crossed by the light rays. If α_0 is the angular dimension of the star observed at the zenith, for an air mass X, we will have:

$$\alpha = \alpha_0 X^{3/5}.$$

For an altitude of 45°, the air mass is 1.41, from which we get a seeing of:

$$\alpha = 2.0'' \times 1.41^{3/5} = 2.5''.$$

The dimension θ of the star at the focus is a function of the focal length F:

$$\theta = F \cdot \tan\alpha = 2100 \cdot \tan(2.5'') = 0.025 \text{ mm.}$$

The useful area of the pixel being $30 \times 19\mu$m, only 80% of the luminous energy is recorded. We define the parameter p as being the fraction of the image spot falling on a pixel (thus here $p = 0.80$).

To obtain the number of photo-electrons created, we integrate the photonic stellar flux (N_λ) into the detector's sensitivity band, taking into account the quantum output (Q_λ), the atmospheric transmission (T_λ). and the transmission of a possible filter F(λ). The flux is proportional to the surface area of the principal mirror. We also have to consider the transmission of the optical system. Transmission of the optical system depends on the obstruction of the principal mirror caused by the secondary mirror and on the reflecting power of the mirrors. If D_1 is the diameter of the primary mirror and D_2 is the diameter of the secondary mirror, we get:

$$\tau_{\text{obstruction}} = \frac{D_1^2 - D_2^2}{D_1^2}.$$

In our case $D_1 = 60$ cm and $D_2 = 17$ cm, giving $\tau_{\text{obstruction}} = 0.92$. We estimate the reflecting power of each mirror at $\tau_{\text{mirror}} = 0.92$. The total transmission is:

$$\tau = \tau_{\text{obstruction}} \times \tau_{\text{mirror}_1} \times \tau_{\text{mirror}_2} = 0.779.$$

Finally, the number N of photoelectrons received by the central pixel of the image spot is:

$$N = \frac{\pi D_1^2 p\tau}{4} \int_{\lambda\text{min}}^{\lambda\text{max}} N_\lambda Q_\lambda T_\lambda F_\lambda d_\lambda.$$

With $p = 0.80$, $\tau = 0.779$, $D_1 = 60$ cm and $F_\lambda = 1$ for all the wavelengths (no filter), in the case of a star of magnitude 0, we find:

$$N_0 = 1.84 \times 10^9 \text{ electrons/s.}$$

Compared to a star of magnitude 0, a star of magnitude 20 is fainter by a factor of $2.512^{20} = 1.10^8$ (Pogson's equation).

We deduce from it the number of electrons produced per second for a star of magnitude 20 by applying the following equation:

$$N_{20} = \frac{1.84 \times 10^9}{1 \times 10^8} = 18\text{e}^-/\text{s}.$$

We now compare this number to the noise to determine if the star is detectable. The CCD's readout noise is of the order of 100e$^-$. To this we add the noise of the dark current. The dark current is about 6000e$^-$ for a temperature of $-40°$C exposing for 300 seconds, or a noise of $\sqrt{6000} = 77$e$^-$. Another noise to consider is that produced by quantification. We evaluate it at 36e$^-$ (see section 1.7.12). From all this we deduce the electronic noise:

$$\sigma_e = \sqrt{100^2 + 77^2 + 36^2} = 130\text{e}^-.$$

The total noise in the image is written:

$$\sigma_t = \sqrt{N_{20} \times t + 2\sigma_e^2 + N_{\text{sky}} \times t}$$

with t the integration time. In this equation, the first term represents the photon noise; the second term is the electronic noise multiplied by 2 since at the moment of processing we subtract from the image a dark exposure that was taken under the same conditions, giving the quadratic sum; the third term is the photon noise caused by the sky background.

To find the contribution of the sky background, we have to calculate the angular surface on the sky covered by 1 pixel. If x and y are the dimensions of the pixel, and F is the focal distance of the telescope, we simply show that this area expressed in square arcseconds is written:

$$\gamma = 4.2510^{10}\frac{x \cdot y}{F^2}.$$

In our configuration, we find: $\gamma = 5.5$ arcsec^2. The number N of photoelectrons produced by the sky background is then:

$$N = \frac{\pi D^2 \gamma t}{4} \int_{\lambda\text{min}}^{\lambda\text{max}} N_{\lambda\text{sky}} Q_\lambda d\lambda = 79e^-/s/\text{pixel}.$$

The value of the total noise is:

$$\sigma_t = \sqrt{18 \times 300 + 2 \times 130^2 + 79 \times 300} = 251e^-.$$

After 300 seconds integration time, the star produces a signal S_{20} on the central pixel:

$$S_{20} = N_{20}.300 = 18 \times 300 = 5400e^-.$$

The signal-to-noise ratio is therefore $5400/251 = 21$. A 20th magnitude star is therefore easily recorded because our calculations show that an object is detectable if it produces a signal higher than three times the RMS noise (3σ). Considering this criterion, we can also calculate that it is possible to observe stars 7 times fainter than magnitude 20, i.e. magnitude 22.

The amount of turbulence limits the faintest stars visible. If the turbulence is strong, the flux is spread over several pixels and the detectivity is proportionally reduced. Careless focusing and poor tracking have the same effect. Our model of flux distribution inside the image spot is therefore a bit simplistic, and in practice the central pixel receives less flux than estimated earlier. In addition, in our example we considered that the radiometric corrections, especially those by the flat-field, were perfect. This is, unfortunately, rarely the case. All this is to say that our calculations are not necessarily accurate. They merely give an approximation of the limiting magnitude.

When we observe extended objects, it is often useful to determine the limiting surface brightness. To do this we can use the preceding calculations. Thus we saw that by exposing for 300 seconds, a star of magnitude 22 is at the limit of detection. In addition we know that the central pixel of the image spot gathers only 80% of the stellar flux. If all the stellar flux would have penetrated into a single photosite, we would have a gain in magnitude of:

$$2.5 \log \frac{1}{0.80} = 0.25.$$

If the luminous spot at the instrument's focus covers exactly the surface of a pixel, we can reach magnitude 22.25. The pixel sustains an angular surface on the sky of 5.5 arcsec^2. The limiting surface magnitude per square arcsecond is therefore:

$$22.25 + 2.5 \log 5.5 = \frac{24.1}{\text{arcsec}^2}.$$

5.5 Deep Sky Imaging

5.5.1 The Quality of the Observing Site

Pointing any telescope equipped with a CCD camera toward the sky is a revelation. All the usual astrophotography objects are taken with amazing ease. A 200 mm telescope under a normal sky almost reaches a stellar magnitude of 19 with an exposure lasting just 15 minutes. At this level of detectivity, the Andromeda galaxy begins to be resolved into stars. With a 300 mm telescope and a 30 minute exposure, stars of magnitude 21 are perceptible. Trying a bit harder, we can reach magnitude 22. All the objects in the famous Palomar Observatory Sky Survey are visible. The sky suddenly appears so immense and rich in objects that we don't know where to begin. This, however, is a fortunate quandary because a CCD camera at the focus of a telescope makes us want to observe very much.

These performances were determined without using filters in front of the CCD. Under these conditions, the term "magnitude" must be used with reservation. Comparison with professional results requires the use of VRI type filters. Roughly speaking, an R filter results in the loss of three magnitudes, in other words, a limiting magnitude 18 when exposing for half an hour with a 300 mm telescope. Meanwhile the CFH telescope on its peak in Hawaii records stars of magnitude 27, that is, 9 magnitudes in difference. Considering that this telescope has a diameter of 3.60 meters, that it is located on a site with exceptional transparency (a gain of two magnitudes with respect to the sky that most amateurs can have access to), and that the CCD and the associated electronics were superbly optimized, the amateur's performance is more than respectable.

An unfiltered CCD can be considered to be operating in the R band, taking into account the curve of spectral sensitivity. This is, of course, a very rough approximation. The quality of the observing site is obviously crucial when we work with such low levels of flux. A very dark sky lessens the problems associated with imperfect correction of non-uniformity by flat-fields. A residual non-uniformity of the order of a percent can be enough to make faint nebulae invisible. We also have to remember that a sky background that is too bright introduces noise into the image (photon noise). The turbulence has to be low. We gain 1.5 magnitudes by concentrating onto one single pixel the flux diluted on four pixels because of turbulence effects. In summary, a black sky background at a site with low turbulence is necessary to reach very faint objects. Professional astronomers do not place their telescopes on high mountains, far from light pollution, for nothing. Moreover, this means that focusing has to be correct in order to concentrate energy in a minimum number of pixels. Typically, with a two meter focal length telescope, FWHM should always be less than one pixel. This parameter must be checked several times a night.

Deep sky images can, however, be made in sites that are not very favorable, such as city suburbs. We will illustrate the possibilities with an example. Let us suppose a bright sky giving a continuous background of 500 quantification units by integrating 60 seconds. The star to be detected produces a stimulus equal to just one quantification unit in 60 seconds. In this case the stellar image will be nearly undetectable because it is submerged in the sky background. By exposing 5 times longer, the star emerges from the sky background by 5 quantification units. The sky background is then at the level 2500, which is tolerable given the large dynamic range of the CCD. Under identical conditions, photographic film would saturate, preventing the detection of faint stars. With a 200 mm telescope, we have reached magnitude 19 (without filters) from a site surrounded by street lights about fifty meters apart, when the naked eye limiting magnitude was only three. Even in the heart of modern cities there is still ample opportunity for significant astronomy.

The deep sky can be observed during periods of full moon if a red W25 type filter is used with the CCD which acts as high-pass type filter that eliminates wavelengths less than 0.65 μm. With this filter, the sky darkens much more quickly than the stars do since the moonlight diffuses with the air molecules (Rayleigh diffusion), making the sky predominantly blue. The filter causes a loss of about 0.7 magnitude. The brightness of the sky background climbs beyond 9500 Å but this is not too important because the CCD's response is limited in this part of the spectrum. Between the Moon's first and last quarters, one should extend the telescope tube with a light shield to avoid stray light.

Figure 5.7 *A five minute exposure of the globular cluster NGC 6544. This object is located in the Milky Way at the edge of a dark cloud. Despite a resolution of 2.2″/pixel, the left part of the image appears as a swarm of stars. This CCD image, taken with the 60 cm telescope at Pic du Midi Observatory, shows stars of magnitude 21.*

5.5.2 The Covered Field and the Resolution

The Achilles heel of the CCD is the smallness of the sensitive surface. Combining a 1200 mm focal length telescope with a 5×5 mm CCD gives an angular sky field of about 15′. Under these conditions it is out of the question to completely capture the extensions of Orion's nebula with a 200 mm Newtonian telescope. A photographic lens with a short focal length can be placed in front of the CCD, but then the resolution becomes a problem. A CCD pixel is usually 20 to 30 μm long while the size of grain of a good photographic emulsion is less than 5 μm. This difference in spatial resolution between film and a CCD is why deep sky imaging with 50 mm focal length optics and a CCD gives an inextricable tangle of stars. However, if CCD's are reserved for observing very faint objects, the loss of wide-field capabilities is less significant. A fifteenth magnitude galaxy is rarely bigger than several minutes of arc! If a CCD must be used to create a wide-field image (imaging a cluster of galaxies, etc.), a mosaic with several successive images of contiguous fields can be made. Images obtained in this way will have to overlap slightly so that they contain stars in common in order that relative adjustments can be made by computer processing.

With the typical focal lengths of amateur telescopes, i.e. 1 to 2 meters, a pixel usually represents 3 or 4 seconds of arc. At this scale light from the stars occupies only one pixel unless they are very bright and because of diffusion. This means that stars of different magnitudes will appear to be the same magnitude on the image, giving it a rather artificial texture. A range of star sizes according to magnitude, characteristic of photographic films, can be simulated either by visualizing the image with a logarithmic scale that emphasizes the diffuse base of the image spot, or by applying a light low-pass Gaussian type filter, or both at the same time.

When stars in an image fill only one or two pixels careful photometric interpretation is essential. Sensitivity inhomogeneities inside the pixel can have unfortunate consequences. Re-read section 5.2.1 which deals with aperture photometry. At professional observatories, deep sky images usually have a scale of about 0.3″/pixel. This fully exploits the superior resolution found at these sites and enables precise photometry by spatially over-sampling the stellar image. At a low altitude site, an excellent compromise between field size and the resolution is a scale of 2″/pixel.

5.5.3 Compositing

Compositing images is a technique that is often used with CCD's. Instead of making a 30 minute image, three 10 minute images are made and summed to reach an *equivalent* 30 minute image. It is, of course, necessary to register the images to one another to within a fraction of a pixel before compositing them. On this subject, re-read sections 4.5.5 and 4.5.9.

Dividing long exposures into shorter ones is a rapid way to detect any abnormality in the image, such as a tracking error, and to correct it with a minimum loss of time. Another advantage is that noise created by cosmic rays stands out when successive images are compared. If the telescope is slightly offset between each of these short exposures, the CCD's local defects can be greatly reduced by averaging after compositing. The ambiguous ghost images produced by reflections on optical surfaces are also eliminated, and more generally all local defects on the CCD's surface. This is the "Random Shift and Add" technique, used systematically when the purpose is to observe very faint objects. In fact, it is essential to remember that in deep sky imagery, the quality of the photometric correction is more important than a high quantum efficiency or a low readout noise. The acquisition of randomly disregistered elementary images and their compositing is a key point to obtain deep images.

The shift between elementary exposures should be worth between 10 and 30 pixels for good performance (to eliminate low frequency defects). Images are then co-registered to a fractional part of a pixel by using the position of one or several stars which are visible on each image (see section 4.5.9). The final useful area of the cleaned picture is the common overlap area of elementary images.

Instead of adding co-registered elementary images, it may be preferable to calculate a median image, pixel after pixel, over the stack of co-registered pictures (see section 4.5.4). The median filter is much longer than a simple arithmetic mean, but it is much more effective in eliminating (completely and automatically) local defects such as cosmic rays.

There is no reason not to composite images taken on different nights. In this way we can build images whose integration times can last several dozen hours or continue an exposure that was unfortunately interrupted by the arrival of clouds.

When we want to observe the faint environment of a bright object without saturating the camera the exposures should be divided into short segments. A typical example is the study of a concentrated globular cluster on which we want to "take out" the faint stars of the halo while resolving the center. Another example is the observation of a planetary nebula surrounding a bright central star. Suppose that the bright part of the object just saturates the camera by exposing T seconds (level 4095 if we code the intensities in 12 bits). Let us sum the two images of the object, each one being acquired by exposing $T/2$ seconds. Although the result is an image in which the brightest point reaches level 4095, the camera will not have saturated during the individual exposures, even when keeping a good safety margin. In fact, we can thus accumulate a greater number of images exposed for $T/2$ seconds. The final intensity will be higher than 4095, but the radiometry will be significant (within the computer the coding of signed integers is done in 15 bits or 32,768 possible levels).

Dividing a long exposure into several short ones has spectacular applications in the observation of comets. The comet is mobile compared to stars and it is absolutely necessary either to track on the core of the comet, or else to calculate its motion in advance and to guide the telescope accordingly. These techniques are not easy to implement because the core can be undetectable, or the telescope may be mechanically unable to accommodate the necessary corrections. With a CCD we can make a series of short exposures using a guide star for tracking so that the comet's motion in relation to the stars is not significant (motion of less than one pixel).

All the exposures acquired in this way will then be composited with a shift corresponding to the comet's motion. In this final image the stars will be streaked, but the comet will be sharp. This technique is both easy and efficient.

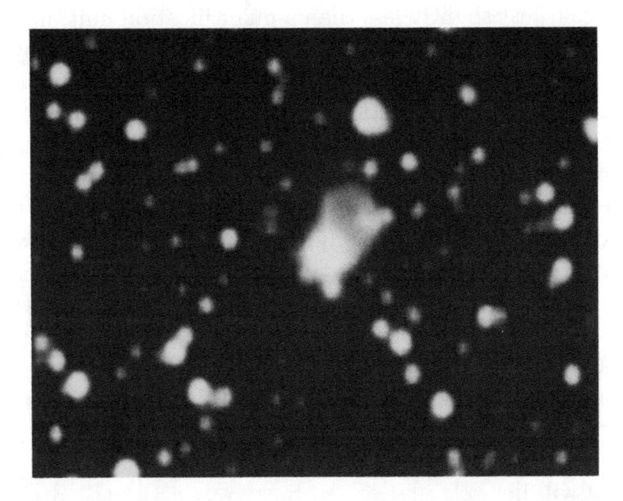

Figure 5.8 *An example of an object that is very difficult to observe: the planetary nebula PK68+1.2 located near a bright star. To get the final image we had to composite 8 images that were exposed only 30 seconds each. The star does not present a blooming streak. This would not have been the case with a single 240 second exposure. Image taken with the T60 telescope at Pic du Midi Observatory with a TH7863 CCD camera.*

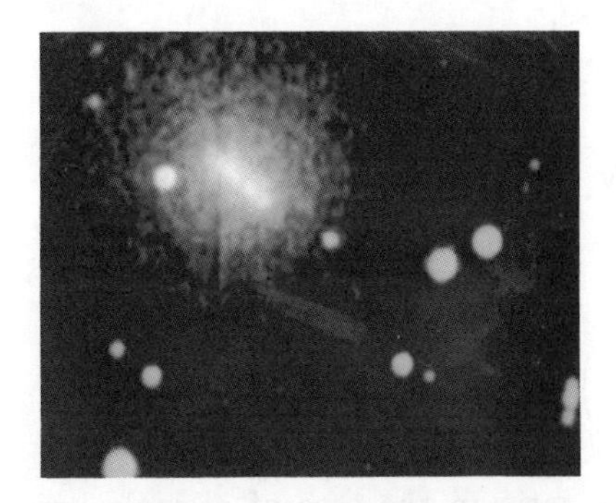

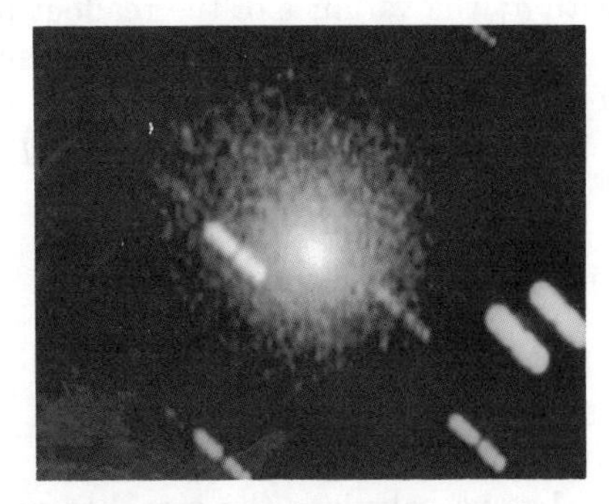

Figure 5.9 *The Brorsen-Metcalf comet observed at the T60 telescope at Pic du Midi Observatory on July 10, 1989 when it had just been rediscovered. On the left, we composited 6 images (each exposed one minute) by recentering them on a field star. Note the significant shift of the comet, justifying the short integration time used for the individual exposures. On the right, the recentering of the 6 images is done on the core of the comet which appears very clearly.*

Compositing is not necessarily better than an equivalent long exposure as far as detectivity is concerned. In fact, the problem is complex because the camera's characteristics and the nature of the objects observed vary. Let us suppose, for example, that we are studying a very faint star that produces a signal just slightly less than a quantification unit, in integration time T. This star is then virtually undetectable. However, by exposing ten times as long, it comes out clearly. What happens if we take 10 individual exposures with integration time T and then sum them digitally during processing? Since the signal produced on an image exposed for time T is less than 1 quantification unit, logically the star should be completely undetectable on the final image because it does not appear in the images taken individually. In practice, the fluctuating character of noise and some statistical principles interact so that the star will very slightly emerge after compositing. However, the star will be very faint and in no way comparable with the result of one long exposure. Let us try to express these relationships mathematically.

Let an image be taken with integration time T, and let N be the number of photocharges and the number of thermal charges produced per unit of time.

The signal will be:

$$S = T \cdot N.$$

The variance of the noise will be:

$$\sigma^2 = T \cdot N + \sigma_1^2,$$

with σ_1^2 the variance of the readout noise.

Now, let us divide our long exposure into n exposures of equal duration. The signal will be:

$$S = \sum_1^n \frac{T}{n} N = T \cdot N,$$

and the noise variance will be:

$$\sigma^2 = \sum_1^n \left(\frac{T}{n} N + \sigma_1^2\right) = T \cdot N + n\sigma_1^2.$$

Let us calculate the signal-to-noise ratio in the case of a single long exposure by the following equation:

$$\frac{S}{B} = \frac{T \cdot N}{\sqrt{T \cdot N} + \sigma_1},$$

and with compositing:

$$\frac{S}{B} = \frac{T \cdot N}{\sqrt{T \cdot N} + \sqrt{n}\sigma_1}.$$

Thus the readout noise increases by a factor of $\sqrt{n}$ in the case of compositing with respect to a unique exposure. In other words, we will have to composite a larger number of exposures than necessary to reach the performance of a single long exposure. In amateur cameras and as long as we expose for longer than a few minutes, the term $\sqrt{T \cdot N}$ is usually greater than the term σ_1, because the CCD is not optimally cooled, which in turn creates thermal charges (remember that the quantity N includes this kind of charge). The effect of the coefficient $\sqrt{n}$ will often be less important than it appears. In addition, the "Random Shift and Add" technique allows us to decrease the spatial noise during compositing, which is impossible in a single exposure. There is an exception when making short exposures to freeze the turbulence in high-resolution imaging. In this case, the thermal noise is low because the integration is short. The readout noise is therefore significant, and for this kind of application it is better to place an image intensifier in front of the CCD. The signal emerging from the intensifier (off the phosphorus surface) is such that readout noise again becomes negligible.

For the compositing technique to be effective the noise should be higher than the quantification unit. Indeed, if the noise is completely included in the value of a unit, it will appear only in a fluctuation of one quantification unit at most, and the compositing will reveal little that is "new." This highlights the importance of having a high-gain amplification of the video signal—while presenting an unpleasant, noisy appearance, the resulting image can contain useful data.

To conclude, it is not easy to decide whether compositing several short exposures or taking one longer unique exposure is better. Our experience teaches that most deep sky imaging should use integration times sufficient to reveal the object in a single exposure. However, it is also beneficial to composite several (2 to 5) of these exposures to reduce the systematic defects of the sensitive surface.

5.5.4 Color Imaging

We know the problems that come up when we want to take deep sky images in color photographs. The different layers that form the emulsion and correspond to the three fundamental colors do not respond in the same way over time to the luminous signal. The result is an image with severe color shift that cannot be used for scientific study.

The optimum solution for obtaining photographic color images is to take three successive exposures on three different black and white plates through three filters: red, green and blue. This is the trichromatic principle. The three negatives are then projected through an enlarger onto color photographic paper. By precisely adjusting exposure times during enlarging, a excellent color match is possible. These operations are complicated and require much experience to master. Seldom is a first attempt successful.

Color imaging is possible with a CCD, but we have to remember that it is out of the question now (1990) to use the common color CCD's made for camcorders (see section 1.6.3). A monochromatic CCD operates much like photography: three images are successively acquired through colored filters. The three images are easily composited by recentering on a star in the field. To restore the color images on a computer screen, re-read section 4.4.3.

During acquisition we will try to have nearly identical density for the three images by adjusting the integration times. If identical density cannot be achieved, we can always make adjustments by balancing the images on the computer (multiplying by a normalization constant). At the moment of acquisition, we should, however, still try not to saturate the stars on a trichromatic component because then the radiometry of the stars in question is lost, and they will appear abnormally colored during restoration.

When choosing filters we can use a combination of BVR filters that will give a good illusion of what the eye would see. However, given the relative low sensitivity of thick CCD's in the blue, we will sometimes have to compensate by using a VRI combination. A VRI combination is not as exact, but the contrast in colors will sometimes be brighter, allowing enhancement of certain morphological details.

Color imaging of gaseous emission nebulae requires special procedures. These objects emit very specific radiation (spectral lines) and images have to be taken through interference filters that isolate them. Filters of a few hundred Angströms will separate the principal spectral lines. We will essentially use the nebular lines OIII at 5007 Å and Hα + NII at 6560 Å. We can complete the trichromy with an infrared filter of the high-pass type (beyond 7500 Å). We can also be satisfied with a two-color image on the two main emission lines. The morphological details, recorded on planetary nebulae in particular, are often spectacular.

5.5.5 Binning

Remember that binning is a technique that consists in gathering information from several adjacent pixels to aid in processing the data contained on the primary pixel. Binning is commonly used to aid in pointing an ob-

Figure 5.10 *Object LBN218 is a pale nebulosity surrounding star 44 Cyg. The image shown here was obtained with a 2×2 binning factor and a camera equipped with a TH7863 CCD mounted on a f/6.1 280 mm telescope. The exposure is only 10 minutes (compositing of 2 images each integrated 5 minutes). Note that 2 × 2 binning is equivalent to a decrease in the focal length of the telescope by a factor 2 (loss of resolution) while keeping the same diameter of telescope. In other words, our instrument was operated at f/3 when we made the image. This performance is close to that of a Schmidt telescope and explains the relative ease with which LBN218 was recorded.*

ject. The pixels are then gathered 4 by 4 by summing the lines and columns in the CCD's registers. Compared to a standard readout of the CCD in full resolution, this allows us to accomplish the following goals:

1. Read the array four times faster (essential when we point to rapidly control the effect of moving the telescope);
2. Display the image four times faster (always the same concern with interaction);
3. Increase the sensitivity fourfold and thereby decrease the integration time needed to detect a given object (more interaction).

The fact that the spatial resolution is degraded by a factor 2 is not generally a problem during pointing. The only exception is locating small planetary nebulae that might be confused with stars.

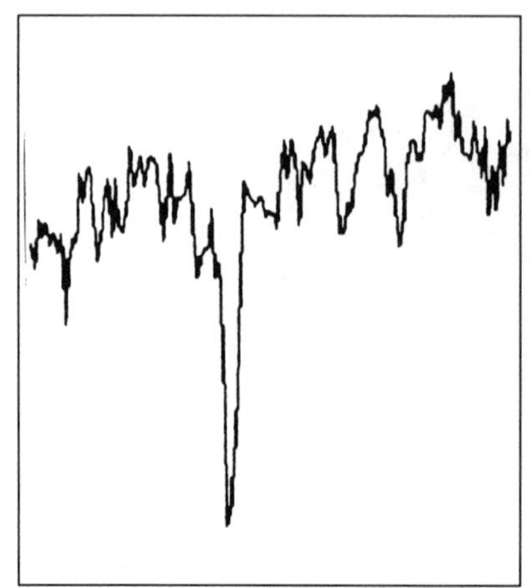

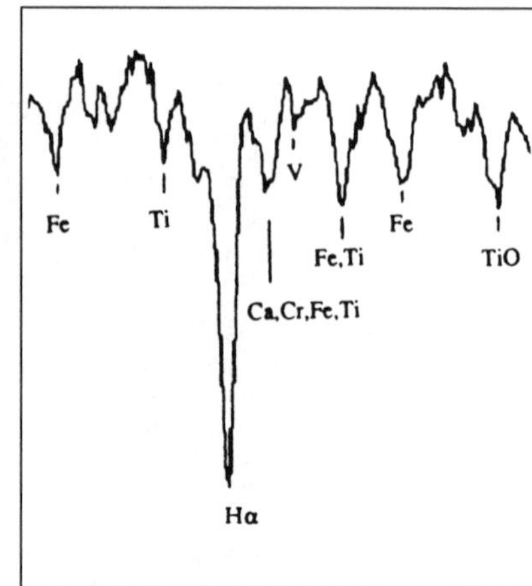

Figure 5.11 *Application of binning in spectrography. On the left, cut in a line along which the spectrum of the star 72 Cyg is spread (magnitude 4.9, K1 type spectrum). The spectrum near the Hα line is quite noisy. In addition, the baseline (average level of the continuum) presents an abnormal inclination as a function of wavelength because of a slight non parallelism of the spectral dispersion with respect to the lines of the CCD. An examination of the image shows that the spectrum is spread over 10 columns. By summing these ten columns we obtain the spectrum on the right with a considerably improved signal-to-noise ratio. Spectrum taken by Alain Klotz with a spectrograph built by Daniel Bardin and with a TH7852 CCD camera installed on the T60 telescope at Pic du Midi Observatory. The sampling is 1 Å per pixel.*

Binning is an important tool for spectrography with a CCD array. In the axis perpendicular to the spectral dispersion, the spectrum always presents a widening because of the star's motion along the input slit or because of poor focusing on the star. All things being equal, spreading of the produced flux reduces the detectivity. Fortunately, thanks to binning, we can gather the signal spread over several lines of the array in one single line and thus obtain a much denser and better quality spectrum.

5.5.6 Photometry

We have already discussed in section 5.2 the CCD's ability to measure flux. We will explore this with three examples. The first concerns photometry with several colors; the second, rapid photometry; and the last, surface photometry.

Star colors immediately come to mind when we compare the relative intensity of stars in images taken over a range from the blue to the near infrared. The high quality of CCD photometric measurements enables

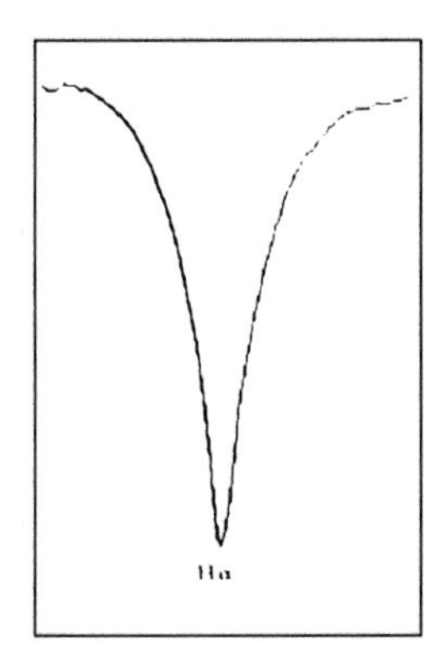

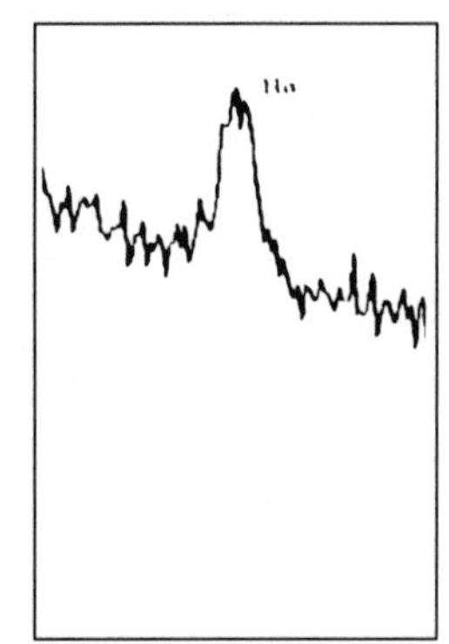

Figure 5.12 *Three examples of very different spectra that show, on the one hand, the power of the CCD technique associated with binning, and on the other, the interest of stellar spectroscopy which thus becomes truly accessible to amateurs. The left spectrum shows the region of the Hα line in the spectrum of Vega. The very characteristic profile of this line in a hot A-type star is clearly shown. The central spectrum is that of a cool star, still in the region of Hα (61 Her of M4 type). The hydrogen line is considerably less evident than in the preceding spectrum. On the right, we have the spectrum of 28 Cyg around Hα. This star is of Be type and presents the hydrogen line in emission. The profile of the line evolves from day to day, and its observation brings precious information about gas motions in the outer layers of the star. Images taken by A. Klotz at the T60 telescope of the Pic du Midi. The sampling is 1 Å per pixel.*

the creation of diagrams like the one shown in Figure 5.13; the HR diagram of an open cluster. The stars were measured by aperture photometry on two images acquired through a green filter (V) and an infrared filter (I).

Another type of observation is shown in Figure 5.14 which plots the light curve produced by Titan occulting 28 Sgr. This rare event happened on July 3, 1989. This light curve is not the result of measurements taken with rapid photometry but is in fact the result of using a CCD. Here is what we did. To be successful, this kind of observation must make the greatest number of photometric measurements within a limited period of time. To optimize data acquisition, only a small part of the array (a TH7863) was digitized, the rest was read rapidly as part of the cleaning sequence. For this particular occasion, we used a window composed of 30×30 pixels. This was large enough to contain the double star plus the satellite and a bit of sky background to provide a zero reference. Only two rapid cleaning cycles were used between two digitizings. Normally there are 10 cleaning cycles. Furthermore, during the cleaning cycles we used a factor of 2 binning in the horizontal register that considerably accelerated operations. All this was possible because between two consecutive readouts of the CCD, very little time went by (several tenths of a second) and the detector did not have time to saturate with thermal charges. This was further helped by the absence of bright objects in the field.

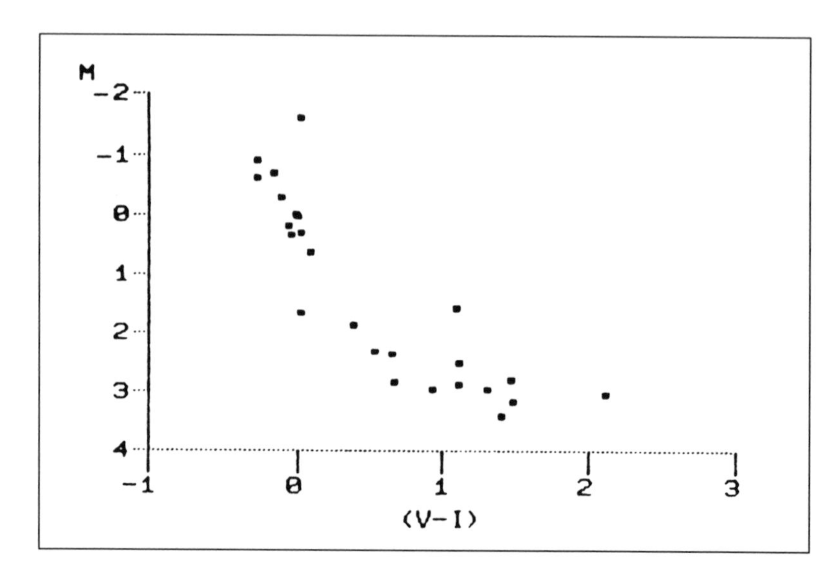

Figure 5.13 *HR diagram of the open cluster M21. We can clearly distinguish the main sequence. The magnitude scale is arbitrary.*

To gain time, we did not display the image on the computer's monitor—only one line of text was displayed and actualized between each acquisition. The text supplied the X, Y coordinates for the brightest point, which was all the information needed to keep the object centered with the telescope's slow motion adjustments. This line also displayed the integrated intensity inside the window (the sum of the intensities of all the pixels), the intensity of the most illuminated pixel (to guard against saturation) and the intensity of the sky background measured in a ring two pixels wide at the edge of the 30×30 window.

The only precaution necessary during this type of acquisition is not to bring the object under study into the sky background measurement zone; the latter would then have an incorrect level. If the tracking precision is poor, the size of the measurement window will have to be increased.

All the calculations on the image are carried out in real time. The results are saved between each acquisition on the hard disk in the form of a sequential file. We thus record elapsed time throughout the observation to within a hundredth of a second, the integrated intensity inside the measurement window and the integrated intensity on the zone covering the sky background at the edge of the window. The time for saving on the disk is almost instantaneous. For processing, the file is read by a second program that carries out the subtraction of the sky background and the display of the light curve.

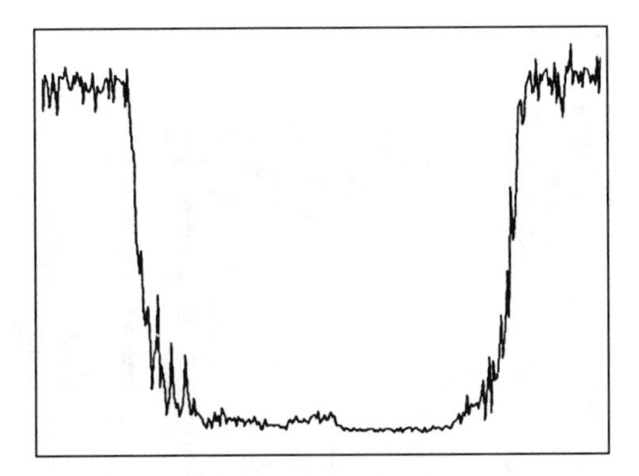

Figure 5.14 *Occultation of star 28 Sgr by Saturn's satellite Titan on July 3, 1989 recorded through a filter centered at 656 nm. The event was recorded with the T60 telescope of Pic du Midi Observatory, but we feel that it could have been studied efficiently with a more modest telescope (200 mm). The occultation lasts a total of 5 minutes. The first and last instants are marked by very important variations in the star's brightness as its light crosses the cloudy layers of Titan's atmosphere. Because of refraction, the star almost never disappears from the observers' view during the event. A central "flash" (rise of the signal in the middle of the occultation) is detected. It is caused by the focusing of the stellar flux on Earth, with Titan's atmosphere acting as a lens.*

If the integration time is minimum, this procedure allows up to four acquisitions per second with a TH7852 or TH7863 CCD and an 8 MHz PC AT compatible micro-computer. This rapid cycle rate is not limited to study of the evolution of a transient event such as the occultation of Titan, but can also be used to observe occultation of stars by asteroids or to study rapid variable stars.

Surface photometry is the study of the distribution of intensities inside extended objects. Figure 5.15 shows work done on a well known elliptical galaxy: M87 in the Virgo cluster. We modeled the form of the galaxy by a system of ellipses. This was done with a very powerful program written by Philippe Prugniel of the Observatoire Midi-Pyrenées and translated to a PC by Alain Klotz (`SPAS` program—Surface Photometry Analysis Software). The program supplies very different parameters: distribution of flux according to the distance from the center, the rotation of the ellipses' axes according to their dimension, the position of the center of these ellipses, etc. All these elements are used for the astrophysical interpretation of what we observe. We should note that just a few years ago it was unheard of that such precise work should be done by an amateur. Today things are different with the availability of CCD's and software like `SPAS`.

Starting with the galaxy model calculated with `SPAS`, we can synthesize an artificial elliptical galaxy that resembles the real image. If we subtract the synthetic galaxy from the original, the elliptical galaxy itself disappears.

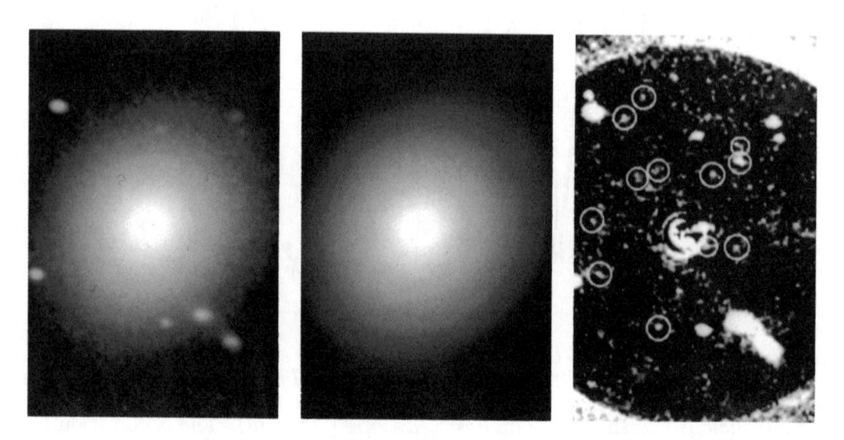

Figure 5.15 *On the left, image of the galaxy M87 taken with a 280 mm telescope in a 20 minute exposure. In the center, a numerical model of the galaxy M87 was made with the* SPAS *program. On the right, we have the result of the subtraction of the model from the original map. The points surrounded by circles are the main globular clusters of the M87 system (the brightest ones are of 19th magnitude).*

In the case of M87, what remain are dozens of globular clusters. It is important to see that the clusters located toward the center would have been completely invisible without the modeling operation because they are considerably fainter than the bright halo of the galaxy.

5.6 Planetary Imaging

In the domain of high spatial resolution the large size of the CCD pixel would at first seem to be a disadvantage when compared to photography. It is true that the focal length of a telescope has to be greatly extended, with the help of oculars or Barlows, to reach a scale of the order of 0.5″/pixel. But at excellent sites like the Pic du Midi a scale less than 0.2″/pixel is needed to process all the information contained in the image.

With such sampling, the planetary disk covers many pixels, reducing the illumination of the focal plane. However, the detector's great sensitivity allows a reasonably short exposure time (less than or equal to one second for Jupiter, about a tenth of a second for Mars).

Selecting images for high resolution can be done just after acquisition thanks to rapid visualization. Hundreds of images may be acquired and examined almost instantly during a session—but only the very best need be retained.

Unsharp masking is the best tool for processing planetary images. This method allows us to enhance very faint details easily. One of the great strengths of the CCD is its large dynamic range. This means that structures that are distinguished by a fraction of a percent with respect to the environment can be emphasized by digital processing. By comparison, photographic film has a much smaller dynamic range, and the processing that has to be done to extract the information is much more painstaking. The CCD's dynamic range must be used judiciously by digitizing on a large number of bits. Experience shows that digitizing on 8 bits (256 levels) is insufficient in planetary imaging to get subtle details. The strict minimum is digitizing on 10 bits (1024 levels), and 12 bits is not too many.

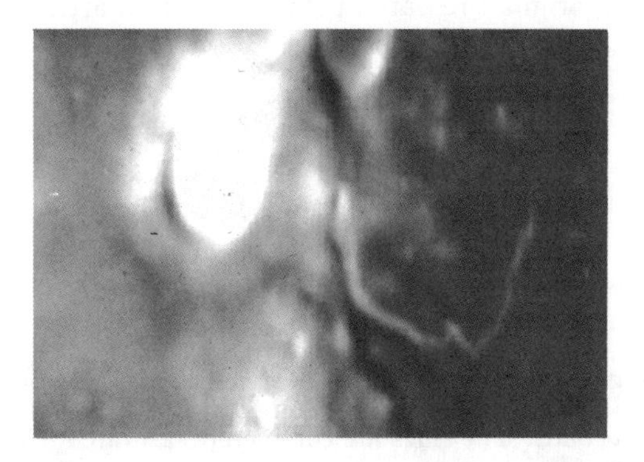

Figure 5.16 *The Schröter crevasse snakes along the moon's surface not far from the Aristarcus crater. The latter is particularly bright toward the period of the full moon (it is the brightest feature of our natural satellite). It was at this moment that our image was taken (280 mm telescope and enlargement of the image by projection with an ocular). This kind of imaging allows one to make very precise albedo maps. Using colored filters shows that the moon has a less neutral color than we usually think. Similar images have been used to make geomorphological maps.*

We absolutely must not forget the CCD's spectral response: we can detect radiation from the blue (0.4 μm) into the near infrared (1 micron). Consequently, you have to consider the chromatic abnormalities that can be caused by the amplifying optics. Chromatism may be limited by using filters. A filter is advisable anyway, given the way the appearance of planetary disks changes with wavelength. For example, Jupiter is an extremely colorful planet. The variation in its structures is striking between images taken in the blue (very contrasted bands, relatively few fine structures) and the near infrared (low contrast but many details). On the planet Mars an infrared filter gives important penetrating power, even when there are dust storms. The clouds of the red planet are easily detectable with a dark blue filter. With filters and a CCD it is possible to make a real climatic study of Mars.

We should also mention gaseous planets, especially Jupiter. A CCD can detect the absorption band of methane at 8860Å. With an interference filter about a hundred Angströms wide, the planet is extremely dark and looks like a piece of coal because of the absorption of radiation by methane. The exposure time we should use when we observe in this band is counted in minutes! Stellar tracking is imperative. The result is worth the trouble: the face of the giant planet is completely changed with a very intense equatorial band, ovals in the temperate zones with no equivalent in the spectral continuum, poles in the shape of bright arcs, strong darkening of the limb, etc. At this wavelength we can observe the Jovian high atmosphere. Planetary observers seeking original research projects might start here.

Figure 5.17 *Negative image of Saturn taken in the methane band with the T1M telescope of the Pic du Midi with a camera equipped with a TH7852 CCD. The exposure is 100 seconds. The rings are completely saturated while the disk shows very contrasted bands.*

In fact, all the gaseous planets can be studied in the methane band. In a short exposure Saturn shows only rings mysteriously floating in space because they reflect all of the sunlight falling on them while the methane on the globe absorbs it. The same principle works for observing Uranus' rings because in the methane band the planet is sufficiently darkened so that it doesn't saturate.

The rotation of a planet or changes in the appearance of its atmosphere can be studied in a gripping way by rapidly displaying images taken at regular intervals one after the other in sequence. The effect produced by this animation is unfortunately difficult to reproduce in this book! Similarly, alternating the display of two images is a good way of detecting a very discreet change in appearance between two images (supernova, asteroid, etc.). This is the electronic version of a *blink* comparator, which has long been used to compare photographic plates. In the `IT` program, a command (`CAPTURE`) copies a part of the video memory into two distinct arrays for the two images. The `BLINK` command carries out rapid display in sequence by successive copies of the contents of the two arrays onto the screen memory (the typical frequency is 5 displays per second). It is even possible to "slide" into real time an image with respect to another to facilitate their superposition.

The ease of digital processing makes it relatively simple to obtain color images. We use the trichromatic method, i.e., three distinct images are acquired through different colored filters and the superposition of these three components during subsequent processing reveals the colored aspect of the planet (it is the same for deep sky imaging). The three filters do not have to correspond to the classic red, green and blue. For Jupiter we commonly use blue (check it for infrared blocking), red and infrared filters. The shades restored are the "true false colors" but the information thus emphasized is considerable. Images in the three filters are not usually obtained simultaneously and we have to make sure that the planet's rotation during the acquisition is not significant for the expected resolution. It is advisable to place the filters on a wheel or sliding mount to allow rapid changes. Be careful, the focusing can be different according to the filter used. As far as possible, the chromatic balance of the three colored components will be selected so that well known details appear as they normally do (e.g., Mars' polar caps are white).

5.7 But Still ...

Often the most common objects that are the most easily observable by traditional means can reveal hidden aspects thanks to the CCD. The sun, a continually boiling surface often with spots, is a case in point. Following the evolution of sunspots is an exciting subject, full of surprises and unknowns. Sun spots can be studied visually or photographically—it's not very complicated. With the CCD, the same type of work is possible but the efficiency is increased. Thus the relative motion of a group of spots will be measured simply and precisely with the geometric properties of the CCD's surface and the ease offered by image processing. Observation over the spectral range of silicon using selective filters will allow segmentation and study of the various levels of the photosphere. The CCD's large dynamic range will be tested in observing discreet bright points hidden inside the shadow of the spots. In short, everything can be read in a new way when a CCD is placed at the focus of an instrument.

Whatever one's specialty, there are not many problems in using a CCD while observing, but the difficulties frequently accompany reduction, especially among amateurs. Digital image treatment can largely be automated, and it is obvious that using a CCD greatly speeds the acquisition of data. For this reason, we tend to acquire more CCD images than photographic images. This multiplicity of images can be a handicap.

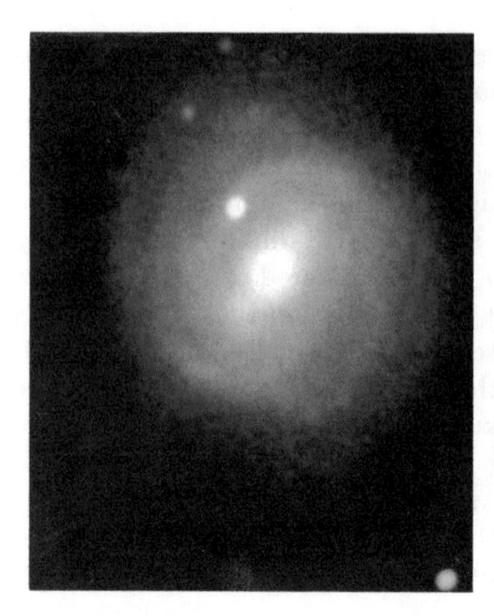
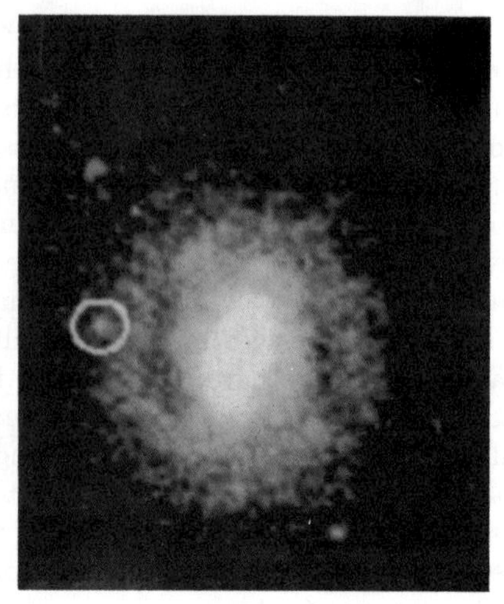

Figure 5.18 *Supernova observations. The picture on the left shows the explosion of a star (SN1989B) in the galaxy M58 (image taken with the T60 telescope at Pic du Midi Observatory in a 20 minute exposure). The picture on the right shows another supernova (SN1989k), with a magnitude of 18.5 in the small galaxy NGC 5375 (composite of 2 images each exposed 5 minutes with a 280 mm telescope). The CCD's high sensitivity is an asset in the discovery of supernova. In fact, by exposing about 5 minutes with a 300 mm telescope, we can detect an additional star down to magnitude 17–18 inside galaxies. The short integration time allows us to survey a large number of galaxies in a given period of time. In addition, it is possible to analyze the image very quickly to detect a possible supernova. The reduction can even be made automatic, using an image bank (such a research program exists in the United States at the professional level).*

The major advantage of a CCD system is precise measurement. All the objects present on a CCD image can be evaluated in many ways: intensity, spectral distribution, size, etc. Much astrophysical data is thus extracted, and that's where the difficulty arises: What do we do with it all? The wisest course of action is surely to join a professional program. Professional astronomers are well placed to choose the most worthwhile observational subjects. They have the information which prevents their wasting their energy by observing an object studied many times before. They use strict, tested reduction techniques and can synthesize work done elsewhere in the world to obtain a better result. However, professional astronomers need us amateurs because, armed with a CCD, we can offer them high-quality data. We need them to advise us on selecting an observing program, to increase our knowledge and sometimes to publish. Therefore, the collaboration between amateurs and professionals can only benefit both communities—this boon to both is a direct effect of the CCD revolution.

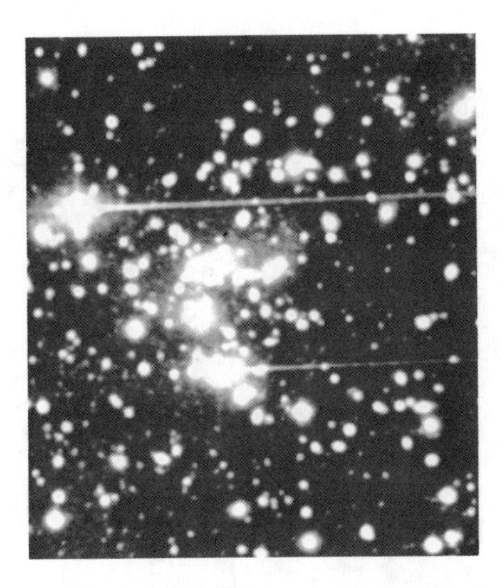

Figure 5.19 *This is the distant galaxy cluster Abell 2065, known as the Boreal Crown. It was taken with the T60 telescope of the Pic du Midi. The cumulated exposure time is relatively long: 45 minutes. In this image (TH7863 array) we can see galaxies down to magnitude 23. In fact, most of the objects are galaxies, not stars. We can count almost 200 galaxies on the image. The brightest ones are included in a pale halo that is very real but absolutely undetectable on the photographic images of the Palomar Observatory Sky Survey.*

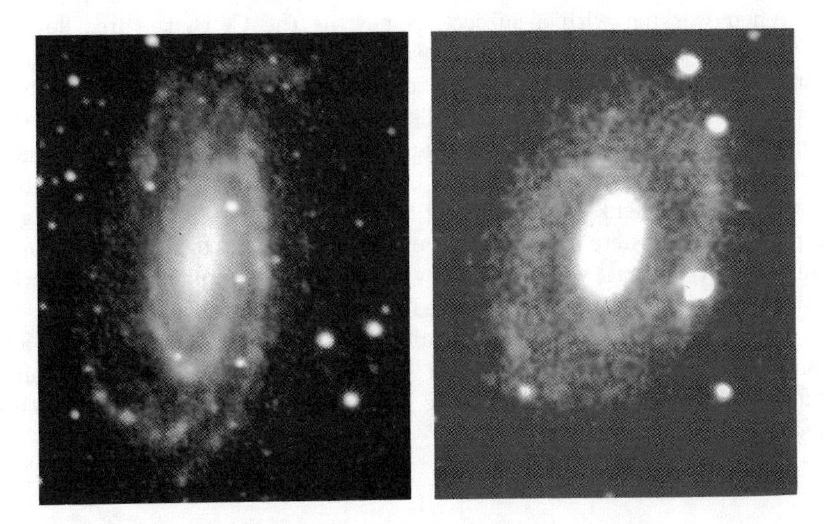

Figure 5.20 *On the left, the galaxy NGC 5033. On the right, the galaxy NGC 210. 10 minute exposures with a 280 mm telescope.*

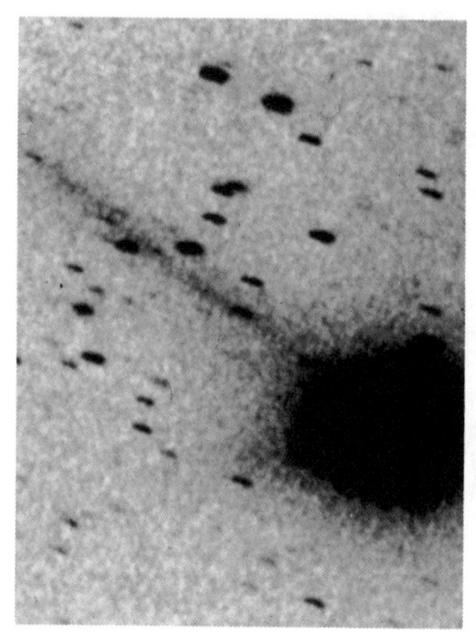

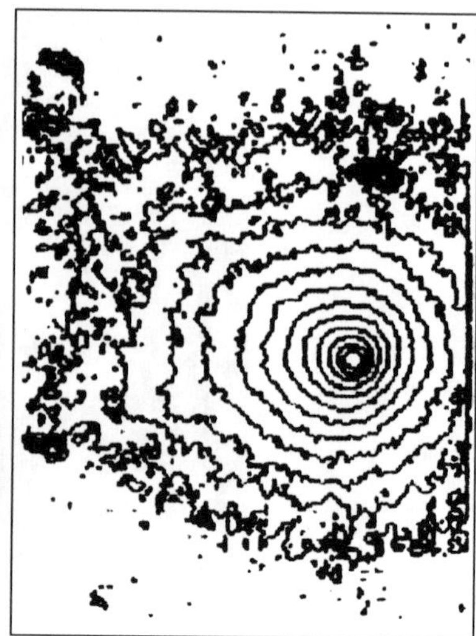

5.21 *Back to Earth after our inter-galactic journey, in the company of the comet Okazaki/Levy/Rudenko (1989r) observed on October 1, 1989, at 19H15 UT. Composite of 6 images each exposed 300 seconds with a 280 mm telescope. We note a very fine plasma tail and an extremely thin dust tail that starts at the core and goes in an approximately horizontal direction. The representation in logarithmic isophote curves clearly shows the asymmetric appearance of the coma.*

When working with a subject as new as the CCD, the free flow of information between amateurs is vital. We can easily imagine an image bank where everyone can bring his latest observations but also take the data required to finish a study. This calls for coordination and discipline (the same filters, same treatment procedures, same image size, same mounts, etc.), but the advantages seem to be so great, to me, that any serious observer would yield. We absolutely have to avoid the anarchy that reigns in photography where hundreds (thousands!) of images of the Orion nebula are taken every year, while not far away there exists a quantity of celestial objects that are completely neglected but deserve systematic study.

Exchanging information also means exchanging one's own technical advances. There is so much to discover about and with a CCD that retaining information is a crime. It's because I believe this very strongly that this book exists.

Figure 5.22 *The galaxy M109. 20 minute exposure with a 280 mm telescope at f/6.1.*

Figure 5.23 *ARP 77 (NGC 1097) is a spectacular galaxy in Fornax. Note the distortion of the spiral arm due to a small companion galaxy. 5 minute integration time using a TH7863 CCD with the 24-inch T60 telescope at Pic du Midi Observatory.*

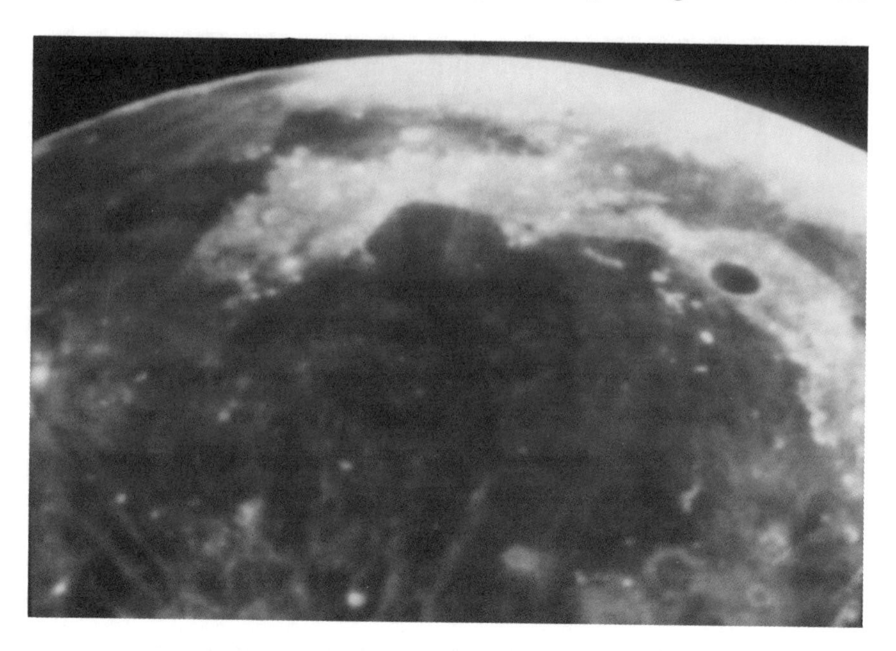

Figures 5.24–26 *Geological studies can be made with a CCD camera and narrow band (typically 500 angstroms wide) filters of atmosphereless bodies, such as the Moon. Figure 5.25 (above) was made at 5500Å and includes the Plato and Mare Imbrium regions of the Moon. The image at the top of the facing page (Figure 5.25) is the result of the* division *of two images taken at 4000Å and 5500Å. The images were co-registered with distortion algorithms before their division to eliminate libration variations between the two exposures which were about one hour apart. This ratio is representative of the Titanium abundance in the Lunar basalts. The image at the bottom of the facing page (Figure 5.26) is another division of two images at 10200Å and 5500Å, which show abundances of Olivine and Pyroxene in the basalts. All these images were sub-second exposures with a TH7863 CCD and the T60 telescope of the Pic du Midi Observatory.*

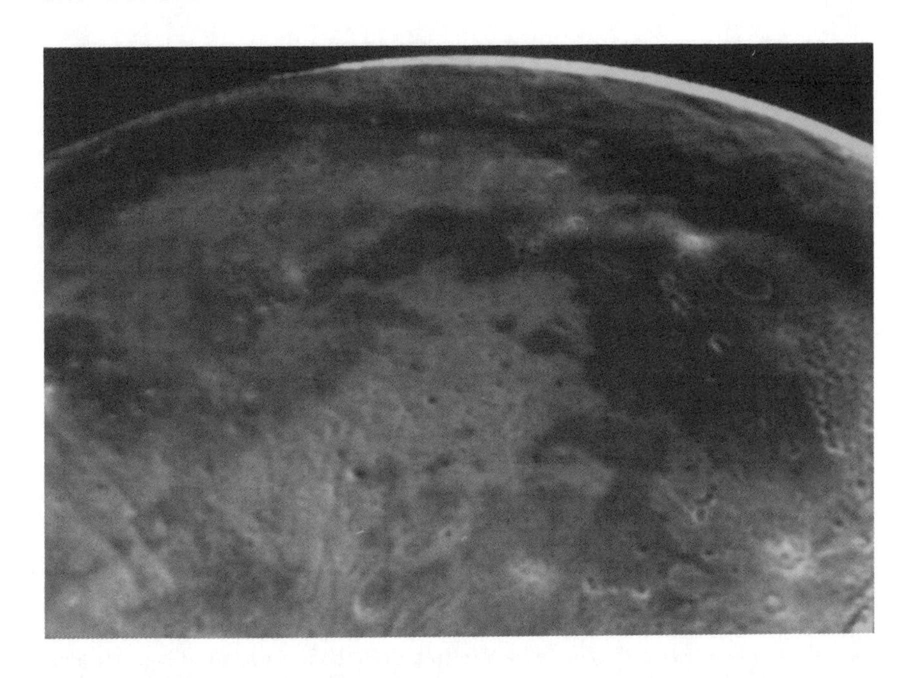

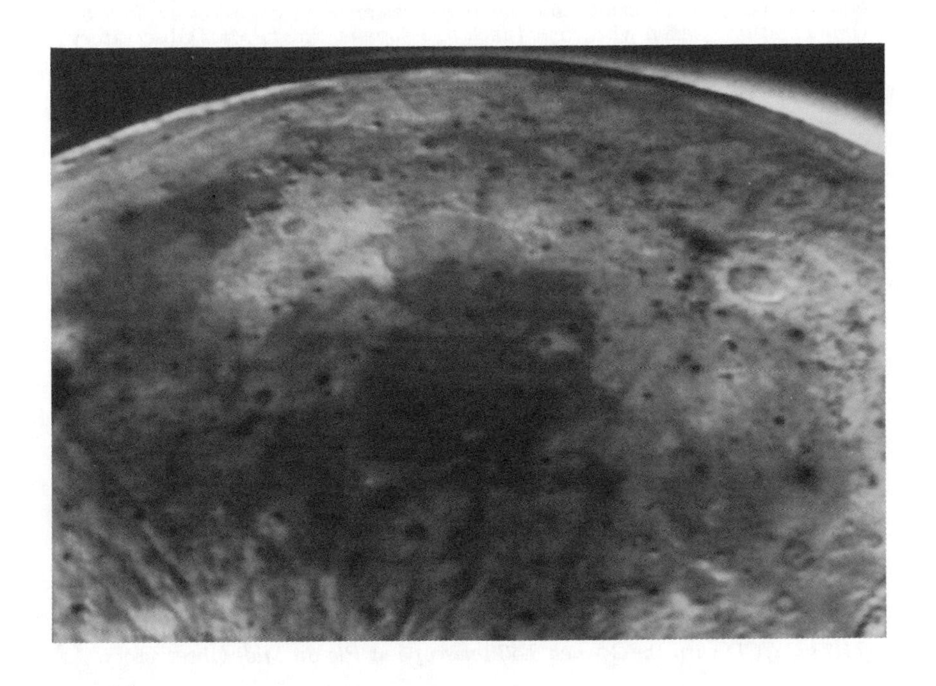

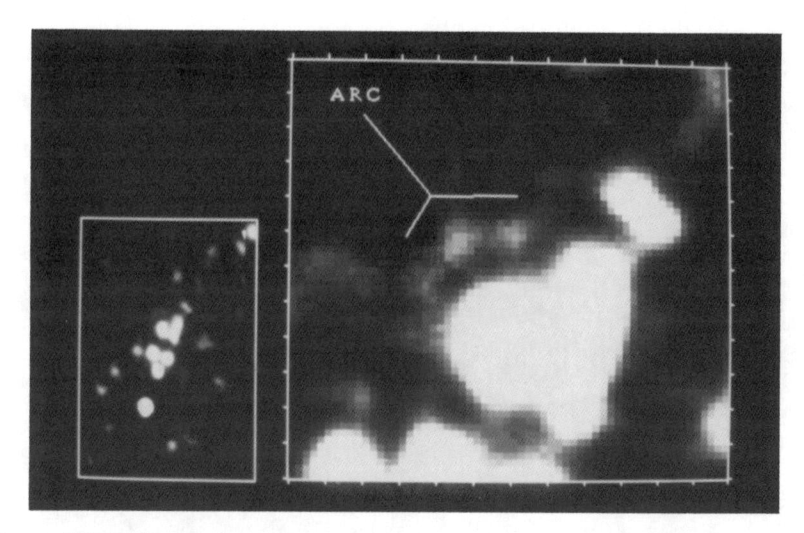

Figure 5.27 *The gravitation arc in the galaxy cluster CL2233-02 is very difficult to observe—the brightest members are barley visible on the Palomar Sky Survey which has a limiting magnitude of 21. At left, a large scale image of the cluster. At right, an enlargement that clearly shows the arc which is about 16″ in length. The spatial sampling was 2.25″/px, which means that the arc was only 7-pixels long. The image was deconvoluted to increase resolution (by about a factor of 2) using methods developed in section 4.5. The image was made by compositing three 600-second exposures taken with the 24-inch telescope at Pic du Midi Observatory.*

Figure 5.28 *ARP 120 (NGC 4435/38) in the Virgo cluster. The interaction between the two galaxies is clearly visible on this 5 minute integration time using a TH7863 CCD with the 24-inch T60 telescope at Pic du Midi Observatory.*

Appendix A

A Linear Array CCD

A.1 Introduction

Linear array CCD's are particularly simple to use. Compared to an area array, a linear array requires fewer charge transfer clocks. In addition, since there are fewer pixels, the demands on the computer are reduced. For anyone beginning to work with CCD's, linear arrays are an excellent testing ground since the technical and financial risks are much lower. Even though the linear array is simpler than the area array, both are similar in the charge transfer mechanism, image processing, and digitizing. We therefore recommend that beginners build their first camera around a linear array. This appendix briefly describes a camera built around a Thomson TH7801 CCD.

A.2 The TH7801 CCD

The Thomson TH7801 CCD has a linear array of 1728 pixels. Each pixel has a square shape 13 μm long, giving a length of about 22 mm. At each end of the photosensitive line, there are 4 elements that are light-proof, acting as dark references, plus several additional stages that are not electrically connected to photo-elements. In all, the video line has 1754 pixels.

This linear array is organized classically (see section 1.6.1) with its two transfer registers on either side of the photosensitive line. One of the registers receives the charges from the pixels in the even row, the other from the pixels in the odd row. A unique feature is that the two registers are multiplexed at the output stage.

Figure A.1 shows the layout of a photosite. The structure is quite different from traditional MOS capacitors found in CCD arrays. These are photodiodes created by the association of an n^+/p junction and a positively biased storage gate. The n^+ zone is defined by insulating tanks in SiO_2 and a p^+ implantation. In the junction, photons produce a photocurrent that charges a classic MOS capacitor (gate, SiO_2 insulation, P-type doped zone), but that is displaced with respect to the photon interaction location. This device has the advantage of not interposing more or less transparent gates in the path of the light. A linear array of this type has a quantum efficiency close to 80% at its peak spectral sensitivity. Seen from the outside, the TH7801 linear array is monophased. It cannot be any simpler.

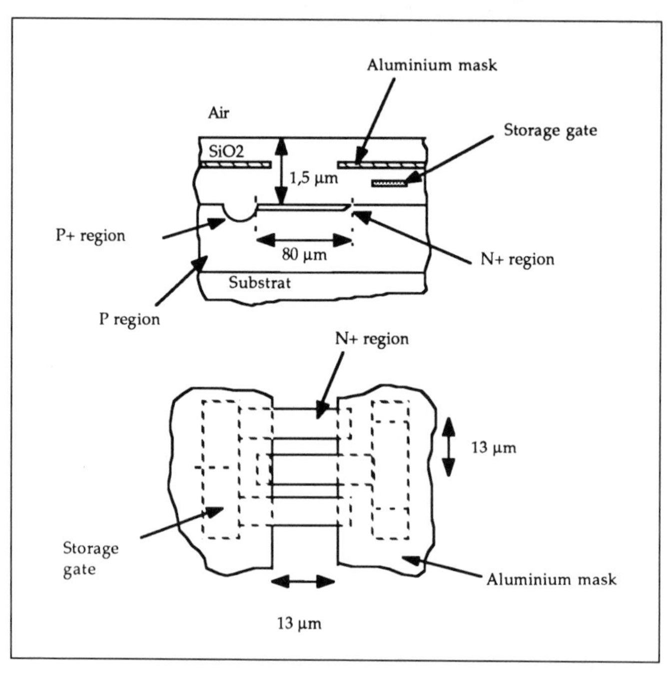

Figure A.1 *Cross-section and top view of a part of the photosensitive line in a TH7801 linear array CCD.*

The TH7801A is recommended rather than the TH7801 because it is twice as sensitive, may internally sum pixels two-by-two from even and odd row, and has an internal sample and hold.

An examination of a video signal coming from a two-register linear array usually shows a shift in level between the even and odd rows. This is called "defect of parity," and is caused because the even and odd row pathways are not identical. In addition, a slight shift of a photo-mask during manufacturing can increase the sensitive surface for some pixels and reduce it for others resulting in a measurable difference in sensitivity. The

TH7801A linear array can sum contiguous pixels two-by-two (binning) in the output diode thus eliminating the parity defect but with a reduction in resolution.

A.3 Setting up the Linear Array TH7801A

At least two phases have to be produced:

1. A phase ØP (called passage) enables the charge transfers from the photosensitive line to the lateral registers;
2. A phase ØT (called transfer) that assures the transfer sequentially to the output stage.

A third clock is optional. This is the ØR phase (Reset) that commands the setting of the output diode before the arrival of a charge packet. This phase is optional because a component functioning mode allows us to produce ØR internally by coupling ØT. However, the generation of ØR by the user is the only way to reduce noise through double sampling. Since our objective was to reduce the electronic circuitry to a minimum, we did not produce the ØR clock externally. A clocking of ØP is sufficient to bring all the charges integrated in the photosites into the transfer registers.

At the leading edge of ØT, an even row register charge packet is transferred to the diode. At the falling edge of ØT, a charge packet from the odd register is transferred to the diode. With this two transfer register device the minimum number of ØT clocks needed to read the linear array will be $1754/2 = 877$.

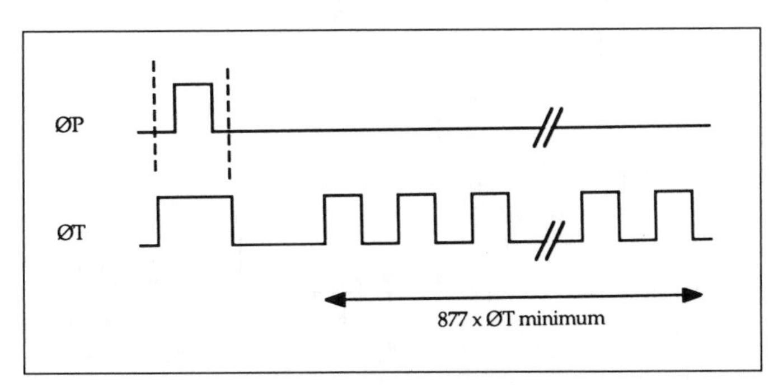

Figure A.2 *Clock timing sequence of the linear array TH801A. Note the relative phasing of the clocks during the transfer from the photosites to the transfer registers.*

A.4 The Electronics

The electronics manage the following functions:

1. Interface with the computer bus through a PIO;
2. Conversion of PIO's TTL level signals to MOS compatible signals with an SN75365 IC (Texas Instruments);

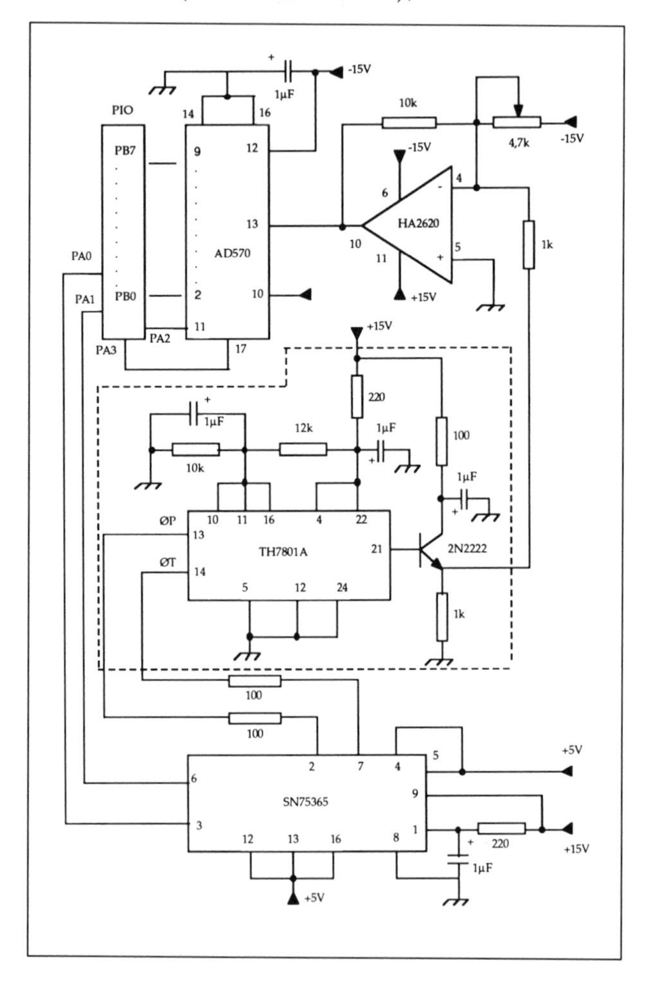

Figure A.3 *Electronic schematic.*

3. Amplification of the video signal with a gain of 10 by an HA2620 (Harris) operational amplifier;

4. Analog-digital conversion of the video signal to 8 bits (Analog Device's AD570 converter).

Figure A.3 diagrams the electronic layout. The framed section delineates the components located in the optical head. The other components can be placed in a small box situated at the foot of the telescope or even mounted on a circuit board that inserts directly into the computer's extension bus. In our case, the electronics were directly powered with DC voltages found on the computer's bus. If the voltages are taken from the computer's bus, additional filtering is required. Usually this can be accomplished by adding capacitors while observing their effect on the signal with an oscilloscope.

A.5 The Software

Phases ØP and ØT are produced respectively on pins PA0 and PA1 of the PIO. At each transition of the ØT clock, a new point on the video line should be digitized. For this, a Start-Convert signal is generated on line PA2. The conversion lasts about 25 micro-seconds. The end of the conversion is signalled by a change on the state of converter pin 17. This signal is scrutinized on line PA3 of the PIO. As soon as the end of the conversion is detected, the binary word is read on port B of the PIO and stored in the memory.

With single-shot functioning of the linear array, several rapid reading cycles must be accomplished before carrying out the integration and the final readout. After integration and just before digitizing, it is good practice to quickly empty the readout registers but not the digitizing ones. Only after having sent the required number of ØT cycles to empty the registers are the integrated photocharges sent into the registers. This procedure significantly reduces the dark signal (a stage of the transfer register would bring two-thirds of the dark current from an image point if this precaution is not taken). The digitized video line is visualized on the screen as a curve. For those willing to experiment, refinements can be added to this basic system (magnifying effects, pointing to measure intensity, etc.).

A.6 Some Application Examples

A two-dimensional image can be obtained by successively storing several digitized lines. To do this, the image formed on the detector is displaced perpendicularly with respect to the photosensitive line while the linear array is continuously read. At the moment of processing, the digitized lines thus acquired are placed side-by-side on the computer screen, restoring the image. This is the familiar scanner technique. Movement of the observed scene can be obtained simply: the driving motor in right ascension is stopped and the observed star moves in front of the linear array naturally. Taking into account the stars' angular velocity, imposed by the laws of nature, and the need to have sufficient spatial resolution, i.e. a long focal length, only bright objects can sufficiently affect the linear array to give a correct image. The moon and Jupiter are two good examples. To record fainter objects, the sweeping speed on the detector should be reduced so that an element of the observed object will be integrated long enough by a photosite. This can be done with stellar tracking, but with a driving speed that is slightly different from the nominal value. The irregularity of the differential movement thus created must be undetectable, otherwise more or less exposed zones will appear in the image and geometric distortions will appear. The image's scale must be identical in all directions, which means carefully setting the sweeping speed, the scale following the direction of the line being set by the telescope's focal length. For this procedure the sweeping speed must be carefully adjusted (setting delays between two successive readings of the linear array).

The scanner technique can digitize photographic images from a negative. The negative is placed in an enlarger and the image to be digitized is projected on the enlarger's baseboard where the linear array is mounted on a translation stage. The stage mounted CCD is moved in a direction perpendicular to the photosensitive line by a stepper motor. The motor is driven by a computer, enabling synchronization between the displacement and the acquisition frequency.

A major problem with this form of digitizing is caused by intensity fluctuations from the enlarger's lamp (powered by AC current—50 or 60 Hertz). In addition, since complete sweeping of an image may last more than a minute, voltage variations will be detectable. The 50/60 Hertz effect disappears if the integration time is relatively long, but the ideal solution is to use a lamp powered by a stable DC power supply.

If instead of a negative we digitize a slide, we have to remember that slides are transparent to the infrared. To capture the image with proper contrast, a cold filter is used to remove the infrared starting at 0.7 μm (or an interference filter working in the visible).

Because of memory limitations an 1728×1728 point image cannot be made. We have to be satisfied with digitizing a part of the line or grouping adjacent pixels (e.g., five by five). The great advantage of digitizing photographic images is that some digital image processing and analysis is then possible.

Linear array CCD's can be used for spectrography. To make a spectrum the photosensitive line is placed in the direction of the dispersion. The large size of the detector's line (compared to area arrays), the ability to record infrared and the sensitivity allowed by temporal integration of the image are advantages in spectroscopy.

In the case of stellar spectrography, the star must be positioned very precisely along the input slit of the spectrograph so that its spectral image falls on the detector. A setting and pointing device is therefore necessary—remember that the linear array is only 13 μm wide. Adjustments should be possible to make the photosensitive line and the dispersion direction parallel. If a diffraction grating is used as a disperser, a red W25-type filter should be placed in the light beam to avoid overlapping of the orders.

To obtain valid astrophysical data the spectra must be radiometrically calibrated. The linear array's output signal is linked to many factors: the spectral distribution of the object, of course, but also atmospheric transmission, the disperser's efficiency, and the detector's sensitivity. The calibration method is to measure a star whose spectrum is known outside the terrestrial atmosphere. The comparison of this spectrum with the ground based signal will allow us to plot a calibration curve that will in turn be applied to other acquisitions.

Figure A.4 shows the relative spectral distribution of an A0 type star (Alpha Lyrae, for example) that can be used to establish the calibration of the spectrograph + atmosphere system (beyond 8500 Å the spectrum of the star in question is very perturbed because of the presence of Paschen series lines).

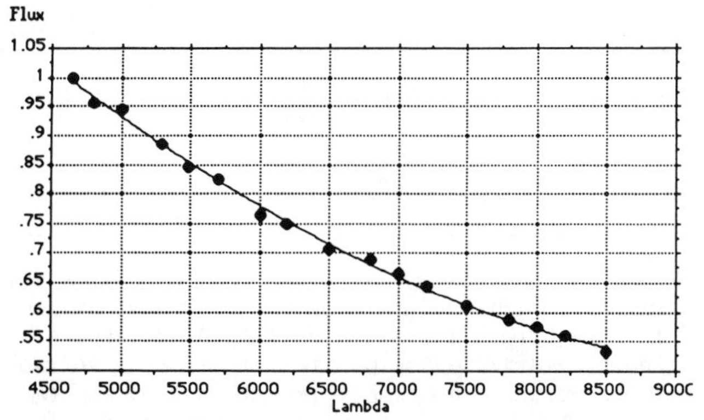

Figure A.4 *Spectrum of a star of A0 type in relative flux.*

Figure A.5 on the next page shows several spectra obtained with a TH7801A linear array and a grating of 300 lines/mm using the 1 meter telescope at the Pic du Midi Observatory. Note the evolution of the shape of the spectra according to spectral type. Note also the presence of lines caused by the terrestrial atmosphere, particularly 7500, 8200 and 9400 Å.

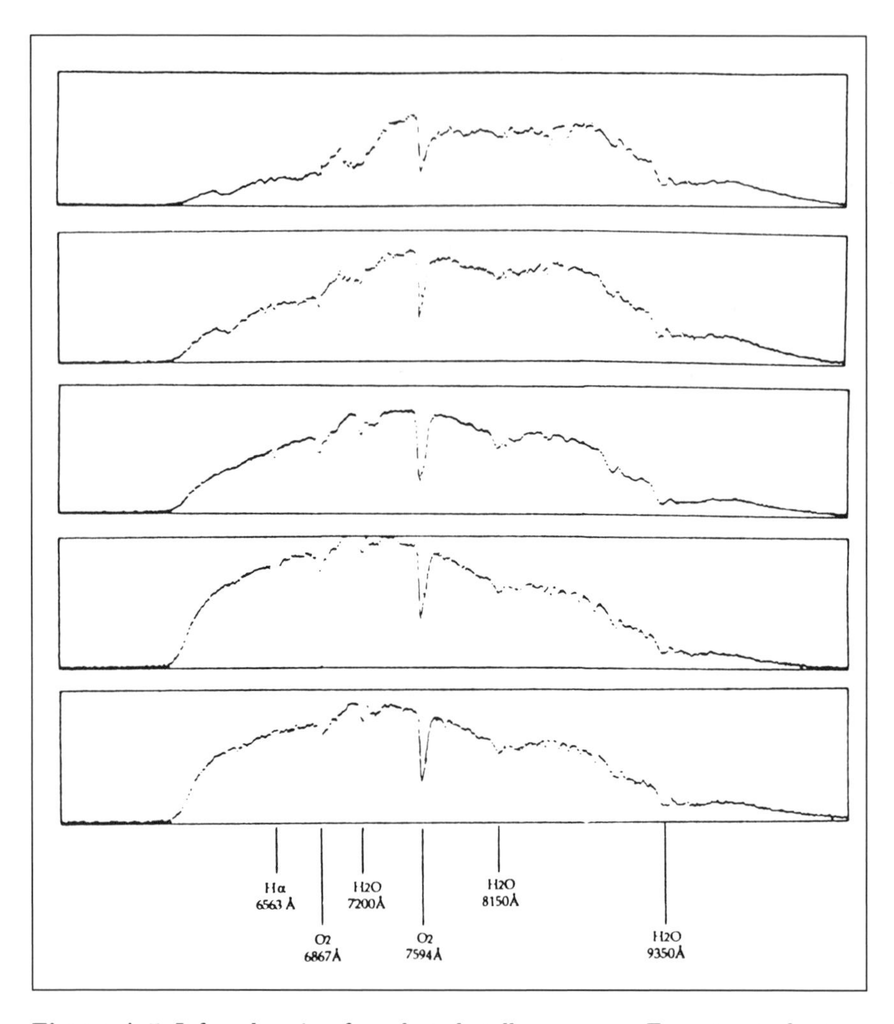

Figure A.5 *Infrared region for selected stellar spectra. From top to bottom: Betelgeuse (type M2), Aldebaran (type K5), the Sun, or more precisely although it is almost the same thing, the Moon (Type G2), Altair (type A7) and Rigel (type B8). The most noticeable indentations in these spectra are caused by the components of the terrestrial atmosphere (essentially H_2O and O_2). We can record the presence of the Paschen series (hydrogen) starting from 8500 Å in the spectrum of stars of A and B type. These spectra have not been radiometrically corrected.*

Appendix B

Sources of Components

B.1 Filters

Edmund Scientific Co., 101 E. Gloucester Pike, Barrington, NJ 08007

Ohara Corp., 50 Columbia Rd., Sommerville, NJ 08876

Oriel Corp, 250 Long Beach Blvd., P.O. Box 872, Stratford, CT 06497

Schott Glass Technologies Inc., 989 York Avenue, Duryea, PA 18642

B.2 CCD's

Eastman Kodak Company, Microelectronics Technology Division, Rochester, NY 14650

EG & G Reticon, Reticon, 345 Potrero Av., Sunnyvale, CA 94086

Fairchild Semiconductor Corp., Digital Unit, 333 Western Av., South Portland, ME 04106

General Electric Co., RCA Government Systems, Rte. 38, Cherry Hill, NJ 08358

Texas Instruments Inc., Semiconductor Group, P.O. Box 655012 M/S 857, Dallas, TX 75265

Thomson Electron Tubes and Devices Corp., Thomson Military and Space Components Division, 40G Commerce Way, P.O. Box 540, Totowa, NJ 07511

B.3 Cooling

Melcor/Materials Electronic Products Corp., 992 Spruce Street, Trenton, NJ 08648

Thermonics, Inc., Route #2, Suttons Bay, MI 49682

Tellurex, 1248 Hastings, Traverse City, MI 49684

B.4 CCD Camera Manufacturers

SpectraSource Instruments, P.O. Box 1045, Agoura Hills, CA 91376

Photometrics Ltd., 3440 East Britannia Dr., Tucson, AZ 85706

Santa Barbara Instrument Group, 1482 East Valley Rd., Suite 601, Santa Barbara, CA 93108

Bibliography

Babey, S., Anger, C. *Digital Charge Coupled Device Camera.* System Architecture. SPIE 570, Solid State Imaging Arrays, 1985.

Barbes, D.F. *Charge-Coupled Devices*, Spriger-Verlag, 1980.

Berger, Blamoutier, Couture, Descure. *Les Dispositifs á Transfert de Charge.* Revue Technique Thomson-CSF 12: (1) 1980.

Besançon, J.F. *Vision par Ordinateur en Deux et Trois Dimensions*, Eyrolles, 1988.

Desvignes, F. *Détection et Détecteurs de Rayonnements Optiques*, Masson, 1987.

European Southern Observatory "The Optimization of the Use of CCD Detectors in Astronomy" Conference and Workshop Proceedings (25), December 1986.

Fauconnier T. *Application des CCD á l'astronomie*, Thesis, Université Paris-Sud, December 1983.

Foley, J., Van Dam, A. *Fundamentals on Interactive Computer Graphics*, Addison-Wesley, 1984.

Fort, B. *Application des CCD á l'imagerie faible flux*, Conference October 1983.

Ghedini, S. *Software for Photometric Astronomy*, Willmann-Bell, 1982.

Gonzalez, R.C., Wintz, P. *Digital Image Processing*, Addison-Wesley, 1977.

Gudehus, D. The Design and Construction of a Charge-Coupled Device Imaging System, *Astronomical Journal* **90**, January 1985.

Hall, D., Genet, R. *Photoelectric Photometry Astronomy of Variable Stars.* Willmann-Bell (1982) 1989.

Howes, M., Morgan, D. *Charge-Coupled Devices and Systems*, J. Wiley and Sons, 1979.

Janesick, J., Elliot, T., "The Future Scientific CCD" *SPIE* 501, 1984.

Janesick, J., Elliott, T., Blouke, M. *Charge-Coupled Device Pinning Technologies*, SPIE 1071-15, Optical Sensors and Electronic Photography, 1989.

Janesick, J., Elliott, T., Bredthauer, R., Burke, B. Fano-noise-limited CCDs.

SPIE, Optical and Optoelectronic Applied Science and Engineering Symposium, 1988.

Kristian, J., Blouke, M., *Charge-Coupled Devices in Astronomy.* Scientific American, 247, 4, 48 October 1982.

Léna, P., *Observational Astrophysics*, Springer Verlag, 1987.

Marion, A. *Introduction aux Techniques de Traitement d'Images*, Eyrolles, 1987.

Melen, R., Buss, D. *Charge-Coupled Devices: Technology and Applications*, IEEE Press, 1977.

Optical Engineering Review, 26, **8**, **9**, **10**, 1987.

Querci, F. (editor) Compte-rendu de la Première Ecole Européenne de Photometrie Photoélectrique, May 1984.

Rosenfeld, A., Kak, A., *Digital Picture Processing*, Academic Press, 1982.

Santinelli, S. Les Dispositifs à Trasfert de Charge. Electronique Applications, (33), (34).

Tyson, J.A. *Progress in Low-Light-Level Charge-Coupled Device Imaging in Astronomy.* J. Opt. Sc. Am., 7, 7 1231, July 1990.

Walker, G., Johnson, R., Christian, C., Waddell, P., Kormendy, J., The CFHT CCD Detector. SPIE 570 State-of-the-Art Imaging Arrays, 1984.

Index

D

E

F

G

Q

R

S

T

U

V

W

Z